**Now available from** MEP

**VALVE USERS' MANUAL** has been compiled by a panel of working engineers from the British valve industry, each a specialist in particular valve types and their applications.

Valve Users' Manual is an invaluable technical reference book for all engaged in the control of fluids – designers, specifiers, contractors, plant engineers, installers, maintenance and service engineers, training officers etc. Many aspects of valve technology are dealt with comprehensively – historical development, valve types, pressure temperature ratings, materials, methods of operation, flow characteristics, installation, maintenance and repair, metrication.

The book is extensively illustrated, with many sectional drawings which show principles and systems, and identifications of parts.

ISBN 0 85298 428 6 /hardcover/A4/103 pages
**U.K. Price £9.00** Elsewhere £13.00 (inclusive of airspeeded despatch)

I wish to order_______________copy/copies of VALVE USERS' MANUAL

☐ Remittance enclosed___________

☐ Please invoice me (orders over £20.00 only)

Signature_______________________________

Name_______________________________

Address_______________________________

_______________________________

**PLEASE RETURN THIS FORM TO: SALES DEPARTMENT,**

# AUTOMOTIVE
## Conference Volumes from 

*Some recent automotive conference volumes include:*

☐ **IMPACT OF VEHICLE DESIGN ON WHOLE LIFE COSTING**

This subject is particularly apt at this time with the struggle to beat inflation and industry in recession. Consumers are looking for durable goods and the more expensive the item, the more durability will influence choice. Manufacturers, in the effort to curb rising prices, are pruning their overheads ruthlessly. However this conference has produced encouraging and constructive ideas which can only be effective.

There are papers on various aspects of corrosion: prevention, testing procedures, and recent research developments; papers on the effect of vehicle design on whole life vehicle costs; depreciation, maintenance, repairs and reliability are discussed; product liability and insurance are also considered. The papers cover all types of vehicle, both passenger and commercial, from the new Ford Escort, buses and trucks to a specialist vehicle for the winter maintenance of motorways.
*ISBN 0 85298 468 5*
UK **£19.00** Elsewhere **£27.00** (inclusive of air-speeded despatch)

☐ **STRATIFIED CHARGE AUTOMOTIVE ENGINES**

Complete contents: Influence on intake swirl on the characteristics of a stratified charge engine with pre-chamber injection; The Deutz stratified charge process; Flow and flame propagation in the Kouchoul engine; Influence of air motion variation on the performance of a direct-injection stratified charge engine – Diagnostic investigation of hydro-carbon emissions from a direct-injection stratified charge engine with early injection; Fuel tolerance tests with the Ford PROCO engine; The anti-knock quality in the stratified charge engine with an auxiliary combustion chamber; Porsche 1-, 6-, and 8-cylinder stratified charge engine with divided combustion chamber (SKS-Engine); Fuel supply characteristics of CVCC carburettor to the auxiliary combustion chamber; High compression stratified charge engines and their suitability for conventional and alternative fuels; A two stroke engine with charge stratification and electronically controlled pressure impact injection; Flow and combustion measurements within a dual chamber stratified charge engine; Development of a mathematical model of flow, heat transfer and combustion in a stratified charge engine; Modelling of drop interactions in thick sprays and a comparison with experiments.
*ISBN 0 85298 469 3*
UK **£25.00** Elsewhere **£35.00** (inclusive of air-speeded despatch)

☐ **SYSTEMS ENGINEERING IN LAND TRANSPORT**

Complete contents: A systems engineering approach to the design and testing of battery vehicle drives; Systems engineering for personal land transport; An optimal control approach to energy minimisation in electric vehicles; Systems engineering in land transport – a computationally efficient approach to the optimisation of traction drive control; A preliminary assessment of controlled cylinder disablement as a potential means of fuel conservation in automobile engines; Derivation of economy strategies for a stepped ratio automatic transmission; Design and performance of the Ratcliffe/Hobbs continuously variable automatic transmission; A design analysis and experimental evaluation of a hybrid transmission system for electric vehicles; Efficiency improvement for engine driven hydrostatic transmissions.
*ISBN 0 85298 457 X*
UK **£11.00** Elsewhere **£15.00** (inclusive of air-speeded despatch)

☐ **TOWARDS SAFER PASSENGER CARS**

This conference deals with a variety of ways and means of reducing accidents for both passenger car users and pedestrians alike.
Complete contents: The dynamics of car impact and occupant protection; The development of impact test procedure for legislation; Safer cars for the pedestrian; Protection of the car driver from steering system induced injuries; The use of computer simulation for the design of safer vehicles; An oblique view of some factors affecting primary vehicle safety; Vehicle safety legislation – a small company's approach; Progress towards safer passenger cars in the United Kingdom; The role of a monitoring organisation; Designing a test dummy for assessing side impact performance; Side impacts – How people are injured in accidents and the use made of information from accidents; Paper on the motor industry code of practice on action concerning vehicle safety defects; The preparation of standards and legislation; Trends in the design of car brakes; Prediction of the determination and movement of the target car in vehicle to vehicle side impact.
*ISBN 0 85298 464 2*
UK **£16.00** Elsewhere **£23.00** (inclusive of air-speeded despatch)

Please tick titles required ☐ Remittance enclosed £ _______________

Please invoice me (orders over £20.00 only)

Name _______________________________

Address _______________________________

_______________________________

*Please return to:*
**SALES DEPARTMENT, MECHANICAL ENGINEERING PUBLICATIONS, LIMITED, PO BOX 24, NORTHGATE AVENUE, BURY ST EDMUNDS, SUFFOLK, IP32 6BW**

# Fourth International Conference on

# PRESSURE VESSEL TECHNOLOGY

Held in London 19–23 May 1980

Organised by The Institution of Mechanical Engineers
The American Society of Mechanical Engineers
The Japan High Pressure Technology Institute

With the constant development of high pressure vessels and the increasing performance expected of them the papers given at the FOURTH INTERNATIONAL CONFERENCE ON PRESSURE VESSEL TECHNOLOGY are particularly significant. The three volumes collect the most up-to-date and learned thoughts on this important subject.

**Volume One** – Materials, fracture and fatigue –53 papers
**Volume Two –** Design, analysis, components, fabrication and inspection –
57 papers
**Volume Three** – Discussions

The first volumes are available now, discussions volume to be published Spring 1981
**The three volumes are sold as a complete set.**

| | | |
|---|---|---|
| Volume One | ISBN 0 85298 459 6 | approx 410 pages |
| Volume Two | ISBN 0 85298 460 X | approx 430 pages |
| Volume Three | ISBN 0 85298 461 8 | approx 150 pages |
| Whole set | ISBN 0 85298 458 8 | |

UK £142.00   Overseas £190.00 (including air-speeded despatch)

I wish to order _______________________________________copy/copies of FOURTH INTERNATIONAL CONFERENCE ON PRESSURE VESSEL TECHNOLOGY.

☐   Please invoice me

Name _______________________________________________________________

Address ____________________________________________________________

___________________________________________________________________

___________________________________________________________________

PLEASE RETURN THIS FORM TO SALES DEPARTMENT,

# MECHANICAL ENGINEERING PUBLICATIONS LTD

PO BOX 24,  NORTHGATE AVENUE,  BURY ST. EDMUNDS  IP32 6BW

# HOWDEN
## Screw Compressors

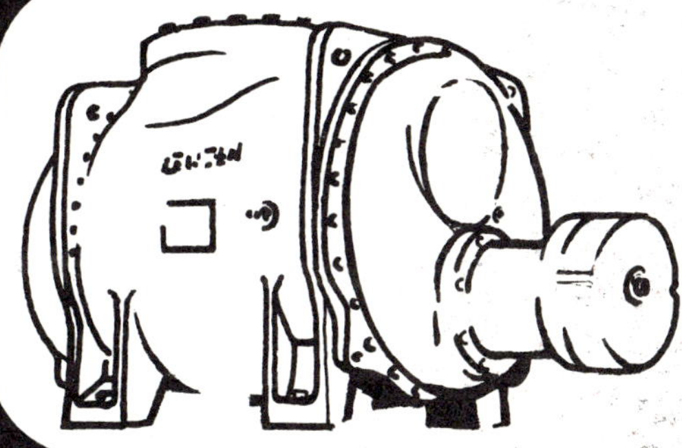

### OIL INJECTED REFRIGERATION COMPRESSORS

These compressors are suitable for use within an evaporating temperature range from approximately $-57^0$C ($-70^0$F) to $+10^0$C ($+50^0$F). They cover a wide range of refrigeration capacities which at an evaporating temperature of $0^0$C ($+32^0$F) is 0.5M Kcal/hr (2M Btu/Hr) to 10M Kcal/hr (40M Btu/Hr)

### OIL INJECTED GAS COMPRESSORS

340 $m^3$/hr to 10,000 $m^3$/hr up to 24.5 kg/$cm^2$ discharge pressure (200 cfm to 6000 cfm up to 350 psig).

### OIL FREE COMPRESSORS

Single stage:  500 cfm to 10,000 cfm up to 50 psig
850 $m^3$/hr to 17,000 $m^3$/hr up to 3.5 Kg/$cm^2$

Two Stage:  1,500 cfm to 10,000 cfm up to 190 psig
2,500 $m^3$/hr to 17,000 $m^3$/hr up to 13.5 Kg/$cm^2$

Three Stage:  5,000 cfm to 10,000 cfm up to 300 psig
8,500 $m^3$/hr to 17,000 $m^3$/hr up to 21 Kg/$cm^2$

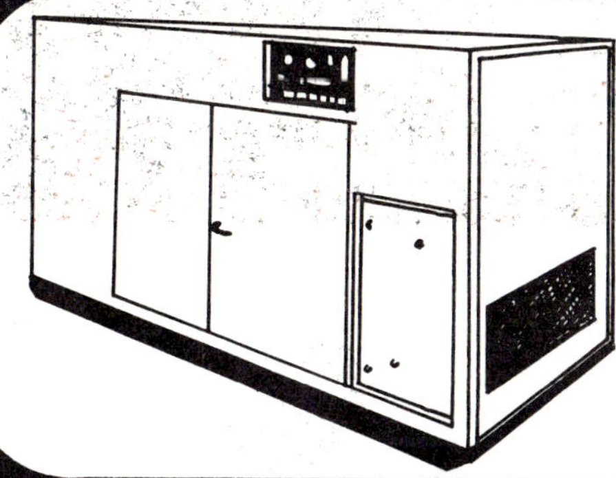

### HOWDENAIR PACKAGED COMPRESSORS

111 $m^3$/hr to 3396 $m^3$/hr up to 13.8 bar g discharge pressure (65 cfm to 2000 cfm up to 200 psig

# HOWDEN
# COMPRESSORS
# LIMITED

133 Barfillan Drive, Glasgow, G52 1BE
**Telephone  041-882 3346**     **Telex  778711**

# Thinking FCCU?

# Then think FW

We've built fluid catalytic cracking units on five different continents, using four different process designs.

In fact, no other contractor has worked on more major licensed designs than Foster Wheeler.

Our experience goes back 30 years or so. This includes the mid-sixties when the advanced designs of today's cat crackers were first being developed. Practically nobody was building new cat crackers then, but we cut our teeth on almost a dozen revamp jobs.

Now in the energy conscious eighties we are well placed to meet rising demand for reliable FCCU-flue-gas power recovery. Already we have successfully engineered four power recovery units around the world.

Experience like this has given us mastery of the complex, detailed mechanical engineering so essential to cat cracker design.

Fluid cat crackers are nothing if not big. Big vessels, big transfer lines. Volume of the total catalyst being moved is upwards of 50 tons per minute. Even the temperatures are big. Anywhere from 900° to 1400°F.

Power recovery expander train engineering is itself a specialist field, requiring know-how of high temperature rotating machinery.

Size, heat and volume of this magnitude multiply the problems of support, stress and expansion, erosion and corrosion.

For example, we've worked on a cat cracker with a reactor and regenerator connected by 100-foot-long risers, which range in diameter from three to four feet.

Despite these dimensions, these lines have to withstand tremendous hostile forces moving and reacting within them. Which calls for some challenging mechanical design detailing, to say the least.

Foster Wheeler Energy Limited
Reading
England

# Bechtel continues to grow in Britain

Since 1952, the professional staff of Bechtel Great Britain Ltd has grown to meet the demands of our increased workload.

From our small beginnings, our London office is now the headquarters of the Petroleum and Chemical-Eastern Division, serving a large geographical area with special emphasis on exports to Bechtel projects worldwide.

To accommodate all its activities at a single location, we are constructing additions which, when completed, will enable us to house more than 2,500 employees.

Bechtel's experience enables us to offer you all aspects of engineering, design, procurement, construction and construction management and cost and schedule control. These capabilities apply not only to petroleum and chemical projects, but to the range of services associated with the construction industry.

For your next project, ask Bechtel to show you how our experience will help you to save money.

# Bechtel

**GREAT BRITAIN LIMITED,**
**BECHTEL HOUSE,**
**245 HAMMERSMITH ROAD,**
**LONDON W6 8DP**
Tel:01-741 5111  Telex:934151

## Bechtel puts its world-wide experience to work in Britain

# FLUID MACHINERY
# FOR THE OIL,
# PETROCHEMICAL AND RELATED
# INDUSTRIES

# FLUID MACHINERY FOR THE OIL, PETROCHEMICAL AND RELATED INDUSTRIES

## I Mech E CONFERENCE PUBLICATIONS 1981-3

European Congress sponsored by
The Power Industries Division of
The Institution of Mechanical Engineers
and
The Division of Mechanical Engineering and Naval Architecture of
The Royal Dutch Institution of Engineers

The Hague 24–26 March 1981

Published by
Mechanical Engineering Publications Limited for
The Institution of Mechanical Engineers
LONDON

Printed by Waveney Print Services Ltd, Beccles, Suffolk

# CONTENTS

# C48/81

# The development of the screw compressor and its application in the petrochemical and related industries

P O'NEILL, BSc, DMS, CEng, FIMechE
Technical Director, Howden Compressors Limited

SYNOPSIS    In the petrochemical and related industries great importance is attached to the ability of machinery to operate reliably over long periods with the minimum of maintenance.

Developments in screw compressors over the last decade have produced units ideally suited to this type of application.  This paper describes the major developments which have taken place in both the screw compressor and its systems.  A number of typical applications in petrochemical industries are described.

## 1. INTRODUCTION

In the petrochemical and related industries processes are often continuous in operation and handle large quantities of chemical materials in one form or another.  In order to ensure economic operation of the plant involved, it is important that all the various items of equipment are as reliable as possible, are easy to repair and overhaul, and it should be possible to add means of predicting problems before actual failure occurs.  It goes without saying that running costs should be the lowest possible also.

In the case of machinery, a number of basic features are known to help to ensure reliability, ease of maintenance and long life.

Firstly there should be the smallest possible number of moving parts.  The less components, the less there is to go wrong.  This may seem to be a 'truism' but it can be mathematically shown that the 'Mean Time Before Failure' of a machine (MTBF) is directly related to the number of moving items involved in its operation.

Secondly, the operation should be entirely rotary or as close to this as possible.  Reciprocating action of necessity involves additional loads and forces in accelerating and decelerating the oscillating masses.

Thirdly, the process should be continuous rather than cyclical as equilibrium conditions can then be reached which are relatively stable in operation.

In this paper one particular class of process equipment is considered, namely compressors and specifically the Screw Compressor.

## 2. INDUSTRIAL COMPRESSORS

Industrial compressors can be categorised by their method of operation and type of action. The screw compressor comes into the category of positive displacement, rotary action, medium volume and pressure.

In the area which could be called medium compression with volumes from 425 to 17 000 $m^3/h$  (250 to 10 000 cfm) and pressures from 1 to 24 bar (15 to 350 psi) the screw compressor is pre-eminent and the various technical features which make it so are discussed below.

## 3. SCREW COMPRESSORS

In the following discussion the principle of operation of the various types of screw compressors are discussed, with comparison against the desirable features mentioned earlier for reliable and efficient operation.

### 3.1  Basic Principles of Operation

The basic principles of operation of the various types of screw compressor are the same whether running 'dry' or 'liquid injected'.  The screw compressor consists basically of two rotors enclosed in a casing.  The rotors differ in shape and are identified as 'male' and 'female'. In Fig. 1 the male rotor is on the left and the female on the right.  If these rotors are rotated, a space is formed (Fig. 1 a ) due to the helix angle of the rotor lobes.  This space is enclosed by the rotors and the casings.  This volume increases as the rotors rotate and gas is drawn in to fill the low pressure area formed (Fig. 1 b ).

A useful analogy in understanding this process is to consider the male rotor lobes as pistons and the female flutes as cylinders.  Rotation of the rotors causes the piston to slide along the cylinder and draw in gas.

Continued rotation results in the lobes remeshing and expelling the gas out of the far end of the casing.  However, if a cover is placed over the discharge end of the machine then the remeshing of the lobes results in the volume of gas which has been drawn into the set being reduced in volume, i.e. compressed between the meshing rotors and the casing.

At a particular point along the length of the casting a discharge port is positioned which will permit the gas to pass out of the compressor. The amount of internal compression which occurs before release is therefore a characteristic which can be varied by the positioning and shaping of the discharge port. This feature is known as the 'Built-in Compression Ratio' of the set and is usually denoted by the symbol $\pi$.

The compressor, therefore, operates on the 'positive displacement' principle. This means that the discharge pressure of the compressor is determined by the resistance to the flow of gas from the compressor and not by a dynamic 'head' developed by the speed of rotation of the rotors. Full discharge pressure can be achieved at any compressor speed. The action of the compressor is, nevertheless, entirely rotary, and dynamically in balance.

As a result of the helical formation of the rotor lobes and the shaping of the castings enclosing them, the gas is drawn in and trapped, internally compressed, and finally discharged, without the need for any form of valve. This has two advantages. Firstly, there are no valve losses, and secondly there are fewer moving components to wear or fail.

## 3.2 Screw Compressor Types

The first type of screw compressor to be developed and marketed was the 'oil free' or 'dry' type. In this design the rotors are arranged, by means of timing gears, to run in mesh but without any contact between the rotors.

Later developments have improved the design of the dry screw compressor and produced liquid injected varieties. The resulting 'family' can be represented as shown in Fig. 2 .

## 4. OIL FREE SCREW COMPRESSORS

### 4.1 Dry Oil Free Compressors

Consider the oil free dry compressor. A typical cross section of such a design is shown in Fig. 3 .

In this design there are only two moving components, the rotors themselves. Air or gas is drawn in at one end of the rotors, is compressed internally and discharged at the other end. This means that one end of the rotor is always in contact with the incoming gas and the other with the discharging gas and as a result one end of the rotor and the casing is hotter than the other. The compressor design has to allow for this effect so that the machine can operate over as wide a range as possible.

The rotors will tend to expand into each other at the hot end and in order to prevent this happening the rotors are manufactured with a cooling chamber down the centre through which oil is passed during normal operation.

In the same way the casings incorporate a cooling jacket surrounding the rotors with particular attention being paid to the coolant flow around the area of the discharge port. Normally the coolant used is water as it is freely available and has excellent heat transfer properties. Other coolants used include oil (although this can restrict possible duty) and glycol/water

mixtures for low temperature applications. It should be said that the screw compressor is a high specific volume machine, that is, it handles a large throughput for its physical size. This means that there is little time for gas cooling to take place and the rotor internal cooling and jacket cooling are primarily for maintaining their geometric shape and dimensions and not really for gas temperature control. Only approximately 7% of heat involved passes to jacket cooling and 7% to the oil system in high temperature applications.

With this type of design maximum operating discharge temperatures of $225^{\circ}C$ are used. What this means in terms of compression duty possible, depends very much on the properties of the gas involved. There are two reasons for this variation.

The first is the theoretical temperature generated by the compression of the gas. The theoretical temperature rise for any given increases in pressure ratio is dependant on the ratio of the specific heats at constant pressure and constant volume, i.e. $C_p/C_v$ which can range from as low as 1.04 with heavy hydrocarbons to 1.6 with helium.

The other important factor is the physical viscosity/density of the gas being compressed. In a dry screw compressor, since the rotors never touch each other or the casings there is a small clearance between the various components through which some of the gas being compressed leaks. This leakage gas has then to be recompressed and since it was heated by the first compression, it increases the discharge temperature by its recompression. The amount of gas which will leak back in this way is determined by the gas properties and a 'thin' gas such as helium has a greater back leakage than say a more dense gas such as propane.

The overall result is that the amount of compression which is possible from a single stage with a fixed maximum discharge temperature depends very much on the gas involved.

Typically air from atmospheric pressure and temperature can be compressed to around 3.5 $kg/cm^2g$ (50 psig). Multiple stages with intercooling are used to achieve higher pressure levels.

These compressors have been called 'chemically oil free' since there is absolutely no contact between the gas being compressed and the lubrication system.

While retaining this feature it is possible to inject a fluid other than oil for one or two purposes, temperature control or cleaning.

### 4.2 Liquid Injected Oil Free Compressors

It has been noted above that the oil free compressor limitation is often maximum discharge temperature. It is possible to reduce the discharge temperature for a given compression duty by adding a liquid to the gas being compressed. The simplest example of this is to add water, although other fluids are of course equally practical. The cooling effect can be due to the heat absorbed in raising the temperature of the fluid, or due to the vaporisation of some or all of the fluid during the compression process.

The application which requires a quantity
of fluid for cleaning is quite different.  In
this case the purpose of the fluid is to remove
some content of the gas being compressed which
tends to be deposited in the compressor.  One
example is gases or gas mixtures which tend to
'polymerise' during compression, that is, form a
gum or plastic like coating on the surfaces within the machine.  In many cases this has the unexpected effect of improving the efficiency of
the compressor during running as it reduces the
clearances.  The rotors by their meshing action
wipe away any tendency for excessive build up.
However, at shut-down these coatings sometimes
bond the rotors together and to the casing
making it difficult if not impossible for the
motor to restart the compressor.

In these cases an appropriate solvent is
sprayed onto the compressor inlet for a few
minutes before shut-down to dissolve and remove
deposits.  Many such installations are operating
very satisfactorily, a typical example being in
the compression of butadiene where the solvent
used is Benzol.  In some cases much larger quantities of fluid are added to wash out dust or
similar products passing into the compressor with
the gas.  As long as the particle sizes are not
excessive or very hard and thus abrasive this is
quite practical.  However, the compressor
design in this case should be arranged to have a
top suction and bottom discharge to ensure complete drainage of both liquid and solids.

Unlike many other types of compressor the
screw can handle large quantities of fluids without any damage, and in fact with many advantages
as will be seen later.

## 4.3  Efficiency in Operation

No matter how reliable any machine might be,
it is of little consequence if it does not
operate efficiently.  Fortunately the screw
compressor is efficient and furthermore, since
the various components do not touch in operation
and therefore do not wear, the efficiency is
maintained.  This basic efficiency is not reduced
by any valve wear or leakage since these items
are not required.

It is interesting to note that, since there
is no 'clearance volume' as occurs with a reciprocating design there is no sudden fall off in
volumetric efficiency due to clearance volume
re-expansion.

At low operating speeds the back leakage
through the operating clearances becomes too high
a percentage of the total throughput and efficiency is reduced.  Hence the fact that there is
a practical operating rotor tip speed range,
nominally 50 to 100 m/s for oil free screw
compressors.

Built in compression ratio is another important variable in screw compressor design which
greatly assists in ensuring high efficiency.  As
described above the discharge port in the casing
is sized to determine how much compression is
achieved within the compressor.  The ideal situation is where the internal compression matches
the overall duty required as shown diagramatic-

ally in  Fig. 4a .

Internal compression lower or higher than this
will result in some losses as shown in  Fig. 4b
& 4c  which reduce the efficiency.

A number of different discharge port sizes
and hence built-in compression ratios may be used
on any given size of machine to enable a high
efficiency to be obtained at any operating pressure ratio within the compressors range.  In
Fig. 5  each of the graphs represents the efficiency of a screw compressor with a particular
built-in compression ratio.  The operating area
of each built-in compression ratio is indicated.

## 4.4  Condition Monitoring

Condition monitoring involves the fitting of
instrumentation to machinery which provides
information on the running conditions of various
parts of the equipment and indicates changes in
any parameter.  The purpose of this information
is to foresee the need for overhaul or adjustment before failure occurs.  The normal parameters involved are temperature and vibration.

### 4.4.1 Temperature

It is normal to fit instruments which record oil supply temperatures, compressor discharge temperatures and where relevant, compressor
suction temperatures and, therefore, these are
not considered as condition monitors other than
that they may indicate such things as fouling of
heat exchangers etc.  Bearing temperature detection can provide a more significant indication
of condition.  In a screw compressor fitted with
solid white metal journal bearings and tilting
pad thrust bearings, temperature detectors can be
placed against the steel backing of the white
metal, as close to the white metal(Babitt metal)
running surface as possible and in the same way
in one of the tilting pads.  Alternatively the
temperatures detectors can be located to measure
the oil drain temperatures at significant positions.

The results obtained have to be interpreted
with care.  A significant change in the running
temperature could be caused by an equivalent rise
in the oil supply temperature or a variation in
the compressor duty and due allowance must always
be made for such factors.

A gradual change in one or more bearing temperature readings can give warning of impending
problems, permitting a planned shut-down to be
organised.

### 4.4.2 Vibration Detection

These devices are available in two basic
forms.  The first is an accelerometer mounted
directly on to the machine external casing and
the second is a series of probes located close
to rotating shafts to detect changes in lateral
and axial movement.

Fitting the casing mounted unit is an economic system which is used quite extensively.  It
is set such that a significant increase in compressor vibration sets off an alarm or shut-down
trip.  An increase in vibration can be caused
by problems within the machine but also can be
caused externally such as by coupling failure or

misalignment. It is sometimes necessary to fit
a time delay during the starting sequence as initial vibration can occur which subsides when
normal running speed is reached. Fig. 6 shows
a typical method of assessment of readings obtained based on the V.D.I.Standard 2056 Group G.

The vibration probe approach is a much more
complex system requiring a fair amount of space
within the compressor and considerable preparation and calibration.

The principle of the system is to mount a
small electric coil close to the rotating shaft.
This coil is fed with a radio frequency voltage.
The current which flows through the coil is affected by the proximity of the shaft to the coil
and hence variation in this distance causes an
equivalent variation in the current. If two
probes are fitted at right angles the combined
signals provide a complete picture of the movement of the shaft relative to the casing.

The system is very sensitive and accuracy of
shaft manufacture is most important as errors in
circularity or concentricity to running centres
will read as shaft movement. These errors are
called 'Mechanical Runout'.

Other factors classed as 'Electrical Runout'
cause a variation in output signal even though
there is no change in shaft-probe gap. Included
here are uneven surface hardness, magnetism in
shaft material, uneven heat treatment and metallurgical segregation. Degaussing, micropeening
and burnishing are techniques used to reduce electrical runout to acceptable levels. Fig.7a
shows a typical shaft as manufactured and Fig.
7b the same shaft after processing as mentioned
above.

The readout from the two probes can be fed
into an oscilloscope for interpretation. An
absolutely perfect set up would produce a 'Lissajous' figure of a true circle. Fig. 8 shows
a satisfactory readout from a processed shaft.

In actual operation each probe is calibrated and the instrumentation is arranged to give
an alarm for a set amount of deviation and to
trip the compressor at a second higher level of
change. When the probes have been 'fingerprint-
ed' with photographs of the 'Lissajous' figure
retained, connecting the probes back into an
oscilloscope can provide a comparison at any time
which can help to give some guide as to the forms
of any change occurring.

A later figure in this paper shows a photograph of a compressor installation incorporating
the equipment described above, both bearing temperature detectors and a full suite of vibration
probes.

## 4.5 Materials of Construction

The standard materials of construction of
screw compressors are steel for the rotors and
cast iron for the casings. However, alternative
materials are often used for both rotors and
casings. Howden have used high nickel content
steel,stainless steel, SG iron and even titanium
for rotors. They have used SG iron, steel, high
nickel low temperature steel,gunmetal and titanium for casings.

## 5. OIL INJECTED SCREW COMPRESSORS

The fundamental difference between an oil
free and an oil injected screw compressor is that
in the oil injected screw compressor, lubricant
(not necessarily mineral oil) is added to the gas
being compressed and removed again after the compression process is complete.

As a matter of interest, the first oil injected screw compressor tests ever carried out,
were by Howden in 1955 and were literally tests
of an oil free compressor with oil sprayed into
the suction port. The results were so impressive that further development was undertaken and
compressors of this new concept were marketed in
1958.

Howden followed this with development of
oil injected compressors for gas type applications in 1959 and they installed the first such
compressor in the world in 1961.

The compressors now manufactured are very
sophisticated derivatives of these early units
which have many additional features to improve
efficiency, give longer life, and incorporate
advanced control and operating systems. The
main areas of development are described below,
covering the compressor, control systems, and
oil separation systems.

## 5.1 Oil Injection

Why is oil injection used?

The most important reason is the enormous
flexibility it gives to the compressor. It was
mentioned earlier that one of the limitations
with all other types of compressor including the
oil free screw, is discharge temperature. When
oil is injected into the screw compressor this is
no longer a limitation.

It can be stated that the power absorbed in
compressing a gas all appears as heat in the system. Normally the vast majority of this heat
appears in the gas itself as increase in temperature, the remainder being absorbed into the
compressor and its cooling systems. However,
in an oil injected compressor, a large part of
the mass flow going through the compressor is
made up of the injected oil and hence this ab-
sorbs the heat. The mass of oil is relatively
large compared to the gas mass flow because the
oil is in the liquid phase, but the volume of oil
relative to the gas is normally less than 1% and,
therefore, the effect of the oil volume on the
gas throughput of the compressor is negligible.

The addition of the appropriate quantity of
oil to the compressor related to the absorbed
power therefore, controls the compressor discharge temperature regardless of the pressure
ratio over which it is operating. As long as
the cooling for the oil is designed to remove the
heat absorbed by it, the system remains under
accurate control with great flexibility.

This injection system cannot be used with
reciprocating or centrifugal compressors as they
are unable to handle liquid, whereas the screw
compressor can and does handle amazingly large
quantities without this even being evident.

The discharge temperature which will result
from any given set of operating conditions can
simply be calculated by the following equation:

Absorbed Power (in heat units) =

$$M_g.C_{pg}.(T_2 - T_1) + M_{oil}.C_{p_{oil}}.(T_2 - T_{oil})$$

Where $M_g$ = mass flow of gas

$C_{pg}$ = Specific heat of gas at constant pressure

$T_2$ = Compressor discharge temperature

$T_1$ = Compressor suction temperature

$M_{oil}$ = Mass flow of oil     (1)

$C_{p_{oil}}$ = Specific heat of oil at constant pressure

$T_{oil}$ = Oil temperature entering compressor

Note: The discharge temperature for the oil and gas are the same.

Associated with the temperature control aspect of
the addition of oil are other advantages such as
efficiency, speed of operation and wider duty
range.

## 5.2 Efficiency in Operation

The presence of the oil in the rotors during
compression forms an effective seal between the
rotors and between the rotors and casing, greatly
reducing the leakage of gas from the high pressure part of the rotors to the low pressure area.
This means that the ability to handle high
pressure ratios and pressure differences due to
the cooling effect of oil injections is com-
plimented by the improved sealing which ensures
high efficiency. A typical performance graph
for an oil injected compressor is shown in
Fig. 9 .

In the same way as with oil free screw compressors the built in compression ratio can be
varied to achieve the optimum efficiency and
Fig. 10 shows the effect on performance of
different values. Also shown on this graph are
the areas for which each built-in compression
ratio would be selected.

Another major effect of the oil injection
is on the operating speed range of the compressor.
It was mentioned earlier that oil free screw
compressors operate over a rotor tip speed range
of 50 - 100 m/s approximately. The sealing
effect of the oil results in a much lower
minimum speed, but the losses in churning the oil
at high speeds also reduces the maximum speed,
again for efficiency reasons. The resulting
operating speed range for oil injected screw
compressors is 15 to 60 m/s approximately. This
speed range confers another advantage in that it
means that the vast majority of compressors are
directly driven by electric motors, either two
or four pole, no speed increasing gearbox being
required.

Oil injected screw compressors also have a
simpler design. As oil is already coming into
contact with the gas being compressed, shaft seals

between the compression chamber and the bearings
are no longer required, also timing gears are
not necessary.

An oil lubricated mechanical face type seal
is fitted on the drive shaft into the compressor
and therefore there is no leakage of gas to atmosphere.

## 5.3 Rotor Profile

Rotor lobe profile is another area in
which development has achieved significant performance improvements in oil injected screw
compressors. The earlier design of circular
arc, symmetric profile has been modified to a
generated arc unsymmetric shape. This has resulted in an improved compression efficiency, at
some duties of as much as 10%.

## 5.4 Capacity Control

Another major feature developed for compressors of this type is fully variable, stepless
capacity control. In many processes and particularly for refrigeration systems some means of
efficiently and simply varying the compressor output is highly desirable. This has been achieved
with the screw compressor by means of an integral
slide valve. The effect of this is to vary the
effective length of the rotor being used. The
whole system is integral, with the compressor incorporating its own hydraulic cylinder which is
actuated by oil from the compressors own lubrication system. A typical part load characteristic is shown in Fig. 11 .

In some installations the discharge pressure
of the compressor falls off with reducing mass
flow, for example, in a refrigeration plant, and
this further enhances the part load characteristic. Fig. 12 shows a typical result.

A cross sectional view of a compressor as
described is shown in Fig.13 .

An oil injected screw compressor can, therefore, be seen to be a machine with a very wide
operating range, suitable for discharge pressures up to 24 bar.g. (350 p s i g ) with discharge temperature accurately controlled regardless of gas or duty, usually predetermined
not to exceed a level around $100^{\circ}C$ ($212^{\circ}F$),
efficient in operation and having the added feature of stepless capacity control from 100% down
to at least 10%. The action of compression is
still entirely rotary although there is the slide
valve mechanism as additional moving parts.

Other lighter duty oil injected screw compressors are also manufactured for pressures up
to 14 bar.g. (200 p s i g ) using anti-friction
bearings because of the lighter loads, some with
and some without the slide valve feature.

## 5.5 Control Systems

The purpose of control systems with this
type of screw compressor is to adjust the output
of the compressor by means of the slide valve to
match the system demand at any given time. This
can mean adjustment to maintain a set suction
pressure or a set discharge pressure. In the
case of a vapour boil off compressor the suction
pressure would be the reference level. With the

compressor feeding a process system, the discharge pressure may be the level to be maintained, likewise on air compressor systems.

The principles of the control systems used are very simple. If the system deviates from the preset level the control system senses this and adjusts the position of the slide valve by operating a solenoid valve to feed oil to the appropriate side of the hydraulic cylinder.

The control system can be electronic or pneumatic, can use a balanced bridge type circuit or a time base.

The simplest system, and one which is used extensively is the time base design. This is a solid state electronic concept incorporating a cycling circuit. Once every cycle the system checks the actual versus the datum condition. If this shows a difference (outwith the 'dead' or acceptable deviation zone) then the appropriate solenoid valve is operated for a period proportional to the deviation. Then the cycling board cuts out till the end of the cycle time and checks again on the next cycle. This system is, therefore, proportional, is solid state and has minimum moving parts for reliability, can sense pressure or temperature, and does not require a position read out from the compressor.

Alternative systems are also used, both pneumatic and electronic, using a slide valve position read out as part of a 'balanced bridge' circuit. A very simple rotary position indictor is incorporated into the hydraulic cylinder of this type of compressor which is used for this purpose.

## 5.6 Materials of Construction

Normal materials are steel rotors and cast iron casings. However, as in the case of oil free versions, casings are made in spheroidal graphite (SG) iron, steel and high nickel content low temperature steel for particular applications.

These compressors are designed to be capable of handling a wide range of gases and gas mixtures and, because of this, no copper or copper alloy components are used anywhere in the machine. This is necessary since the casings of the compressor are at suction conditions and the gas, therefore, comes in contact with all parts of the compressor. The shaft seal is in this category also and it therefore seals the casing from suction pressure to the outside atmosphere.

## 5.7 Selection of Lubricants

In selecting suitable lubricants, two separate areas have to be considered, chemical compatibility and physical suitability. Some gases chemically interact with lubricants and as such are not suitable for handling with this type of compressor. An example would be chlorine gas. This interacts with mineral oils forming sludges and breaks down the oil chemically. There are no commercially available alternatives of appropriate viscosity and as a result this gas is handled by oil free screw compressors but not by oil injected.

The oils selected for many gases including a number of hydrocarbons such as methane ($CH_4$) are standard paraffinic base mineral oils, often referred to as 'turbine oils'.

Where lower temperatures are involved, particularly in industrial refrigeration systems where waxing of the oils on low temperature heat exchangers can reduce their efficiency, mineral refrigeration oils, usually of naphthenic base, sometimes dewaxed paraffinic, are used. These would be used when operating on ammonia ($NH_3$) or chlorodifluoromethane (R22).

When gases to be compressed have a high solubility in the lubricant, consideration has to be given to the dilution effect on viscosity.

The amount in solution is related to pressure and temperature. Henry's law states that the amount of gas which will dissolve in a fluid is proportional to the absolute partial pressure. Since the lubrication system in an oil injected screw set is at discharge pressure, pressure effects are therefore important. The temperature effect is related to the vapour pressure of the constituents, which results in a lower percentage in solution at increasing temperatures. The typical effect for a given gas/oil combination is shown on  Fig. 14 .

The effect on viscosity of various levels of dilution is graphically represented on a 'Refutas' chart as shown on  Fig.15 .

Study of these various relationships enables appropriate selections to be made. Typical examples are mineral oil for methane and ammonia, polyglycol for propane and R12, silicon synthetics for hydrogen chloride etc.

## 5.8 Oil Separation

Oil separation is a very important part of an oil injected screw compressor installation, partly because it is undesirable to have contaminants in the gas after compression and partly because constant loss of oil would add to the cost of operation of the plant.

Fortunately considerable research has been undertaken in this field in the last decade or so and systems have been developed with remarkably high efficiencies.

It has been found that approximately 99.7% of the oil in the gas stream can be recovered via gravity simply by reducing the velocity of the mixture and with appropriate changes in direction of flow.

The 'impingement' pack will achieve oil separation down to about 35 parts per million by weight (p p m ) with most gases with a minimum pressure drop.

Higher levels of efficiency are achieved by using 'coalescers'. A problem that exists with oil separation from a gas is that a proportion of the oil is present in very finely divided form, rather like a very fine mist. Some of these droplets are so small that they are around molecular size, and, as such, tend to act like a gas. As an example they can be passed through a cloth without wetting the fibres. The coalescer is an element made from material with carefully controlled pore sizes. These pores and the various changes in direction in flowing through the elements cause the fine particles to combine and eventually to flow out of the downstream side of the element in normal liquid form.

The thickness of the element is a compromise between ultimate efficiency and pressure drop, since the basic equation for flow through a porous material is:

$$P = C \frac{b.V.Y.}{A} \quad \text{where}$$

P = pressure drop

C = constant

b = thickness of material $\qquad (2)$

V = volume flowing through material

Y = coefficient of fluid viscosity

A = cross sectional area

Elements are now fitted to standard air compressors which guarantee oil separation of 5 p p m. Higher efficiency elements are also available for more stringent applications.

Vapour pressure of the lubricant is important as oil in the vapour phase is a gas and therefore mechanical separation is not successful in dealing with this aspect. When the ultimate in separation is desirable, therefore, cooling of the gas before final separation is used and, again using air as the reference gas, oil separation down to 0.003 p p m can be achieved, i.e. virtually oil free.

## 5.9 Condition Monitoring

It is possible to fit to oil injected screw compressors the same condition monitoring equipment already discussed with regard to oil free screw compressors. However, there is one major area of difference. In the case of the oil injected screw design the outer casing of the compressor is a pressure wall as the casing is subjected to suction pressure. This means that all connections or probes for bearing temperature or vibration detection must be sealed as they pass through the casing and must be suitable for the pressure difference.

Externally mounted vibration detectors do not require any special consideration and many are in use. The chart shown in Fig. 7 earlier is equally applicable to this type of compressor.

## 6. APPLICATIONS

The following figures are illustrations of screw compressor plants of the various types discussed in the text above.

Fig. 16 shows a combined oil free/oil injected screw compressor installation. These units are used on liquid gas tankers for the purpose of re-liquefying the gas which boils off due to heat leakage into the cargo while it is in transit. Many such plants have been installed by Kvaerner Kulde A/S, Oslo, Norway.

The plants operate on a cascade cycle. The first stage oil free compressor compresses the cargo boil-off gas and passes it through a low temperature cascade heat exchanger for re-liquefaction. The cascade heat exchanger is the evaporator for the high stage part of the cycle. The high stage cycle can operate on any suitable refrigerant, and in this case R22 is used.

Fig. 17 shows a Mollier chart representation of the above system and Fig. 18 a diagrammatic representation of the plant involved.

In Fig. 19 the oil free propane compressors operate with suction temperatures down to -40°C (-40°F). The two compressor packages are located at the Sullom Voe oil terminal in the U K.

The compressors were manufactured with high nickel content steel casings, jacket cooling is with glycol and they are fitted with bearing temperature detectors and a full suite of vibration probes. The total installation is flameproof.

## 7. CONCLUSIONS

The text of this paper has described the various type of screw compressor which have been developed over the last 10 years or so. It has been shown that these compressors are entirely rotary in operation, have very few moving parts, are robust, can handle liquids, and have a high operating efficiency which is also maintained in service due to the lack of wearing parts. The compressors have been progressively developed over the years for improved efficiency, longer life and ease of maintenance.

In the context of the process and petro-chemical application, appropriate condition monitoring equipment can be fitted and many such plants are in operation.

Some of the various types of applications described are shown in the installation photographs, demonstrating that it is not just claimed that the compressors are suited to this type of application but have been proved to be so by actual running experience.

## 8. ACKNOWLEDGEMENT

The author would like to record his thanks to Howden Compressors Limited for permission to publish this paper.

<u>REFERENCES</u>

1. Lyshom A.J.   A new rotary compressor.
   Proceedings of the Institution of Mechanical
   Engineers, Volume 150 No. 1. 1943.

2. Perry E.J., Laing P.D.   Positive displacement rotary compressors as applied to
   refrigeration.   Proceedings of the
   Institute of Refrigeration 1960/61.

3. Laing P.D., Perry E.J.   The development of
   oil injected screw compressors for
   refrigeration.   Proceedings of the
   Institute of Refrigeration 1963/64.

4. O'Neill P., Beatts W.   The oil free screw
   compressor.   Proceedings of the Institution
   of Mechanical Engineers 1970.

5. O'Neill P., Briley G.C.   Design of the oil
   injected refrigeration screw compressor and
   its effect on reliability and maintenance.
   Proceedings of Purdue University Compressor
   Technology Conference 1972.

6. O'Neill P.   The screw compressor.  A short
   history of its development and its application to the fields of air conditioning
   and mine cooling.   Proceedings of South
   African Institute of Mechanical Engineers
   1977.


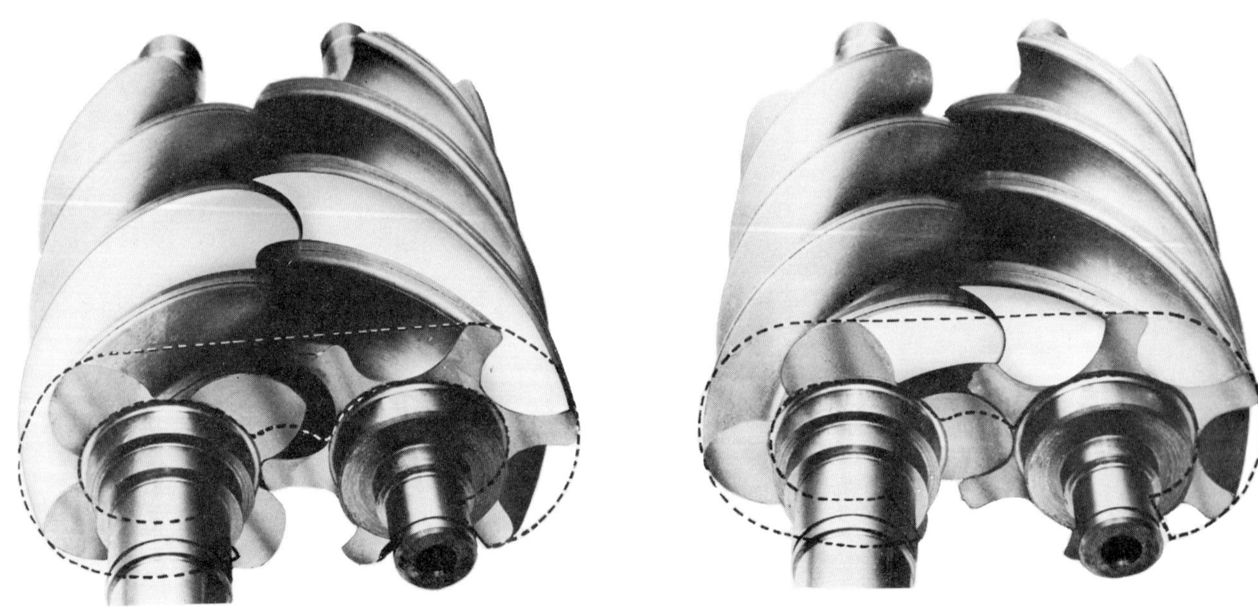

Fig 1   An illustration of the operation of a screw compressor

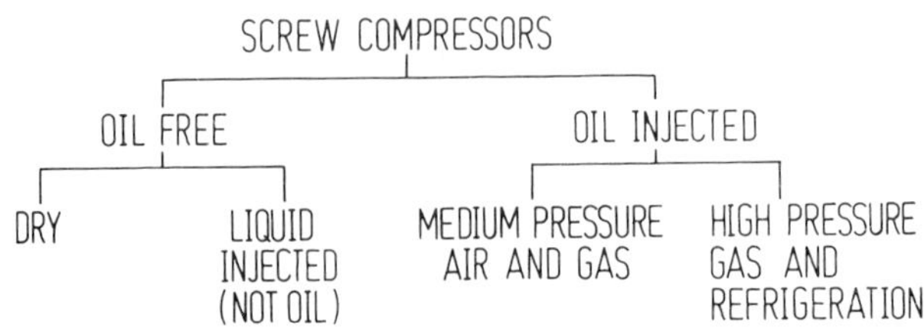


Fig 2   The relationship between different types of  screw compressor

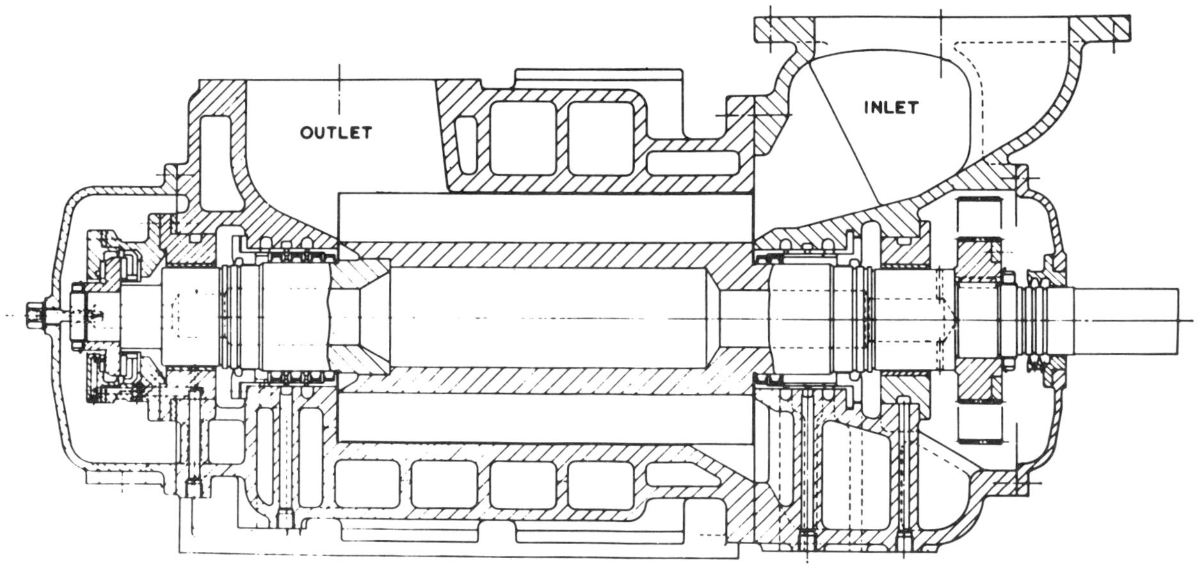

Fig 3   Cross section of Howden oil free screw compressor

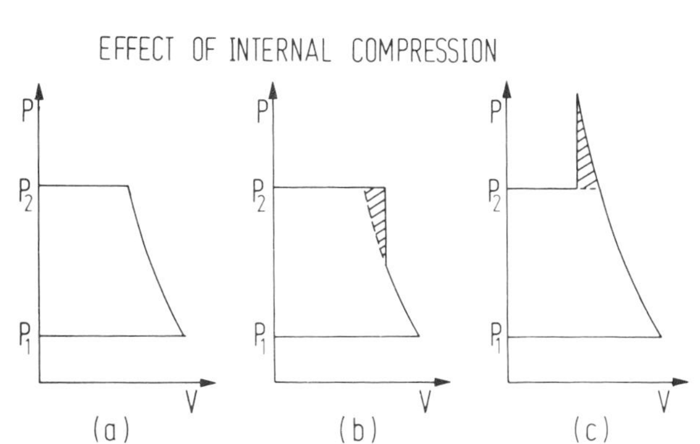


Fig 4   *p, v* diagrams of internal compression

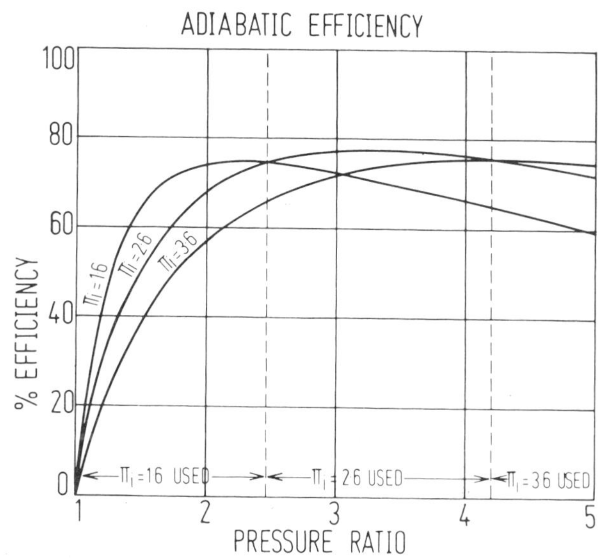


Fig 5   The effect of different built-in compression ratios

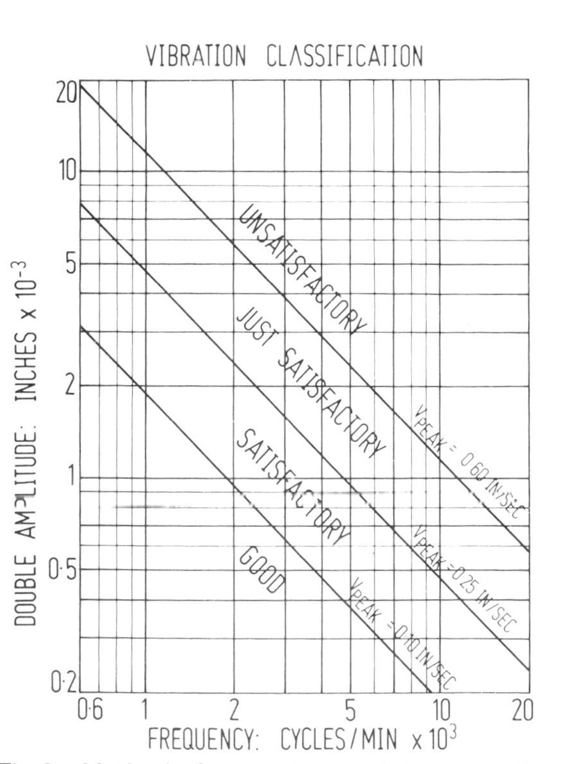


Fig 6   Method of assessing peak to peak vibration measurements

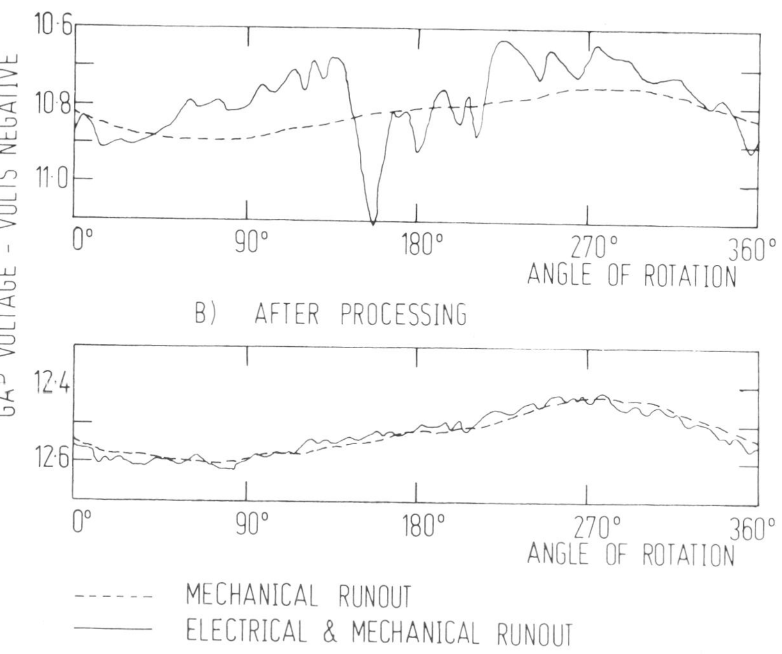


Fig 7   Vibration probe read outs of compressor shafts

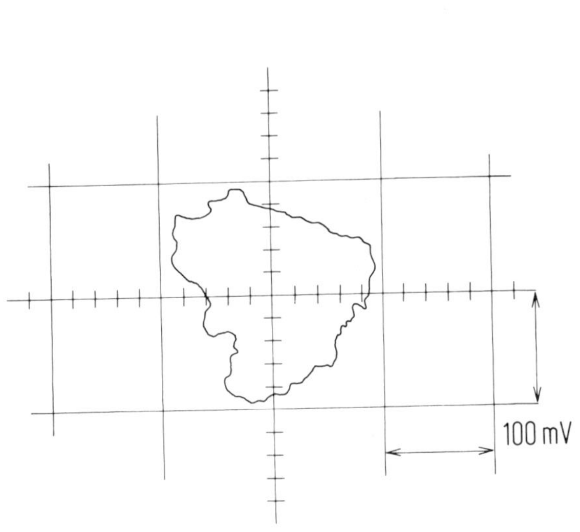

Fig 8    Lissajous figure from vibration probes of a processed shaft

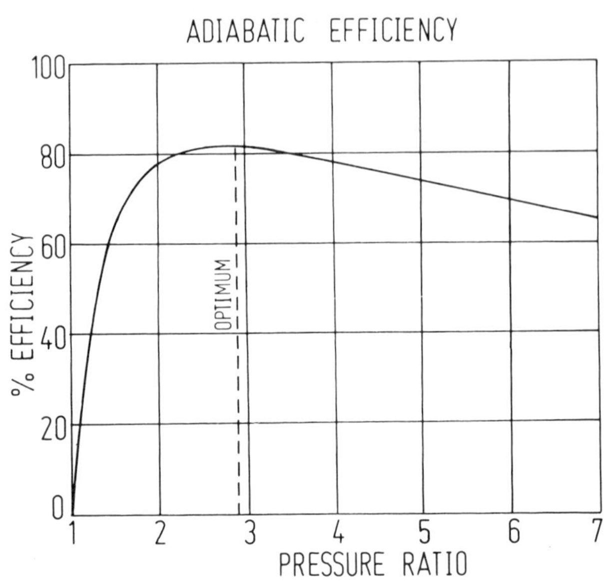


Fig 9    Typical efficiency graph of oil injected screw compressor

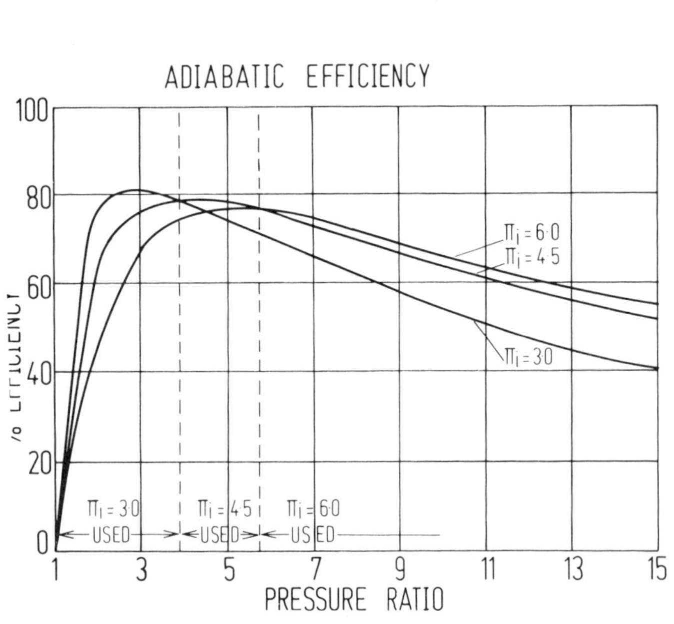


Fig 10    Typical efficiency graphs showing effect of built-in compression ratios in oil injected screw compressors

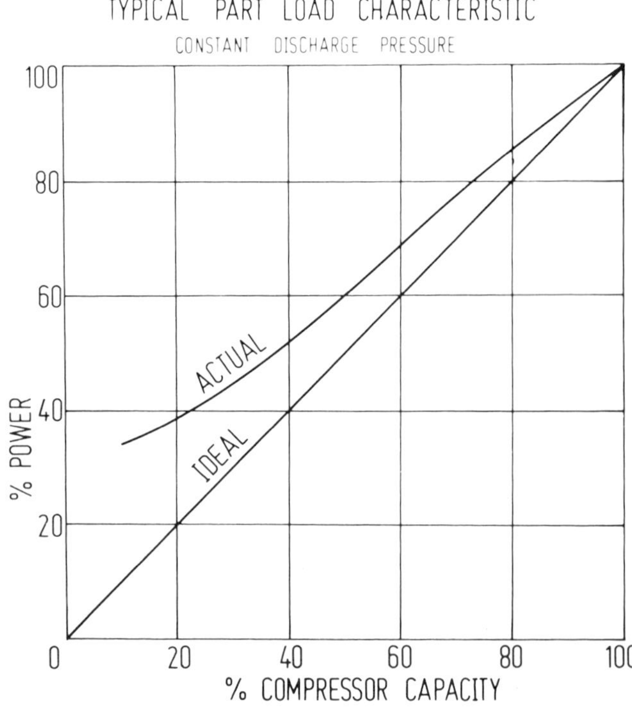


Fig 11    Typical slide valve part load characteristic with maintained discharge pressure

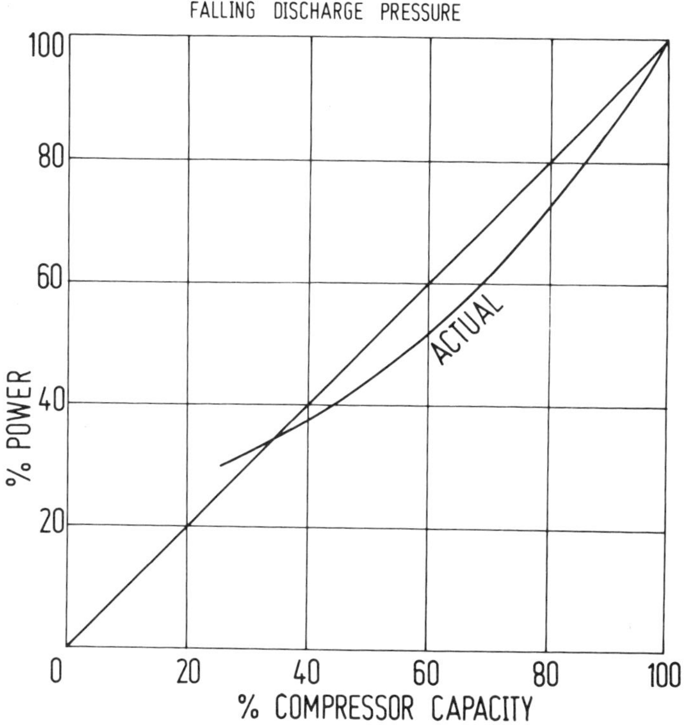



Fig 12   Typical slide valve part load characteristic
allowing for some fall in discharge pressure

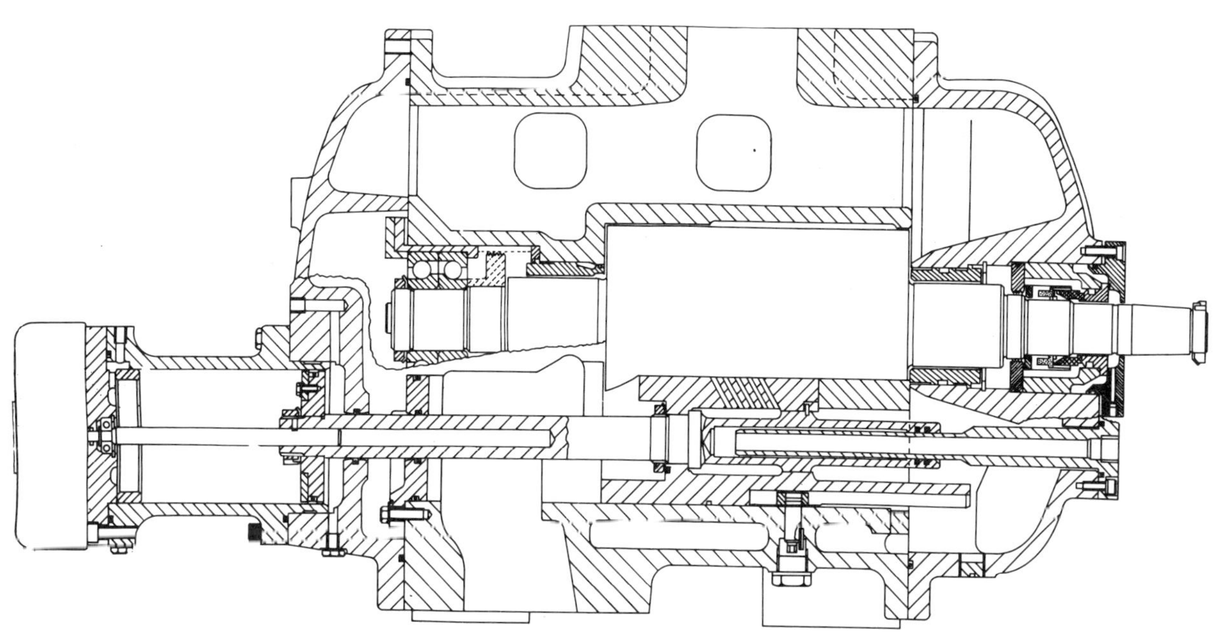

Fig 13   A vertical cross section of a Howden oil injected screw compressor incorporating slide
valve capacity control

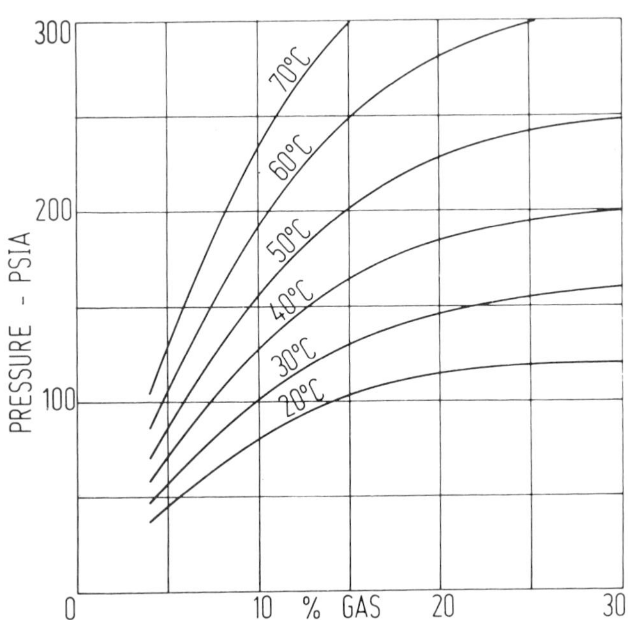


Fig 14    Typical hydrocarbon gas/oil mixture relationship for various temperatures

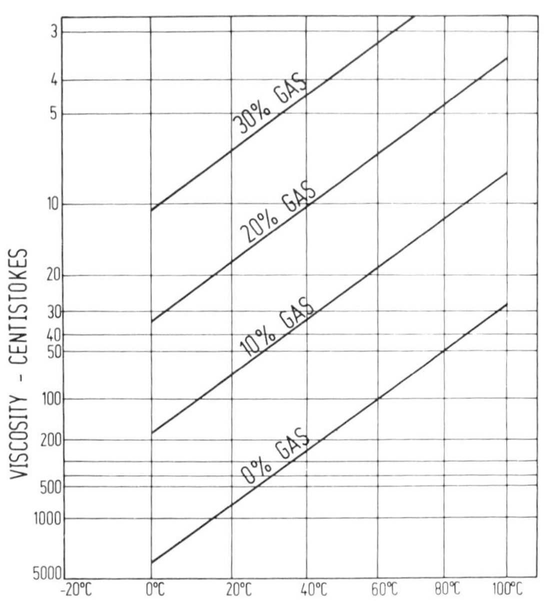


Fig 15    Typical effect of dilution on viscosity of hydrocarbon gas/oil mixture shown in Fig 14

Fig 16    A combined oil free/oil injected Howden screw compressor installation in a hydrocarbon gas reliquefaction plant

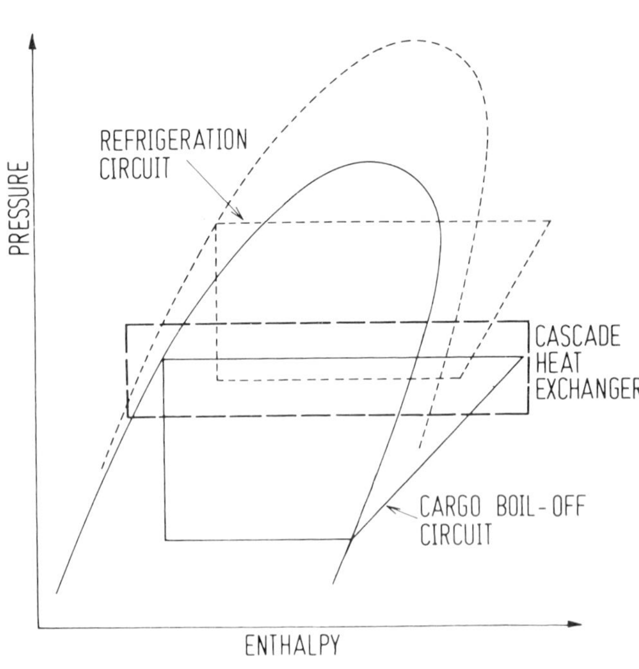


Fig 17    Mollier chart representation of cascade reliquefaction system

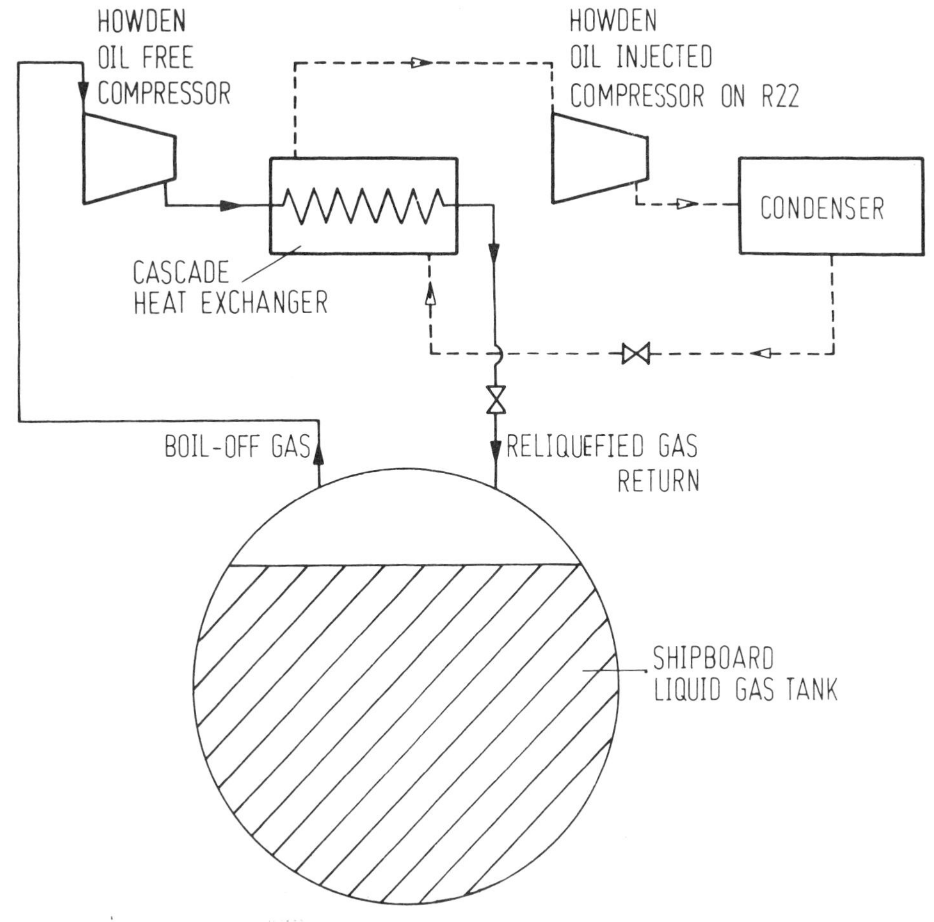


Fig 18    Diagrammatic representation of reliquefaction system

Fig 19    A Howden oil free size H408 screw compressor skid mounted package handling propane

# C56/81

# A comparison of the characteristics of screw and centrifugal compressors and a method of comparative assessment for maximum energy saving


P A O'NEILL, BSc, DMS, CEng, FIMechE
Howden Compressors Limited, Glasgow



SYNOPSIS
In the present world situation where energy is both short in supply and increasingly expensive, it is important that industrial equipment is selected with minimum energy consumption as one of the most important selection criteria.

This paper compares the operating characteristics of oil injected screw and centrifugal compressors and proposes a method of comparative assessment taking account of operating conditions over a full year. The results show a significant difference from those based on methods generally used at present.


## 1. INTRODUCTION

In selecting compression equipment for new installations it is often necessary to compare centrifugal compressors with screw compressors. The characteristics of these two machines are quite different, and as a result, the effects of changes in operating conditions are also quite different. This paper reviews these differences with respect to centrifugal compressors and oil injected screw compressors with slide valves at full and part loads and suggests methods of equipment selection which take into account the effects of changes in running conditions and their effect on operating costs.

## 2. OPERATING CHARACTERISTICS

### 2.1 Screw Compressors at Full Load

The screw compressor is a positive displacement machine. This means that it draws in gas, traps it and compresses it within the compressor (the built-in compression ratio effect) and pushes it out of the discharge. The pressure developed at the discharge is that pressure necessary to move the gas out of the compressor and no more. Therefore, if the discharge of the compressor is simply an open pipe, discharge pressure will be atmospheric pressure. If the discharge is into a pipeline at a given pressure, or into a condenser, the discharge pressure will only be the pressure necessary to transfer the gas into that system. The significance of this characteristic is that the power absorbed by the compressor is proportional to the discharge pressure.

The action of drawing in the gas, trapping, compressing, and discharging it, is carried out by the physical components of the compressor. Even though the action is entirely rotary there is no question of developing a pressure or head due to operating speed. Only the pressure necessary to discharge the gas is developed regardless of the compressor speed, and in fact regardless of the diameter of the rotors or the density of the gases at suction either.

Figure 1 shows a graph of power and volume of a screw compressor operating at fixed speed with a fixed suction pressure on a basis of discharge pressure.

These characteristics give the screw compressor considerable flexibility and are most important in operational power saving.

It should be noted that there is no 'clearance volume' in a screw compressor as occurs in a reciprocating compressor which is also a positive displacement design. The screw compressor draws in, compresses and discharges the gas by virtue of the helical shape of the meshing rotors (Fig.2) and the shape of the rotor casings and gas ports. There are no suction or discharge valves necessary and, therefore, no gas volume left in the machine at the end of the compression cycle which would be associated with any such valves. This ensures that limitations due to residual clearance volumes and associated re-expansion and re-compression do not occur and very high operating pressure ratios are quite feasible and practical. A more detailed explanation of these features is provided in other technical papers on this subject. The volume handled by a screw compressor is directly proportional to rotor speed. Figure 3 shows a graph of power and volume of a screw compressor on a basis of speed. The effect of speed was mentioned briefly earlier.

### 2.2 The Screw Compressor at Part-Load

Screw compressors can be operated at part-load in various ways. It should also be mentioned that there are different types of screw compressor, oil free, and oil injected. The comments in the previous section are true of both types, but the part-load situations differ. In this discussion the part-load characteristics are those of the oil injected design with a slide valve as this is the type most normally coming into competition with the centrifugal compressor.

#### Variable Speed

The ideal method of operating a screw compressor at part-load is by speed variation. Since the compressor is positive displacement and can, therefore, develop full discharge pressure regardless of speed, there are no limitations due to pressure considerations. Also, volume and power are directly proportional to speed.

Figure 4 shows a typical part-load characteristic with variable speed drive where suction and discharge pressures remain constant during off loading as would occur in a process compressor application.

The other extreme is where both suction and discharge pressures alter as the compressor is off-loaded. This occurs in the case of a refrigeration process such as a brine or water chiller. Here the condensing pressure falls as the condenser, with its fixed heat exchange surface area and constant water flow, becomes more effective at the reduced gas mass flow at part load. Similarly the suction pressure increases slightly as the heat exchanger (evaporator) on the suction also becomes more effective for the same reasons. Figure 5 shows the characteristics for this latter case, based on refrigeration capacity versus power. This is excellent where a suitable variable speed drive unit is available such as a steam or gas turbine, gas or petrol engine. Variable speed electric motor drivers are also available but are not normally viable alternatives on economic considerations.

### Constant Speed

In the case of constant speed drives the capacity of the screw compressor is controlled by means of an internal slide valve. This gives stepless volume variation from 100% down to 10%. Full discharge pressure can be maintained at all times.

The performances achieved during part-load operation depend on a number of factors. The simplest case is where both suction and discharge pressure remain absolutely constant as occurs in many process type applications. Figure 6 shows the off-load characteristics for this case. The other extreme as mentioned earlier is where both suction and discharge pressures alter as the duty changes as in refrigeration plant. Figure 7 shows the off-load characteristics applicable in this case based on refrigeration capacity.

Other applications will lie somewhere between these limits as suction and discharge conditions dictate.

### 2.3 Centrifugal Compressors at Full Load

The centrifugal compressor is an aerodynamic machine. It carries out the process of compression by generating a kinetic head or 'momentum' in the gas as it passes through the spinning impeller and by 'slowing down' the gas after it leaves the impeller converts the kinetic energy in the gas to pressure energy or in other words 'head'.

The discharge pressure capability of a centrifugal compressor is, therefore, directly related to the diameter of the impeller, its speed of rotation, and the density of the gas it is handling.

As there is a clear gas connection through the compressor from discharge, to the suction, via the impeller, changes in discharge conditions have a direct effect on the suction volume. It should be noted that there are no suction or discharge valves involved, just as there are no such valves in a screw compressor either.

Aerodynamic compressors can suffer from a phenomenon known as 'surge' and this is an important consideration with regard to selection of the compressor operating point.

It has been mentioned that a centrifugal compressor develops a 'head' or discharge pressure related to gas density and impeller diameter and speed. If, for any reason, the pressure of the system into which the machine is discharging is above the head developed, a back flow occurs from the system into the impeller. This has the effect of increasing the density in the impeller, the head capability then increases and there is an appropriate discharge into the system again. However, as soon as the density in the impeller reverts to normal, the head falls and back flow occurs again. This pressure and flow oscillation occurs at high speed and causes severe vibration and high temperature and the compressor cannot be permitted to operate under these conditions. Surge can be caused by increased back pressure above compressor 'head', or reduction of head developed due to reduction in suction gas density or suction throttling, or reduction of impeller speed.

Centrifugal compressor characteristics tend to be drawn on a different basis from those for positive displacement machines, due to the differing characteristics noted above. A typical centrifugal compressor characteristic is shown in Fig.8. The effect of varying discharge pressure (head) on suction volume is evident and the area of surge also. The angle of the characteristic line varies with the angle of impellor blading but the principle remains the same.

The power absorbed by a given centrifugal compressor can be seen not to fall in line with a reduction in discharge pressure. The simple graph shown in Figure 8 indicates that both volume and power would increase with a reduction in discharge pressure. However, an increase in volume is often not acceptable. A process unit may be designed to handle the volume generated by a plant and an increase in compressor suction volume might reduce the pressure of the process unacceptably and might cause problems with surge in extreme cases due to reduced head capability. In a refrigeration plant the suction conditions are the controlled area and the compressor throughput is adjusted to maintain them regardless of variation in condensing conditions. Figure 9 shows a centrifugal compressor performance with varying discharge pressure and suction volume controlled to design conditions. The relatively flat power characteristic results from the compressor generating a discharge pressure or head to suit the design conditions which (at other duties) is in excess of requirements and the extra power is, therefore, wasted.

### 2.4 Centrifugal Compressors at Part Load

There are several ways of off loading centrifugal compressors. However, care has to be taken to avoid moving the operating point into the area of surge and flexibility is limited for this reason.

### Variable Speed

Consider firstly speed variation and the limiting cases already considered for screw compressors, namely suction and discharge pressures remaining constant when capacity is varied (process compressor) and suction pressure increases and discharge pressure falls with capacity change (refrigeration process such as brine or water chilled).

Figure 10 shows a typical characteristic of a centrifugal compressor with variable speed. This indicates that with constant suction and discharge pressure the compressor could be off loaded by speed reduction to approximately 80% volume. Of course, this does not mean that further off loading is not possible. Gas recirculation can be used to give control down to 0% flow but with no power saving associated with it. Also some further reduction might be possible with variable inlet guide vanes used at the reduced speed but very little improvement could be expected.

At the other end of the spectrum where both suction and discharge pressures vary during off loading (refrigeration and air conditioning applications) the situation is much better. The critical area is the reduction of discharge pressure. If this occurs to a sufficient degree, off loading down to 10% is possible without surge. Figure 11 shows a typical characteristic under such optimum conditions.

Constant Speed

Where a constant speed drive is being used such as an electric motor, other methods of off loading have been developed. Variable Inlet Guide Vanes are used to control the gas flow into the eye of the impeller. These have the dual effect of throttling the incoming gas into the impeller and imparting an angle of swirl to it to give a clean flow into the impeller. Using this system a typical characteristic would be as shown in Figure 12. With this system applied to the two limiting applications, the following results are achieved. With fixed suction and discharge pressure, off loading by V I G Vs would be limited to around 70% volume to avoid surge. Further off loading would require gas recycling which is fully variable with regard to gas flow but without power saving. Figure 13 shows the resulting characteristic.

In the other limiting case with both suction and discharge conditions varying as the compressor is off loaded it is possible to off load to as low as 10% as shown in Figure 14. This is the best situation that can be achieved. It must be stressed that in many instances it is not possible to off load to anything like as
a percentage without moving into the surge area and it is necessary to resort to recycling to achieve the flexibility required.

3. COMPARISON BETWEEN SCREW & CENTRIFUGAL COMPRESSOR CHARACTERISTICS

As noted earlier the characteristics of these two types of compressor tend to be drawn on a different basis. In the following comparisons the graphs have been redrawn on common basis to make them directly comparable.

3.1 Compressors Operating at Full Load Fixed Suction & Discharge Pressures

Consider firstly the application where the compressors are operating at their full load condition and under what could be called 'process conditions' where both suction and discharge pressures remain constant. Under these conditions comparison is simple as the absorbed powers are constant at all times and selection of equipment can simply be based on first cost and running costs with some allowance being made for complexity, ease of maintenance etc. It will be found that the performances of screw compressors and multistage centrifugal compressors in the areas where both are suitable will be of similar levels Some actual examples are discussed later.

3.2 Variable Discharge Pressure

Now let us consider the application where the compressor suction pressures are constant but there is variation in the discharge pressure. This is a typical case in refrigeration plants. The process requires cooling to a certain temperature e.g. a chemical cooling load or a brine or water chilling duty, and the suction duty is, therefore, constant and automatically controlled to match the design operating temperature. The condensing part of the plant where the plant heat is rejected, must be designed to cope with the worst set of circumstances foreseen to ensure that the plant can operate at all times of the year. This means allowing for the hottest time of the year, and perhaps also the highest ambient humidity also, where cooling towers or evaporating condensers are involved.

An example of a plant of this nature is that of underground mine cooling where the compression plant is located on the surface. These plants are usually required to operate at full design suction conditions to cool water to the specified temperature for passing underground to provide temperature control underground. However the condensing systems experience a wide range of conditions from the heat of the day to the cold of the night, and from summer to winter. Figure 15 shows the characteristics of the two types of compressor under these same conditions. This shows very important differences between the 'fixed head' pressure generation of the centrifugal and the 'discharge pressure only as high as necessary' operation of the screw. (Note: the centrifugal characteristic is valid for variable and constant speed. Change of speed or change of IGV angle both reduce efficiency by similar amounts as shown on figures 10 and 12).

3.3 The Effect of Fouling Factors

This is an area almost always ignored in comparative assessments but which can have surprisingly significant effects on absorbed power. Fouling factors are, of course, the allowances built into heat exchanger designs so that, when fouling to the level allowed for occurs, the heat exchanger will operate at its design heat transfer rate. Significantly from the time of installation up to this 'fouled' condition it will be more effective than at design conditions.

Table 1.

Example of Effect of Fouling Factors

|  | Design Conditions | Earlier Running Conditions |
|---|---|---|
| Evap. F.F. | (0.001 ) | (0.0005 ) |
| Cond. F.F. | (0.002 ) | (0.0005 ) |

Refrig. Cap.    7480kWR (2125TR)    (7480kWR (2125TR)
Evap.Temp.     -0.7°C (30.8°F)     1.2°C (34.1°F)
Cond. Temp.    45.2°C (113.4°F)    38°C (100.4°F)
Cent. Comp.Pr.1996kW (2675 hp)     1711kW (2293 HP)
Screw Comp.Pr.1989kW (2666 hp)     1538kW (2061 hp)

While the fouling factors used are on the high
side, they are intended to demonstrate the importance of this rarely considered factor.   The
screw compressor shows to advantage in energy
saving.

### 3.4 Variable Suction Pressures

The remaining possibility of fluctuation
in suction pressures and/or suction densities
which can occur in some process type plants,
particularly where the gas is a mixture of hydrocarbons is more readily handled by the positive
displacement screw machine.   A change in suction
density or pressure causes a centrifugal machine
to develop a different 'head' (discharge pressure)
with associated power increase or decrease surge
limits permitting even though the system discharge pressure does not change.   The screw is
not affected in this way and the effect of power
under these circumstances would be negligible.

### 3.5 Compressors Operating at Part Load
### Fixed Suction and Discharge Pressures

Using the variable speed characteristics
of the two types of compressor, Figure 16 has
been prepared.   In order to achieve the full
range of off loading with the centrifugal design
recycle has been used when surge prevents further
speed reduction.   This figure shows a very large
factor in favour of the screw design under these
conditions.

Figure 17 is the equivalent graph, this
time using the fixed speed characteristics.   As
using the integral slide valve is not as efficient as speed variation the screw compressor curve
is less favourable but is still very much better
than that of the centrifugal design.

### 3.6 Variable Suction and Discharge Pressures

Figure 14 and 19 show the comparison
graphs for variable speed and fixed speed drives
respectively.   It must be emphasised again that
the centrifugal characteristic is the optimum
assuming that the fall in system discharge pressure during off loading is sufficient to keep the
operating point out of the surge area.   It is
often not possible to do this, in which case recycling becomes necessary with consequent reduction in power saving.

### 4.0 RANGE OF COMPRESSORS AVAILABLE

Before any attempt is made to compare the
relative merits of screw and centrifugal compressors for any application it is worth while to
consider the range of sizes available.   There is
a considerable overlap between the ranges of the
two types on the market.  The screw compressors
available from different manufacturers differ in
sizes and the following comments apply to those
manufactured by the Authors Company.   These
screw compressors cover a range of suction volumes up to 10,000 m³/h (6000cfm) and discharge
pressures up to 24 bar g (35 psig).  This
is equivalent to water chilling package sizes up

to approximately 10 000 kWR (3000 TR).

Centrifugal compressors are available in
sizes from perhaps a quarter of this up to very
much larger.

Where the required plant size is outwith
the screw compressor range, plant selection is
made on comparison between different centrifugal
designs.  It is still worthwhile carrying out a
careful check of the various operating duties involved and the characteristics of the different
designs offered as there may be significant gains
to be had at conditions other than the design
maximum.

### 5.   SUGGESTED METHOD OF PLANT ASSESSMENT

### 5.1 Present Methods of Assessing Competitive
### Equipment

It has been shown above that two plants may
have virtually identical absorbed powers at one
condition and quite different ones when these
conditions alter.

It is, therefore, desirable to have some
method of taking this circumstance into account.
In plant selection the two main factors are
purchase price and running cost.   There are many
other important features such as reliability,
ease of maintenance, availability of spares and
service, size, weight, country of origin etc. but
once these have been reviewed the major two mentioned become the deciding factors.   In the present climate of ever increasing energy prices, the
running costs of a plant are becoming a very
major part of plant selection.  One method of
tying together purchase and running costs, which
is gaining popularity, is as follows.  The various
acceptable bids are compared on power absorption.
A value is given to each kilowatt of power used,
based on cost per kilowatt hour over a number of
years operation, sometimes 3 years sometimes 5
years or more.  Each bid is compared against the
lowest power absorbed bid and the kilowatt power
difference multiplied by the value given to each
kilowatt hour and added to the bid price.  The
totals thus obtained are then used for bid selection.

This is a sound basis for best equipment
selection but as normally used suffers a very
serious flaw which tends to make the results of
very doubtful validity.   This flaw is that the
bid comparison is based on the powers absorbed at
the design point for the plant.

In sizing any plant and its drive it is
obviously necessary to rate them both to be capable of coping with the most arduous conditions
foreseen.  If we consider the case of a refrigeration plant this will mean  that the plant must
be able to operate against the maximum condensing
conditions which will normally occur on the day
with the maximum wet bulb temperature.  This then
decides the size etc. of the compressor and the
power of the driver.  This is the design point.
However, the plant will run at lesser conditions
for the majority of its operating life.  As we
have seen above, the absorbed power characteristics of screw and centrifugal compressors
operating away from their maximum design point
differ greatly.  It is, therefore, necessary to
use a slightly more sophisticated method to
obtain more genuine operating costs.   The final

assessment would still use the cost per kilowatt hour usage but would be much more accurate.

### 5.2 The Inaccuracy of Averaging Operating Conditions

It is quite inaccurate to use simple averaging of operating conditions, whether pressure or temperature.

Consider the simplified case of a plant operating at, say, 43°C (110°F) condensing temperature for half of its operating time and 24°C (75°F) for the other half with suction pressure constant in both instances. On a discharge pressure basis the first duty would be 100% and the second would then be 60%. A simple average of discharge pressures would give the following results:

Table 2.

Average Method

| | |
|---|---|
| Average Dis.Pressure relative to design condition | 80% |
| Centrif. Power on this basis(fig.15) | 86% |
| Screw power on this basis (fig.15) | 81% |

Actual Method

| | | |
|---|---|---|
| Actual Dis.Pressure for 50% time | = | 100% |
| ∴ Power usage of Centrif. over this period | | 50% |
| Power usage of Screw over this period | | 50% |
| Actual Dis.Pressure for other 50% | = | 60% |
| ∴ Power usage of Centrif.for this time (fig.15) | = | 80% |
| Power usage of Centrif. over this period | = | 50% of 80%=40% |
| Power usage of Screw for this time (Fig.15) | = | 63% |
| Power usage of Screw over this period | = | 50% of 63%=31.5% |
| Total power usage of Centrif.= 50 + 40 | | =90% |
| Total power usage of Screw = 50 + 31.5 | | =81.5% |

The accurate method of working shows up almost twice the power saving compared to the averaging approach. The same errors occur if averaging is carried out on a temperature rather than pressure basis and also on part load operation.

It is also important to note that the more running that occurs at conditions other than 100% the greater the discrepancy will be.

### 5.3 'Actual Operating Review' Method

In order to obtain a reasonable order of accuracy a method which encompasses the total range of operating conditions should be used but which does not involve the use of averages and is not excessively complex. The method described below is one solution to this problem.

A city in the southern hemisphere which has a fairly wide range of climatic changes over the year has been used for this example (Johannesburg, South Africa). The application chosen is a nominal 8800 kWR (2500 TR) water chilling plant.

The condensing conditions will be dictated by the wet bulb ambient temperature at any given time. Figure 20 shows an annual record of the maximum and minimum wet bulb temperatures over a year. We have already seen that to use average values is invalid. It is necessary, therefore, to take account of the fact that the plant will operate against the maximum wet bulb temperatures during some part of the day and the minimum during some part of the night. Also the temperatures in both cases vary with the time of year as is clearly seen in Figure 20.

Analysing the chart (Fig.20) it has been possible to produce the following approximations. The year can be split into six roughly equal periods during which the wet bulb temperature could be said to lie in the listed bands. Typical equivalent condensing temperatures are shown also:

Table 3

| Wet Bulb Temperature | Condensing Temperature |
|---|---|
| 18 to 21°C | 37°C (98.4°F) |
| 14 to 18°C | 30°C (85°F) |
| 9 to 14°C | 23°C (73.4°F) |
| 6 to 10°C | 18°C (64.4°F) |
| −1 to 6°C | 14°C (57.2°F) |
| −6 to −1°C | 12°C (53.6°F) |

The plant must be rated to give 8800 kWR (2500TR) at 37°C (98.4°F) condensing temperature. The condensing temperatures are conservatively selected and could be reduced by increased condenser sizing. The conditions are, however, the same for both types of machine.

Using the proposed 'Actual Operating Review' method of assessment the following results are obtained:

From Figure 15

Table 4

| Condensing Temperature | Percentage of Design Duty Power | |
|---|---|---|
| | Centrif. | Screw |
| 37°C | 100% | 100% |
| 30°C | 86% | 83% |
| 23°C | 82% | 71% |
| 18°C | 80% | 62% |
| 14°C | 78.5% | 57% |
| 12°C | 78% | 54% |

The full comparisons are shown in Figure 21.

As the values on Figure 21 show the more detailed analysis of actual running condition yields significantly different results from the simple full load comparison. Very substantial financial benefits can be obtained as a result of the small amount of extra work required in analysis.

### 5.4 Principles of 'Actual Operating Review' Method

A summary of the 'Actual Operating Review'

method of assessment is as follows:

a) Obtain information on the range of climatic conditions over which the plant will be called to operate. (This applies predominantly to refrigeration/air condition plant. If process plant will experience fluctuations for any other reason a similar review of pressure changes should be undertaken). Details of maximum and minimum wet bulb temperatures which occur over a year can usually be obtained from local weather centres or airports, similar to the information shown in Figure 20.

b) Break down the temperature record into periods of approximately equal duration over certain temperature bands. It may be found that the logical break down suggests that over some temperature bands the time period should be approximately twice other bands. In this case, for simplicity in evaluation, they would be considered as two periods at the same temperature level.

It is important to remember that banks at maximum and minimum must be selected, average temperature values must NOT be used.

As the high temperatures occur in daytime and the lows at night (in general) it is satisfactory to assume that when the appropriate bands are chosen, half of the time is spent at maximum conditions (daytime) and half at minimum (night-time). If the plant concerned only runs during the day or for less than the full 24 hours/day, the relevant periods of operation can be balanced to allow for this. Daytime only running would eliminate the need to use the minimum conditions in the study replacing these with minimum daytime temperatures.

A cross check of the selected time periods is that the sum of hours run at the various conditions should be equal to the planned annual total running hours.

The total number of periods used in the review should, if possible, not exceed 6 to 8 in number to avoid excessive complexity.

c) Obtain a performance quotation from the various compressor/plant suppliers for a plant to meet the maximum design specification. Obtain the performance of this plant, offered to meet the maximum condition, at the other operating bands identified in (b). This may be either on the basis of the same plant operating against the other wet bulb temperatures, or discharge pressures.

d) Prepare relative assessments on 'Actual Operating Review' method which gives true average power usage over a years running. This simply involves adding up the various actual operating powers over each period and dividing by the number of periods. This is the reason for selecting periods of equal duration.

Where performances at the various reduced condensing temperature conditions are not readily obtainable the graphs included in this paper may be used with a fair degree of accuracy.

## 6. OTHER AREAS OF COMPARISON

### 6.1 Operating Speed

Screw compressors, being positive displacement, have a fairly wide range of efficient operating speeds. The type considered in this paper, the gas and refrigeration compressors, operate within a rotor tip speed range of approximately 15 to 60 metres per second. They can, of course, be driven by variable speed drive units such as gas turbines, engines etc., but normally they are directly coupled to electric motor drivers. No speed increasing gearbox is normally necessary.

Centrifugal compressors are operated at a speed which enables the combination of impeller diameter, suction gas density and speed to develop the necessary head or discharge pressure required for the maximum design condition. In most cases this is considerably greater than that available from electric motor drivers and speed increasing gearboxes are used to achieve the necessary impellor speed.

### 6.2 Gases and Refrigerants Handled.

Screw compressors being positive displacement can handle almost any gas, ranging from the lightest, e.g., helium to the heaviest, e.g. propane.

Where a choice is available as occurs in a refrigeration plant, the heavier high pressure gases are adopted as more refrigeration output is obtained from any given screw compressor size. Typical refrigerants are, therefore, Ammonia ($NH_3$) R22, R12, propane, and propylene but many others are used for special applications.

Centrifugal compressors can handle a wide range of gases also but where a choice is available will tend to use the lighter low pressure gases such as R11, R12, R21, etc. Very light gases such as helium are difficult for centrifugal compressors since they rely on density as one of the major means of developing pressure and the density of the Helium is remarkably low.

A result of this 'natural selection' in refrigeration plants is that with screw compressors the refrigerant pipes are often of smaller diameter, higher pressure whereas those of a centrifugal are larger in size but low pressure.

Where pressures are sub-atmospheric any leaks will be inwards and difficult to locate, where the pressures are above atmospheric leaks are outwards and contamination of the system with 'non-condensibles' does not occur.

### 6.3 Surging

The phenomenon of 'Surging' is described in the text earlier. It is an occurrence which can only happen with aerodynamic compressors.

It is impossible for this to occur in any positive displacement compressor design, which includes screw compressors.

### 6.4 Mechanical Simplicity

Both types of compressor considered in this paper are relatively simple mechanically compared to many other types available in the market place

The screw compressor comprises basically two rotating meshing rotors, rigid in construction and dynamically in balance. The only other moving component is the slide valve which moves axially under automatic control to steplessly adjust throughput. The compressor is normally directly coupled to its driver which may be an electric motor or other prime mover.

A centrifugal compressor normally operates at much higher speed than the driver and a speed increasing gearbox is therefore necessary.

In order to be competitive, most applications use multiple staging. This can be carried out either by using several impellers mounted on the one shaft, the gas sometimes being intercooled between stages, or by several wheels on their own shafts rotating at different speeds all driven by gears from the one prime mover. Movable guide vanes are often incorporated for capacity control as are gas recycle systems to avoid surge problems.

Overall, therefore, a screw compressor installation has considerably fewer moving components than an equivalent centrifugal compressor plant.

## 6.5 Liquid Carry Over into Compressor

It occasionally happens that some liquid is passed directly into the compressor either due to a fault developing in the overall system or due to plant maloperation. This is more likely to happen in a refrigeration plant.

In the case of a screw compressor, a slug of liquid will cause no damage to the compressor.

A centrifugal compressor will suffer damage which may be severe.

## 7. CONCLUSIONS

1) Screw and Centrifugal compressors, although both entirely rotary in operation have quite different operating characteristics.

2) The range of sizes available of each type is different. Small screw compressors are marketed and large centrifugal, but in the middle range both types are in direct competition.

3) At the present time when energy costs are expensive and rising continually, it is important that plant selection takes energy usage into account in the most accurate manner practicable.

4) The 'Actual Operating Review' method proposed in this paper supplies a means of assessing annual running costs taking into account the most important variables involved. The example worked out in the text shows that simple comparison of running costs at maximum design conditions can be totally misleading. The more sophisticated method proposed provides a more accurate management tool, allowing full allowance to be made for current and increasing energy costs.

## 8. ACKNOWLEDGEMENT

The author would like to record his thanks to Howden Compressors Ltd., Glasgow, for permission to publish this paper.


## REFERENCES

1. Lyshom A.J. A new rotary compressor. Proceedings of the Institution of Mechanical Engineers, Volume 150 No.1. 1943.

2. Perry E.J., Laing P.D. Positive displacement rotary compressors as applied to refrigeration. Proceedings of the Institute of Refrigeration 1960/61.

3. Laing P.D., Perry E.J. The development of oil injected screw compressors for refrigeration. Proceedings of the Institute of Refrigeration 1963/64.

4. O'Neill P., Beatts W. The oil free screw compressor. Proceedings of the Institution of Mechanical Engineers 1970.

5. O'Neill P., Briley G.C. Design of the oil injected refrigeration screw compressor and its effect on reliability and maintenance. Proceedings of Purdue University Compressor Technology Conference 1972.

6. Aloi W.C. What you always wanted to know about Centrifugal Refrigeration Systems. Proceedings of Purdue University Central Chilled Water Conference 1976.

7. O'Neill P. The screw compressor. A short history of its development and its application to the fields of air conditioning and mine cooling. Proceedings of South African Institute of Mechanical Engineers 1977.

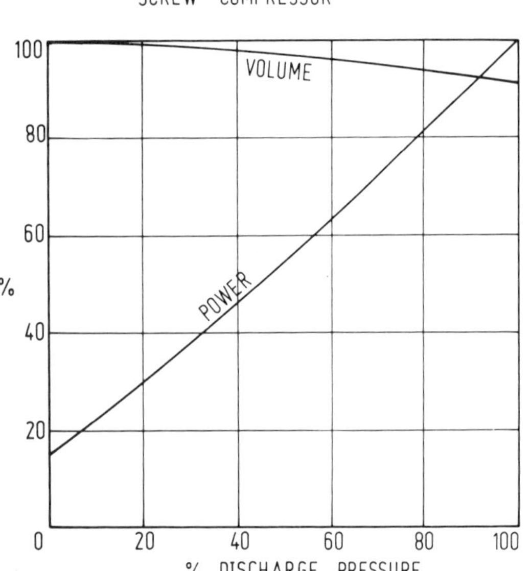


Fig 1   Power and volume characteristic at constant speed with fixed suction pressure

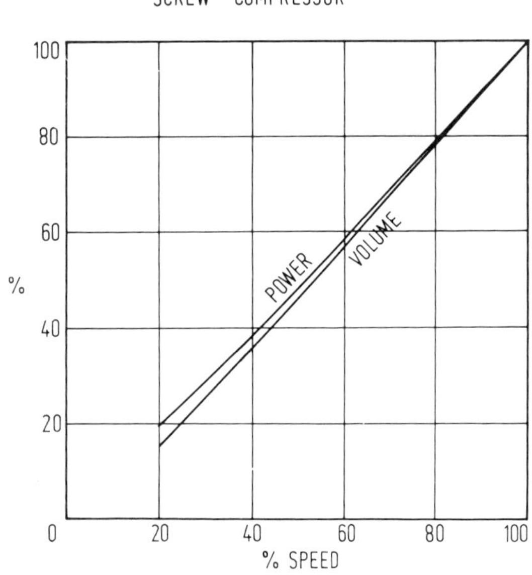


Fig 3   Power and volume characteristic at varying speed

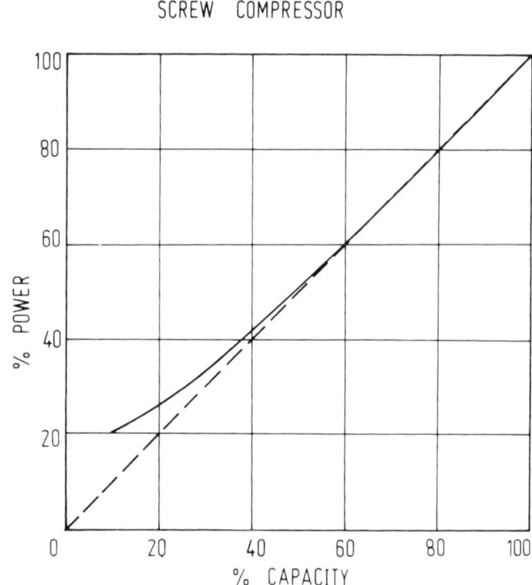


Fig 4   Part load characteristic with variable speed, suction and discharge pressures constant

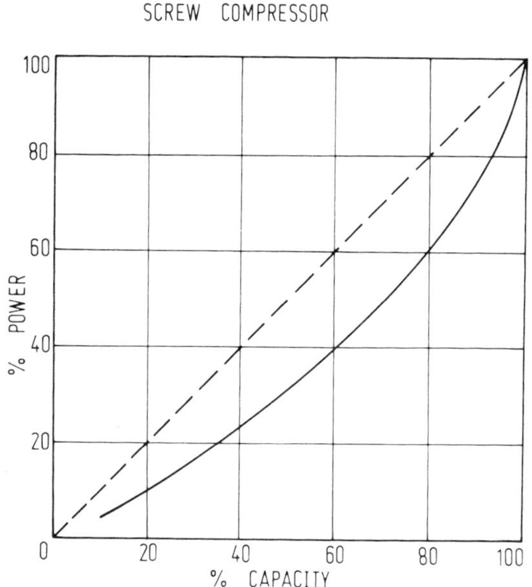


Fig 5 Part load characteristic at varying speed: suction and discharge pressures varying (see text)

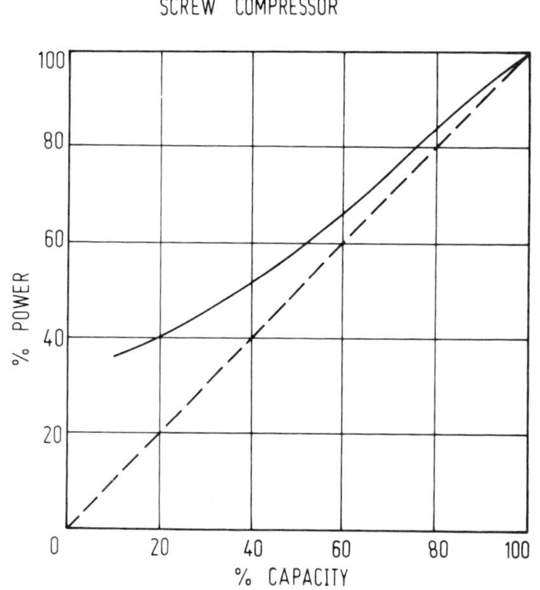


Fig 6 Part load characteristic at constant speed: suction and discharge pressures also constant

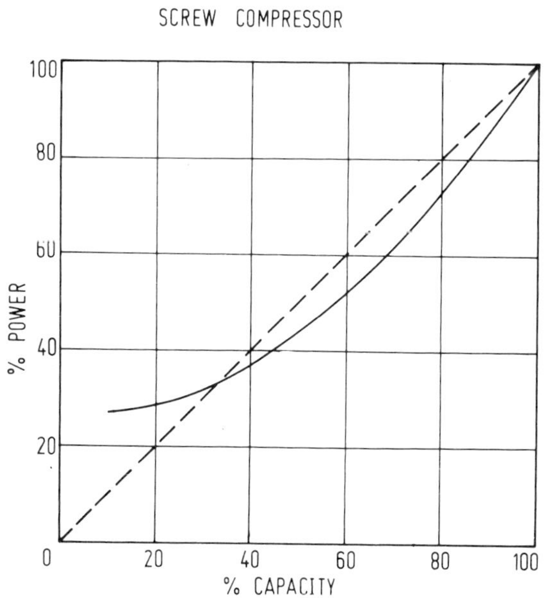


Fig 7 Part load characteristic at constant speed: suction and discharge pressures varying (see text)

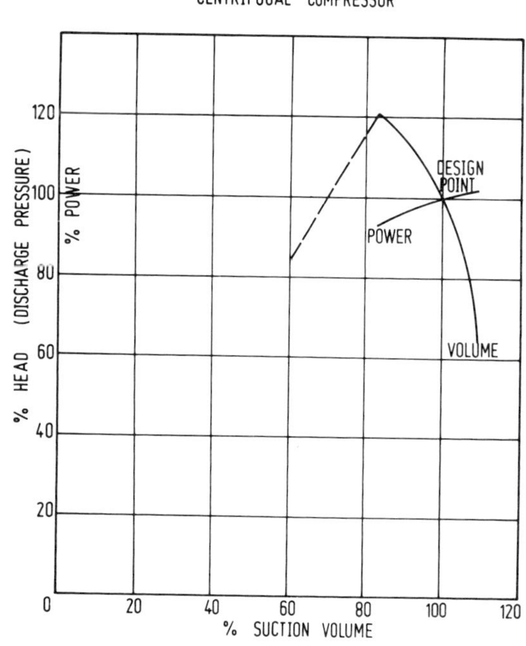


Fig 8 Pressure, volume and power relationship at constant speed

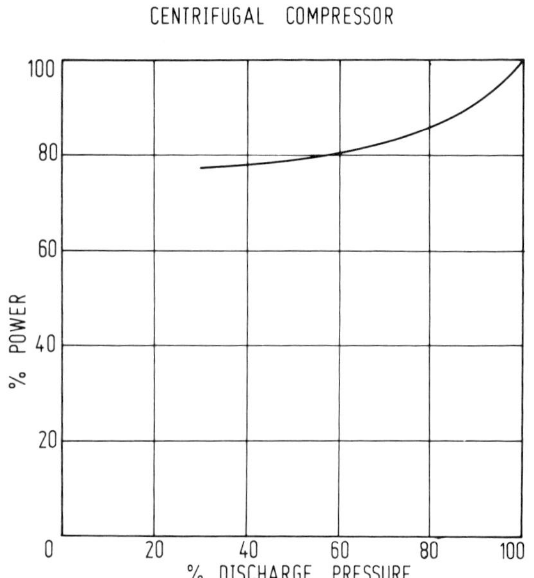


Fig 9    Power, discharge pressure relationship at constant speed with fixed suction pressure

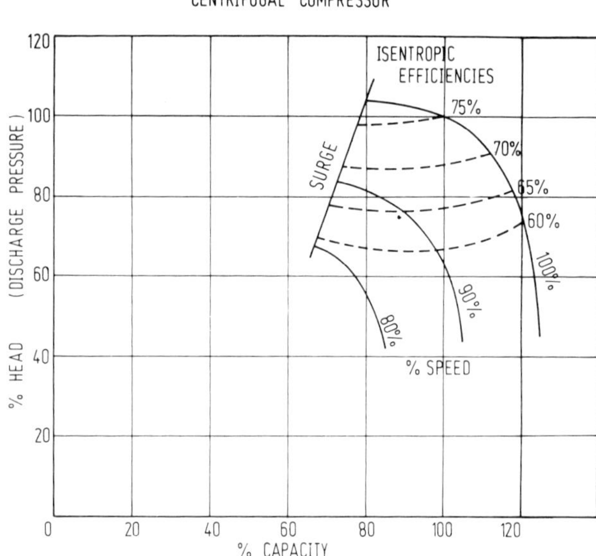


Fig 10    Variation in capacity with speed control

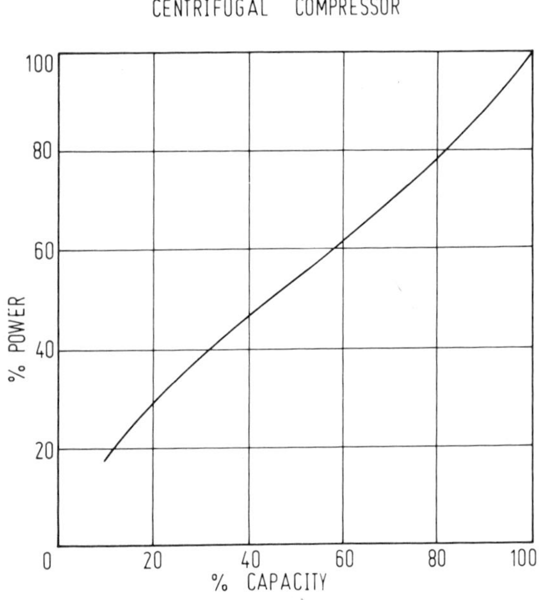


Fig 11    Part load characteristic at varying speed: suction and discharge pressures varying (see text)

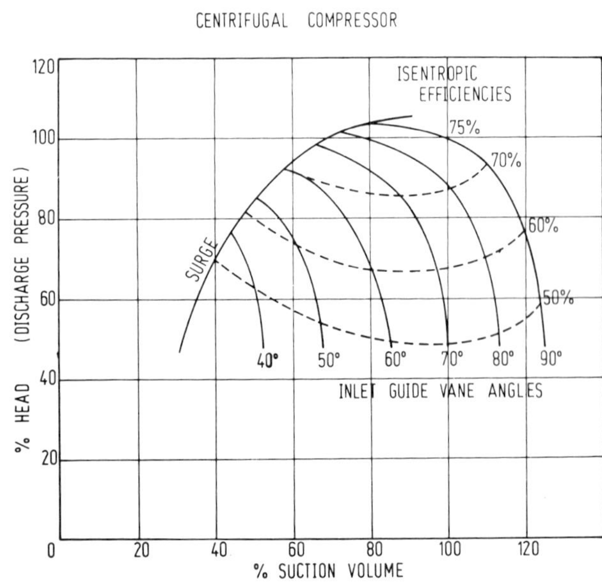


Fig 12    Variation in capacity with inlet guide vane control

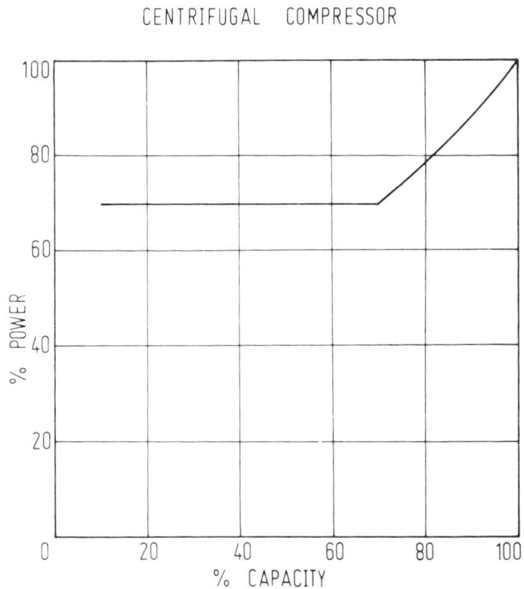


Fig 13    Part load characteristic at constant speed: suction and discharge pressures also constant

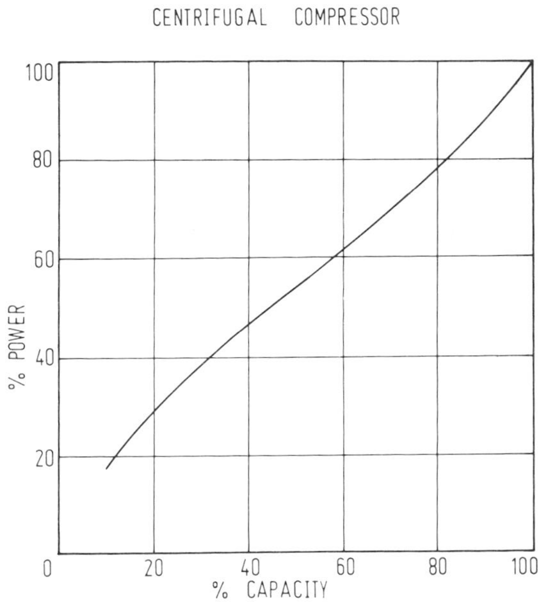


Fig 14    Part load characteristic at constant speed: suction and discharge pressures varying (see text)

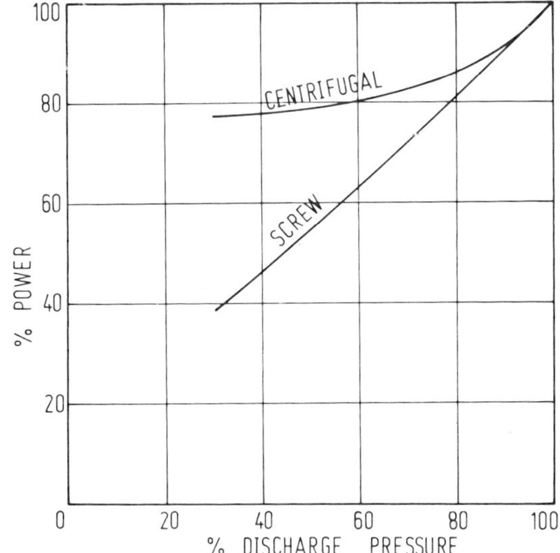


Fig 15    Comparison of power and discharge pressure characteristics at constant speed with fixed suction pressure

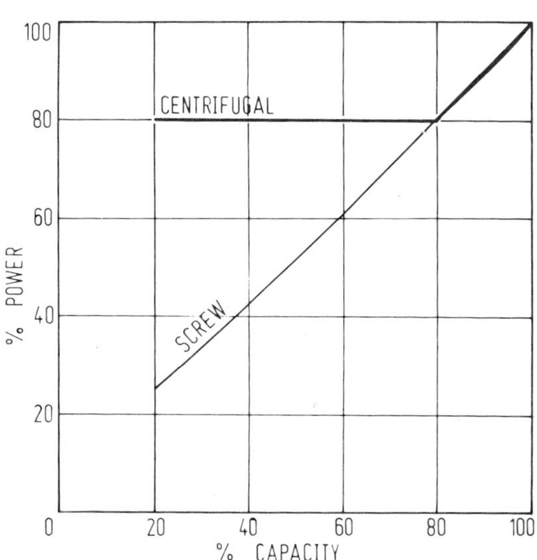


Fig 16    Comparison of part load power characteristics at variable speed with suction and discharge pressures constant

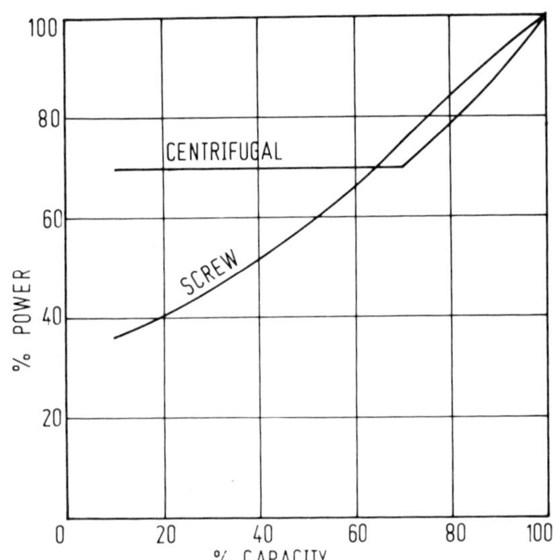


Fig 17    Comparison of part load power characteristics at constant speed with suction and discharge pressures also constant

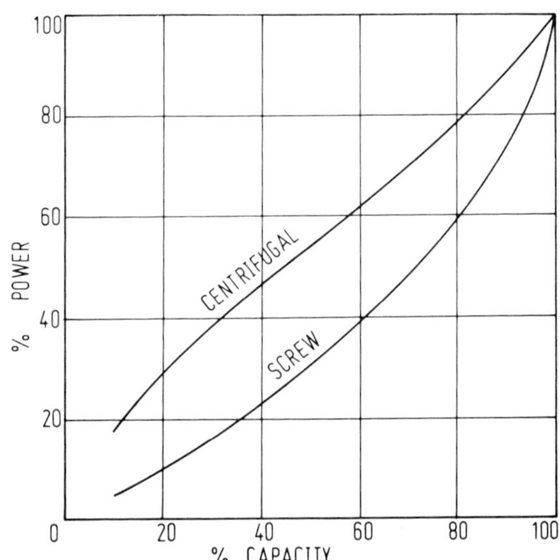


Fig 18    Comparison of part load power characteristics at variable speed: suction and discharge pressure also varying (see text)

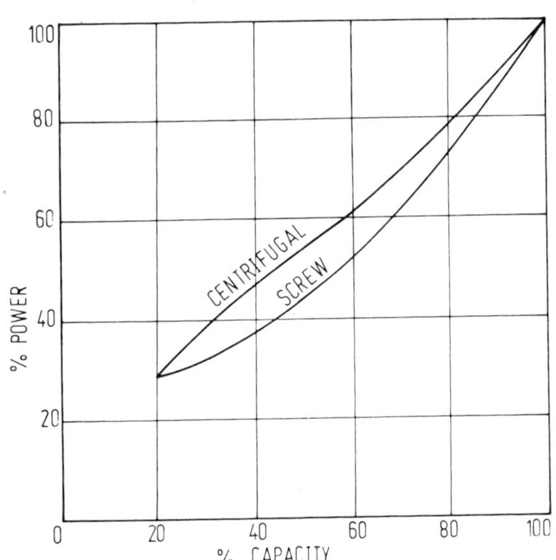


Fig 19    Comparison of part load power characteristics at constant speed: suction and discharge pressures varying (see text)

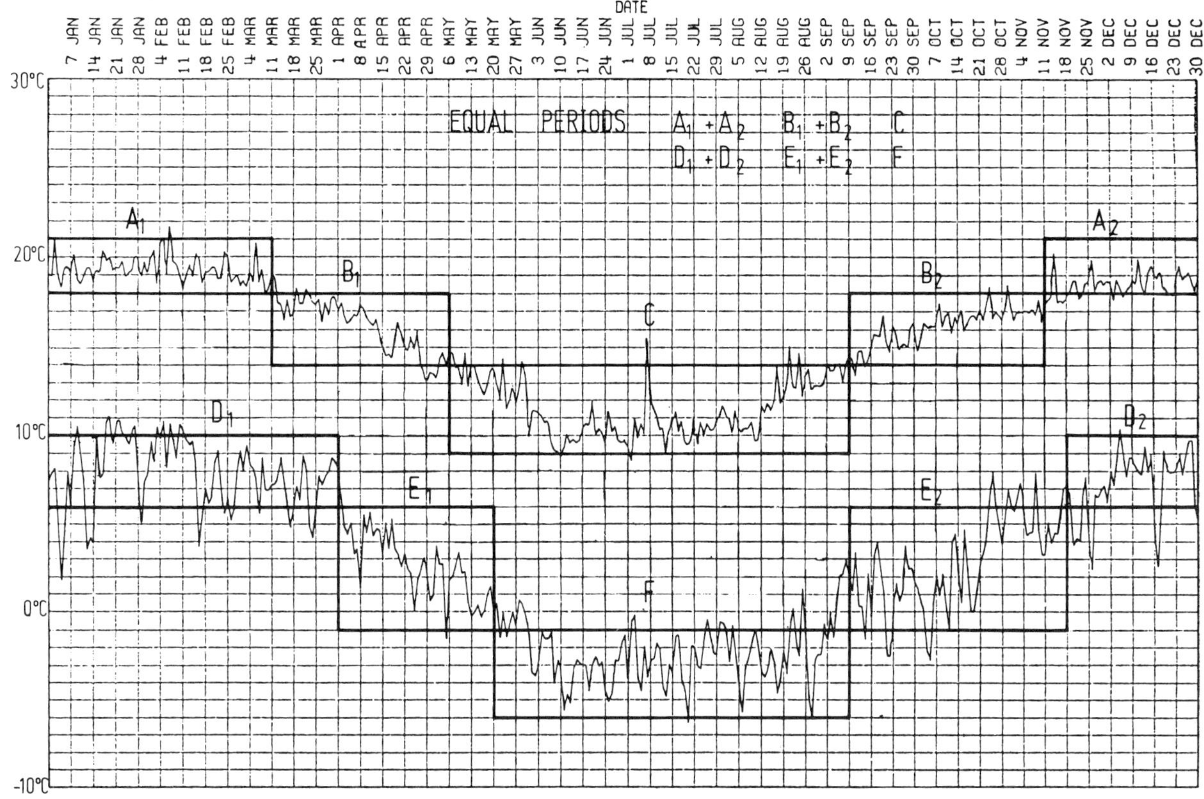


Fig 20  Maximum and minimum wet bulb temperatures occurring over a 12 month period at Jan Smuts Airport, South Africa

COMPARISON OF POWER USAGE ASSESSMENTS BETWEEN CENTRIFUGAL AND SCREW COMPRESSORS

| Plant Duty | Refrigeration Capacity | 8,800 KWR (2500 TR) |
|---|---|---|
| | Evaporation Temperature | +1.7°C (35°F) |
| | Design Maximum Condensing Temperature | 37°C (98.4°F) |

**Assessment Based on Maximum Design Conditions**

Centrifugal Power for Rated Duty    1640 Kw  (2200 hp)

Screw Compressor Power for Rated Duty 1665 Kw (2231 hp)

**Assessment Based "Actual Operating Review" Method**

Centrifugal Power for Rated Duty    1640 Kw

Screw Compressor Power for Rated Duty  1665 Kw

Actual Condensing Conditions for equal periods of year:-

Actual Power from Table

| | Centrifugal | Screw |
|---|---|---|
| 37°C | 1640 Kw | 1665 Kw |
| 30°C | 1410 | 1382 |
| 23°C | 1345 | 1182 |
| 18°C | 1312 | 1032 |
| 14°C | 1287 | 949 |
| 12°C | 1279 | 899 |

True Average
$$= \frac{Total}{6} = \frac{8273}{6} \qquad \frac{7109}{6}$$
$$= 1379 \text{ Kw (1848 hp)} \qquad 1185\text{Kw (1588hp)}$$

Direct Comparison of Powers at Rated Conditions
Shows Power Advantage in Favour of

Centrifugal Compressor of    1.5%

Comparison on Basis of Actual Operating Conditions
Shows Power Advantage in Favour of

Screw Compressor of    16.4%

Fig 21  Results of 'actual operating review method' assessment

# Mathematical simulation of turbo-compressors in chemical processes

H VOSS, Dipl-Ing
MAN (Maschinenfabrik Augsburg-Nürnburg) GHH Sterkrade, Germany

SYNOPSIS   The application of mathematical models to the design of compression systems allows machinery to be correctly sized and controlled to give optimum performance when operating within a process network. The procedure presented in this paper makes it possible to determine the dynamic behaviour of a turbo-compressor within a process system by mathematical simulation. Compression systems under examination are constructed from mathematical models of individual components with the aid of an indexing system.

The operating conditions for the individual components are determined as a function of time for a given interval. By reference to various examples from chemical processes studies of start-up and shut-down procedures of compressor trains are shown.

NOTATION

Symbols

| | |
|---|---|
| A | Area |
| i | Number |
| K | Damping coefficient |
| M | Torque |
| m | Mass |
| n | Speed |
| p | Pressure |
| t | Time |
| $\alpha$ | Flow coefficient |
| $\gamma$ | Angle |
| $\zeta$ | Loss coefficient |
| $\varkappa$ | Isentropic exponent |
| $\rho$ | Density |
| $\psi$ | Flow function |
| $\theta$ | Mass moment of inertia |
| $\omega$ | Angular velocity |

Indices

| | |
|---|---|
| a | Driver |
| b | Acceleration |
| d | Valve |
| g | Closing mechanism |
| k | Flap |
| o | Basis |
| r | Friction |
| s | Flow |
| v | Compressor |

1 INTRODUCTION

Problems can occur if a correctly surged compressor is placed within a process system without sufficient investigation into the interaction of machine and plant dynamics leading to poor utilisation of the installations.

Modern calculation methods permit the design engineer to perform a careful and exact rating of a turbomachine that will translate into a correctly dimensioned machine to give the required operating behaviour. Final evaluation of the turbomachine design and its safe operation within the stable operating range can only be ensured if all operating requirements with respect to the overall plant dynamic behaviour, in which there may be rapid changes in the operating conditions of various plant components, must also consider the storage capacity of the system (Ref. 1), the time response of the shut-off valves, and the inertia of the machine itself.

To ensure the operational reliability of the plant as a whole and to establish exactly what equipment is required to achieve this, it is recommended that extensive studies are performed as early as possible in the project planning stage. To permit a complete system to be simulated in an economic way, a particularly flexible digital calculation method has been developed with wide application potentials.

2 MATHEMATICAL SIMULATION

Previous methods for the calculation of dynamic response were predominantly designed for specific types of plant, such as Blast Furnace air supply systems (Ref. 2) or gas turbines (Ref. 3,4,5). It is felt, however, that any calculation method should not be confined to a particular type of system, but rather enable the user to assemble any required loop from standard components (Ref. 6). The simulated system is built up using the indexing system described, which, once assembled, establishes a so-called loop matrix. This matrix includes all

components according to type and number
with consecutive numbering of each of the
components' battery limits. The data  -
available from the loop matrix are used to
link the components concerned and to control the mathematical simulation. Fig. 1
shows the structure of the programme and
the sequence of the calculation.

## 2.1  Indexing of loop

Fig. 2 is a flow diagram of a compressor
set with two intercoolers and includes
indexing. The first digit designates the
type of component as follows:

| | |
|---|---|
| Compressor (axial or centrifugal) | 0, |
| Pipework and vessels | 1, |
| Coolers | 2, |
| Non-return valves | 3, |
| Valves | 4. |

The second digit indicates the number of
identical components. To permit consecutive numbering of the limits establishing
the junction points with the next component, a third digit was introduced to
which a plus/minus sign rule applies to
indicate the flow direction of the circuit fluid at the start of the calculation, i.e. mass inlet 'positive', mass
outlet 'negative'.

Where studies are to be performed on turbo
-compressors in large systems, limitation must be made with respect to the
other system components that cannot be
contained within the calculation.

In such a case the system is assumed to
have its limit at those points where despite incoming or outgoing flows only
insignificant changes are to be expected
in the gas data during the examined time
interval.

## 2.2  Calculation method

The dynamic behaviour of the plant under
consideration is to a large extent determined by the dynamic reaction of the turbomachinery. The calculations can either
be based on measured performance characteristics or on performance curves converted from measurements in a polynominal
form which describe the behaviour of the
machine within its stable operating range.
The actual duty point of a compressor is
defined by the prevailing pressure ratio
between discharge and suction nozzle and
by the speed, and, if adjustable stator
blades are provided, also by the stator
blade position. The control elements on
the compressor, such as suction throttling devices or adjustable stator blades,
are not treated as independent components.
Compressors are driven either by motors
or by turbines, or, as often occurs in
the chemical industry, by a combination
of these. In steady-state operation of
any compressor set there is an equilibrium of power between the driver and
the compressor; and disturbance in this

respect causes the speed of the shaft
system to be changed, apart from initiating other dynamic response reactions.

To study the balance of power, the torque
curve of the unit must be pre-determined
as a function of speed or of time. When
preparing the torque curve, it may also
be necessary to consider the setting
times of trip devices, residual expansion,
or the windage losses of turbines.

Changes in the speed of the shaft system
result from the acceleration torque $M_b$
which is derived from the difference
between the driving torque $M_a$ and the
sum of the stationary compressor torques
$M_v$ including their friction component $M_r$:

$$M_b = M_a - \sum_i (M_{v,i} + M_{r,i}) \qquad (1)$$

The change of speed then is as follows:

$$\frac{dn}{dt} = \frac{30}{\pi \cdot \theta} M_b \qquad (2)$$

where $\theta$ is the mass moment of inertia
of the entire shaft system.

The influence of the pipework and vessels
on the dynamic behaviour of the system is
considered in the form of storage capacity and in terms of the possible mixing
of various gas flows. The total of all
incoming and outgoing mass flows $\dot{m}_i$ of
the system equals the change of the
stored gas volume, m

$$\frac{dm}{dt} = \sum_i \dot{m}_i \qquad (3)$$

The pressure and the temperature in the
system under consideration change as a
function of the difference in the mass
balance.

With regard to heat exchangers, their
dynamic behaviour involves several problems and requires much expenditure for
accurate modelling. Therefore, an approx-
imate calculation method is adopted which
mainly considers the function of the
cooler as a flow-conducting component.

The heat exchange rate is governed by

a) control of the cooling-water flow
   rate to give constant gas exit temperature, and

b) a constant cooling-water flow rate.

In case b) the gas exit temperature is
determined using equation (Ref. 7) via
an energy balance at the cooler. This
presupposes that the differential temperature between the gas outlet and the
water inlet (approach temperature) is
known for the design point.

Non-return valves within a pipework system are designed to prevent the fluid from flowing in the reverse direction.

The valve is opened by the flow-medium pressure and under steady-state conditions is set at an angle where the opening torque arising from the change of impulse and the differential pressure at the flap equals the torque provided by the closing mechanism.

The hydraulic properties of the flap depend upon the flow coefficient as a function of the pressure ratio as well as of the flap angle and the loss coefficient as a function of the residual area and the flow velocity (Ref. 8).

The initial position of the flap is determined by the equilibrium of the flow and closing torque and is found by means of an iterative loop at the end of which the following equations must be fulfilled for $\gamma$ , i.e.

$$M_g - M_s = 0 \qquad (4)$$

and

$$p = f ( \varsigma , \dot{m}_o ) \qquad (5)$$

Any change of the mass flow rate produces a torque at the flap

$$M_k = M_g - M_s , \qquad (6)$$

which is partly required to overcome the resistance provided by the hydraulic braking equipment, and for acceleration of the flap.

$$M_k = \theta_k \frac{d\omega}{dt} + K \omega \qquad (7)$$

where K is the damping constant for the braking cylinder.

If there is a flow reversal, closure of the flap is accelerated by the hydraulic forces, thus reducing the effective closing time by a considerable margin.

All valve settings for mathematical simulation are modelled by means of the active valve area as a function of time.

As part of the analysis, the density and the pressure at the valve inlets and the pressures prevailing in the subsequent systems are determined.

The mass flow rate through any one valve is defined by:

$$\dot{m}_d = \alpha A \Psi \sqrt{\frac{2 \ae}{\ae - 1} \varsigma_1 P_1} \qquad (8)$$

where $\alpha$ is the flow coefficient.

$\Psi$ represents a flow function which is dependent only on the valve pressure ratio and on the isentropic exponent $\ae$ .

3 TYPICAL APPLICATIONS

3.1  <u>Start-up of a 36-MW compressor set</u>

The start-up of a compressor which is used in an air separation plant for the production of 2250 tons per day oxygen was analysed and compared to measurements taken on the test stand.

Starting a large-capacity axial compressor set with intercoolers produces operating conditions in the stage groups which are far away from their design point. This type of installation is particularly suitable for checking the calculation method here presented due to the occurrence of steep gradients in the variables of state as a function of speed. Figs. 3 and 4 show the machine in the works test facility of M.A.N. - GHH STERKRADE. The compressor comprises a low-pressure section, a medium-pressure section and a high-pressure section with intercooling after each stage group, and was driven by a steam turbine. In order to reduce the absorbed power during the test runs, the compressor was throttled to such an extent that the pressure downstream of the LP section was still below ambient pressure. To improve flow conditions in the medium speed range, an intermediate vent was connected to a by-pass line leading to the compressor inlet.

For start-up purposes the suction throttle and the four rows of adjustable stator blades of the LP section were set to the 'closed' position where the first four rows of stator blades are adjusted to such an angle that the pressure is continually decreased, minimizing the stage power input.

The variables measured during start-up were continuously recorded by means of a process computer. Shown in Fig. 5 as a function of speed are the suction pressure, the pressure upstream of the fifth stage (where the pressure decrease within the compressor reaches a minimum value) and the pressure downstream of the 1st intercooler.

Likewise the discharge pressure of the MP section was not recorded and Fig. 6 indicates the suction and discharge pressure curves vs. speed for the HP section.

When taking torque measurements attempts were made to exclude the acceleration torques for short periods during start-up by step-wise increase of speed.

When operating at its design speed of 3000 rpm the torque measured on test was only 30% of that during normal operation.

## 3.2 Emergency shut-down of a nitric acid plant

The nitric acid plant is a good example of the interaction between a complex turbomachine set and large storage volumes within the plant. Fig. 7 shows a schematic of such a plant in which the machine train comprises a steam turbine, an air compressor, an NO-gas compressor and a tail-gas expansion turbine.

In the event of a plant trip or if the ammonia combustion process fails, the machine set is immediately shut down via the trip devices on the steam and tail-gas expansion turbines. At the same time, the compressor blow-off valves are opened in order to prevent surging. Non-return valves arranged between the compressors and the plant prevent unloading of the plant by flow reversals, which would, due to the nitrous gases, give the inherent risk of corrosion. The mass moments of inertia of the non-return valves and the response time of the blow-off valves exert considerable influence on the operating mode of the compressors. As long as the non-return valve remains open, the storage volume of the vessels permits only a minor decrease of the back-pressure prevailing at the compressor. As the **driv**er has been tripped, the speed of the machine set decreases so rapidly that under certain specific conditions the compressors may leave their stable operating range.

The study was aimed at determining the minimum necessary blow-off valve opening time to give stable compressor run-down after an emergency trip.

Fig. 8 shows the duty point curve (run down after trip) for different blow-off valve opening times and it will be seen that an opening time of 1.6 sec. or above cannot prevent surging. Further calculation results are also plotted in Fig. 8. Conditioned by the response time of the trip devices and by the residual expansion of the steam through the turbine casing and gas through the expander casing, the drive power quickly reduces to zero, so that the shaft speed decays after only a short dwell time. The middle curve shows the flow velocity at the non-return valve in relation to the nominal diameter. Velocities which are negative indicate reverse flow, which on account of the increasing differential pressure at the valve approaches a maximum value even though closing is progressive. The bottom graph shows the non-return valve angle as a function of time. The effect of flow reversals clearly reflected by the acceleration of the valve movement. Only after an angle of approx. 15° has been reached, does the second stage of the braking cylinder come into action with a considerable increase in damping rate.

## 3.3 Trip functions in the machine sets of a terephthalic acid plant

The increase in the number of casings in a machine train as well as more machine components has made it difficult to assess the effects of different operating modes without resorting to mathematical simulation.

For start-up and run-down of a single machine train, in the event of a failure, or during the shut-down of one or several machine trains, instantaneous changes in the operating conditions of the compressors are liable to arise which may make it necessary to protect the machine train.

By reference to a two-shaft compressor unit for terephthalic acid production plant with an annual capacity of 450 000 tons, a trip study was made with the aim to prevent surging of the turbo compressors.

Fig. 9 shows the flow diagram of the compressor set. The first shaft system comprises a motor, an axial-flow compressor with interstage cooling and with variable stator blades and an expander. The second shaft system which is installed downstream of another intercooler, consists of a motor, a centrifugal compressor with interstage cooling, and a steam turbine.

At the time of measurement, both compressor systems were driven only by their motors, with the expander and steam turbine being windmilled. The machine train was isolated from the process by a non-return valve, and was operated with opened blow-off valves downstream the centrifugal compressor.

The duty point selected for the trip tests featured a pressure ratio of $\pi = 29$ and a stator blade position of the LP axial compressor of 50%. Both shaft systems were running at rated speed. At the moment of shut-down (T = 0 sec.) both motors were isolated from the supply grid, while issuing a 'quick-opening' signal to the valves between the axial and centrifugal compressors and to the blow-off valves which were already partly open.

In Fig. 10 the results from the simulation are plotted vs. the measured variables recorded during run-down. They show decrease in speed, and falling discharge pressure curves for both shaft systems, shown as a function of time.

Calculations and measurements indicated that the LP sections of the axial and centrifugal compressors went into surge after less than 2 seconds.

In line with current development of the calculation method, the duty points determined at the point of surge for each speed-related characteristic curve are fixed in the programme.

Despite this limitation, the simulated values compared well with the measured curves. The main reason for this is the comparatively large buffer volumes contained in the piping and coolers which in conjunction with the LP compressor may form a contained oscillating system.

Further studies were concerned principally with protecting the axial compressor, and the tripping of one shaft system: protection of both shaft systems was also considered in the event of frequent trips.

Figs. 11 and 12 represent the results of this simulation, which are shown vs. the changes the actual conditions measured during the shut-down tests. Starting at T = 0 sec., the illustrations indicate for both compressor casings the calculated transient duty points within the individual performance characteristics for decreasing speed.

To prevent surging of the machines, the intermediate relief valves downstream the LP sections of both shafting systems must be incorporated in the control loop, and the stroking times of the blow-off valves downstream the centrifugal compressor must be considerably shortened.

In Fig. 13 we see a comparison of the stroking times and capacity of the valves concerned.


4 SUMMARY

. Manufacturers of complex turbomachinery trains are often confronted with specific questions which can only be answered if the interaction of the various system components and their dynamic behaviour are known.

It is only then that the turbomachines can be adapted to the downstream plant in an optimum way, thus further improving the efficiency of the process as a whole.

To solve this problem, a calculation method was developed which could be applied to a multitude of applications, thus permitting any required compressor loop to be simulated. Comparison between the theoretical results and actual measurements taken in the field has proved the calculation method to be reliable. The programme structure is such that further development of the system can be made.

REFERENCES

1. KÜPER, K.D., LANGHANS, J.    Ein Programm zur dynamischen Berechnung des Druckverlaufs in gekoppelten Volumina. Atomenergie 17 (1971) No. 3 P. 163 - 166.

2. FASOL, K.H., JÖRGL, M.P., TUIS, L. Untersuchung der Regeleinrichtungen von Hochofen-Windversorgungssystemen am Analogrechner. Brennstoff-Wärme-Kraft 28 (1976) No. 2, P. 43 - 48.

3. HARMS, A.    Über das Zusammenwirken der Anlagenteile einer Gasturbinenanlage beim Regelvorgang. Energie und Technik 18 (1966) No. 4, P. 133 - 139 and No. 5, P. 190 - 193.

4. BAUERFEIND, K.    Die Berechnung des Übertragungsverhaltens von Turbo-Strahltriebwerken unter Berücksichtigung des instationären Verhaltens der Komponenten. Luftfahrttechnik-Raumfahrttechnik 14 (1968) No. 5, P. 117 - 124 and No. 6, P. 143 - 151.

5. KREY, G.    Das dynamische Verhalten von einwelligen geschlossenen Gasturbinen. Doctoral thesis TU Hannover (1974).

6. VOSS, H.    Contribution to the Calculation of the Dynamic Behaviour of Industrial Compressor Circuits. Von Karman Institute for Fluid Dynamics, Lecture Series 2 (1978).

7. BAMMERT, K., TWARDZIOK, W.    Das stationäre Betriebsverhalten von konventionell- und nuklearbeheizten Gasturbinen. Atomenergie 11 (1966) No. 5/6 P. 185 - 204.

8. TRAUPEL, W.    Thermische Turbomaschinen. Vol. 1 and 2, 2nd edition, Springer-Verlag (1966).

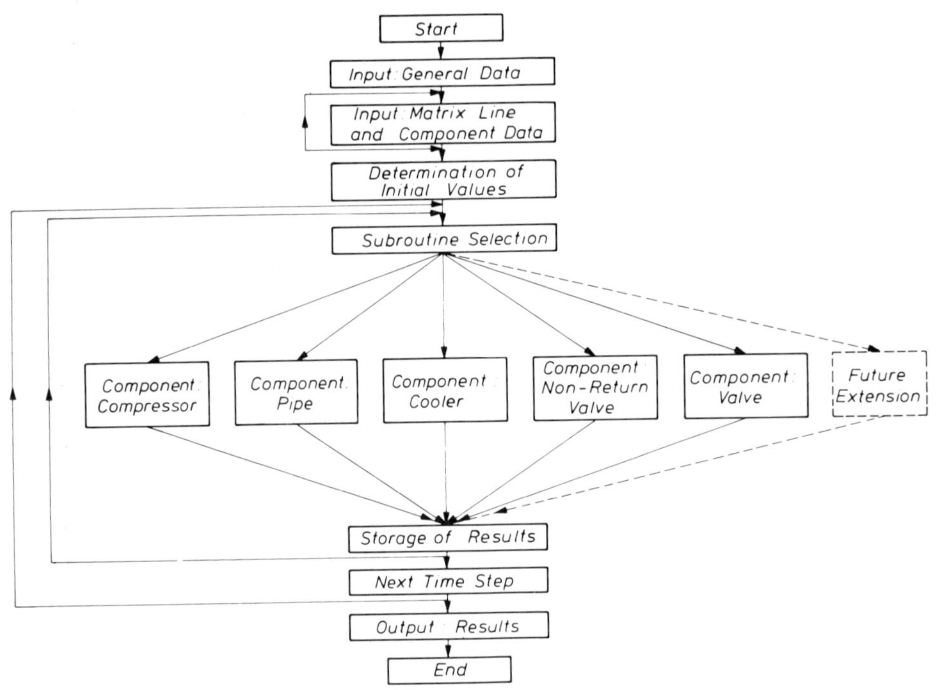

Fig 1    Block diagram

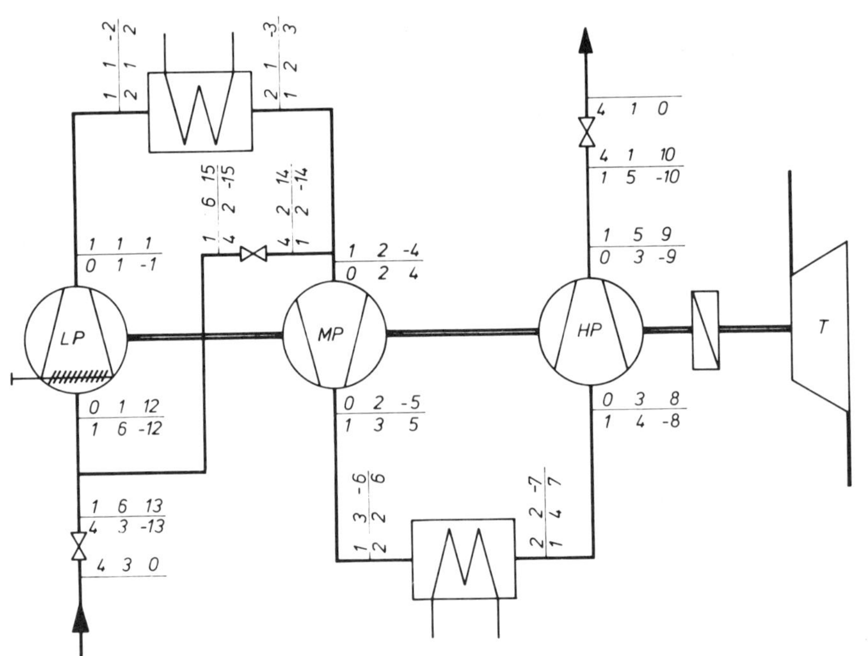

Fig 2    Calculation model

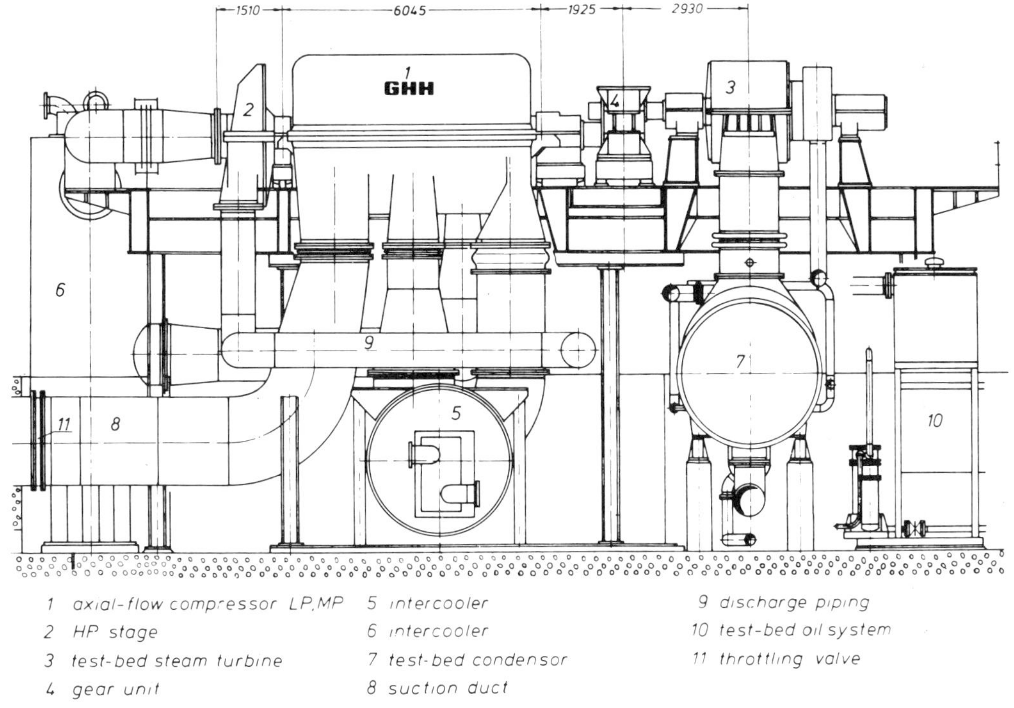



1 axial-flow compressor LP,MP
2 HP stage
3 test-bed steam turbine
4 gear unit
5 intercooler
6 intercooler
7 test-bed condensor
8 suction duct
9 discharge piping
10 test-bed oil system
11 throttling valve

Fig 3  Axial-flow compressor on test bench

Fig 4  Final inspection after testing

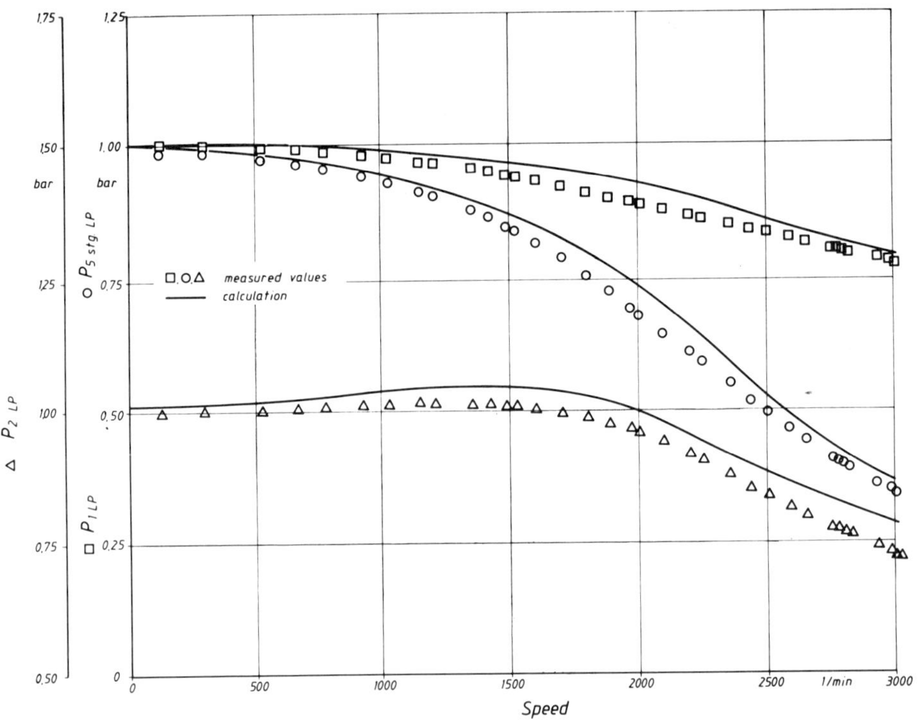

Fig 5    Course of calculated and measured operating conditions for the LP section

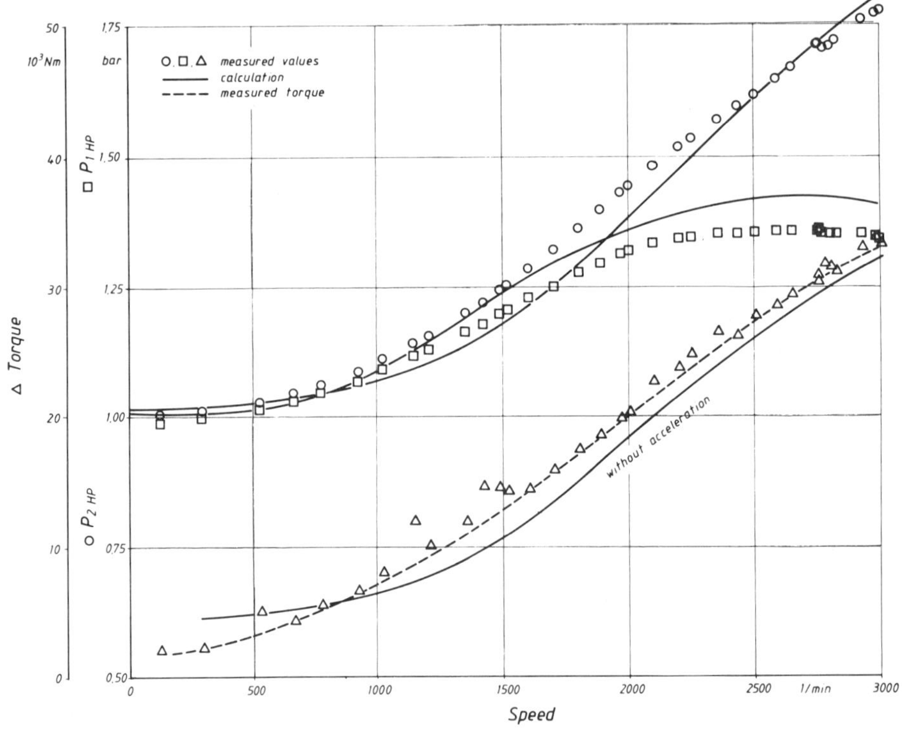

Fig 6    Course of calculated and measured operating conditions

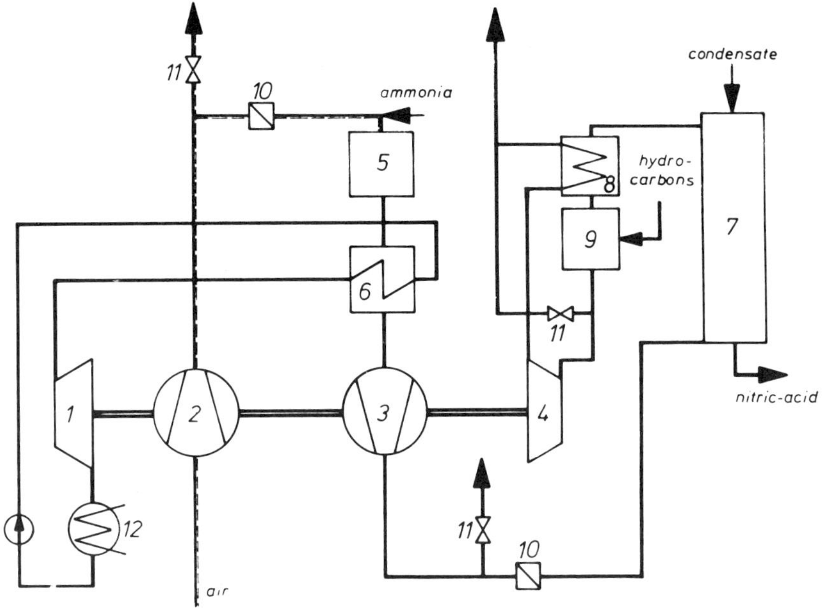


Fig 7    Diagram of a nitric acid plant

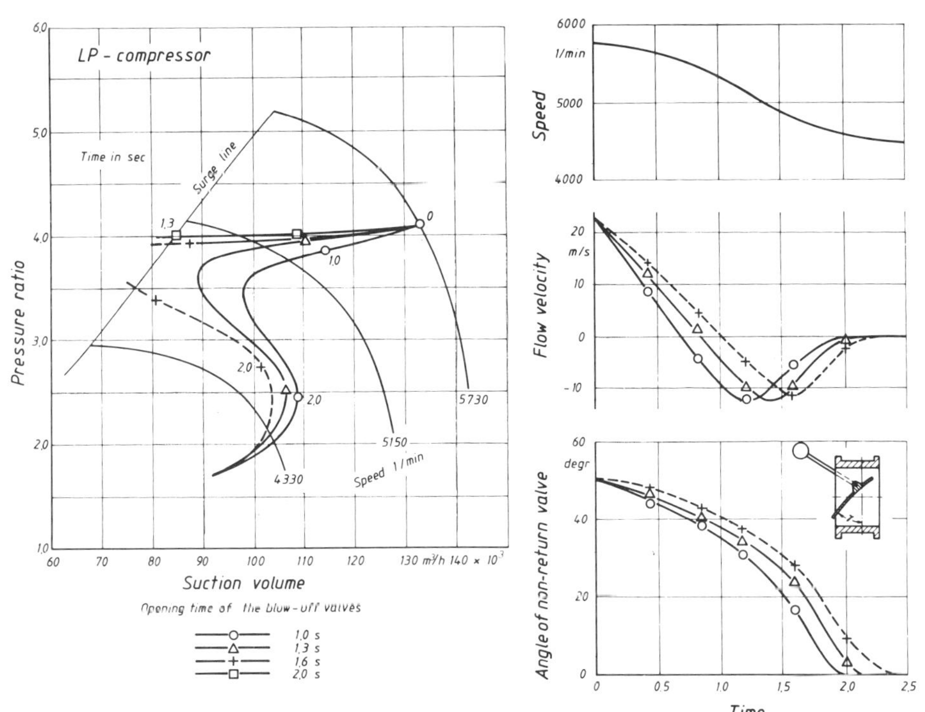


Fig 8    Emergency shut-down of the machine train

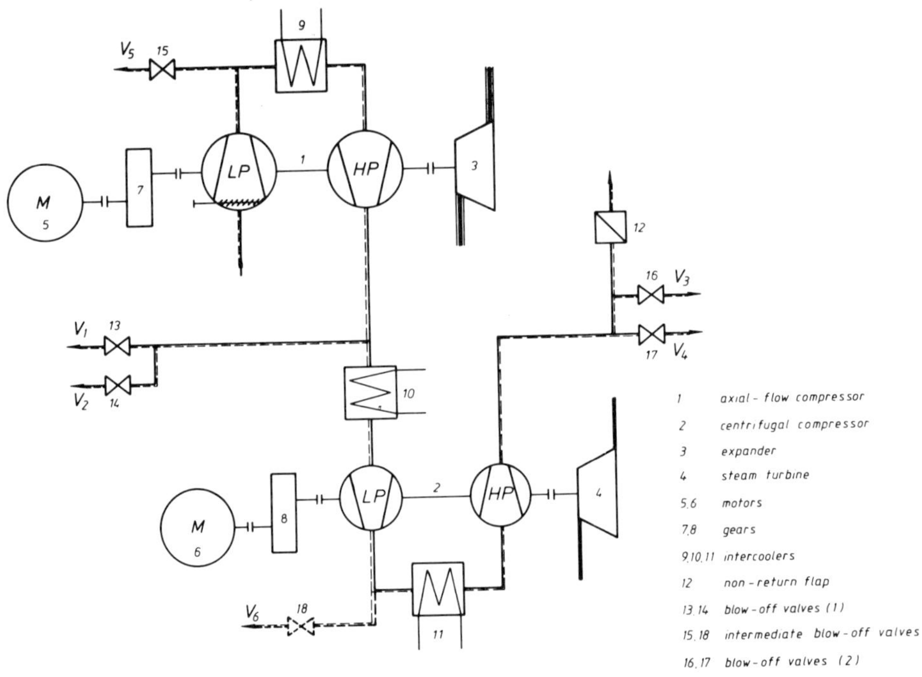


Fig 9   Flow diagram of a two-shaft compressor unit

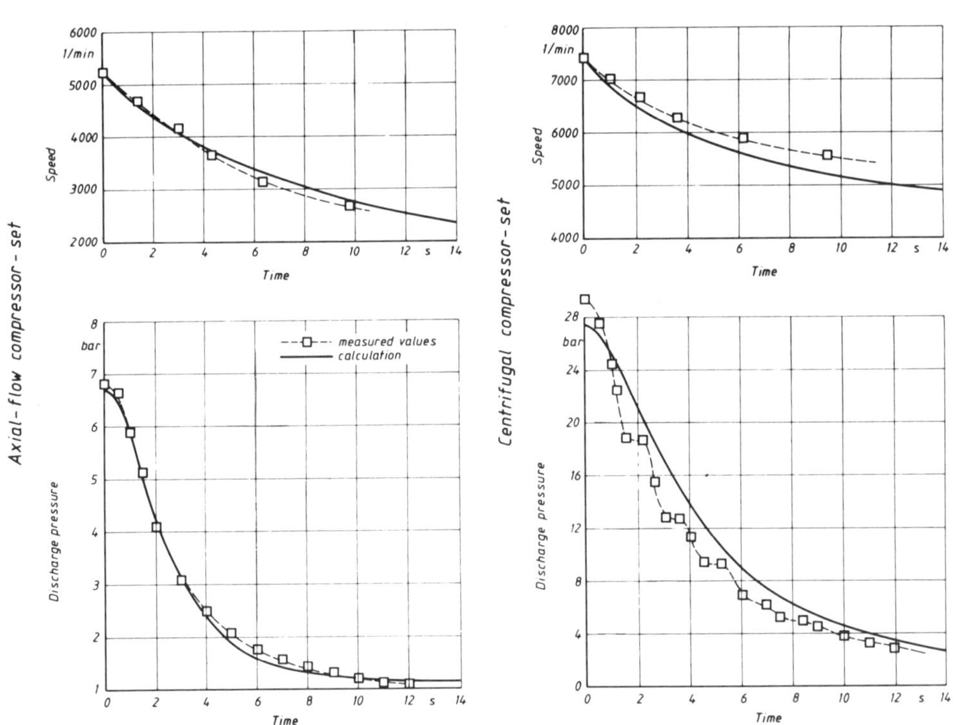


Fig 10   Comparison of simulated and measured values at shut-down

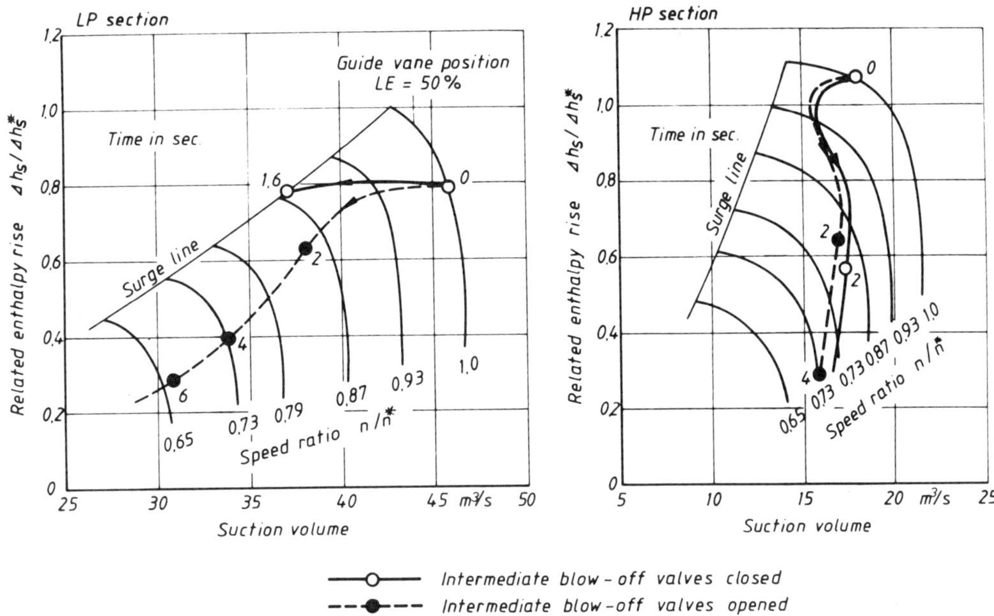



Fig 11   Simulation results for the axial-flow compressor set

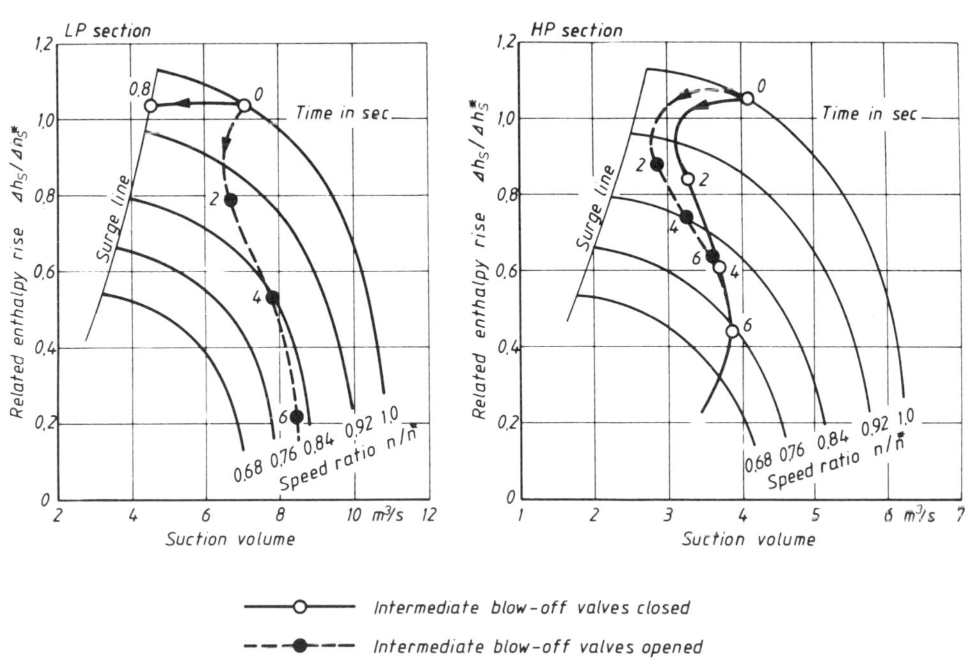



Fig 12   Simulation results for the centrifugal compressor set

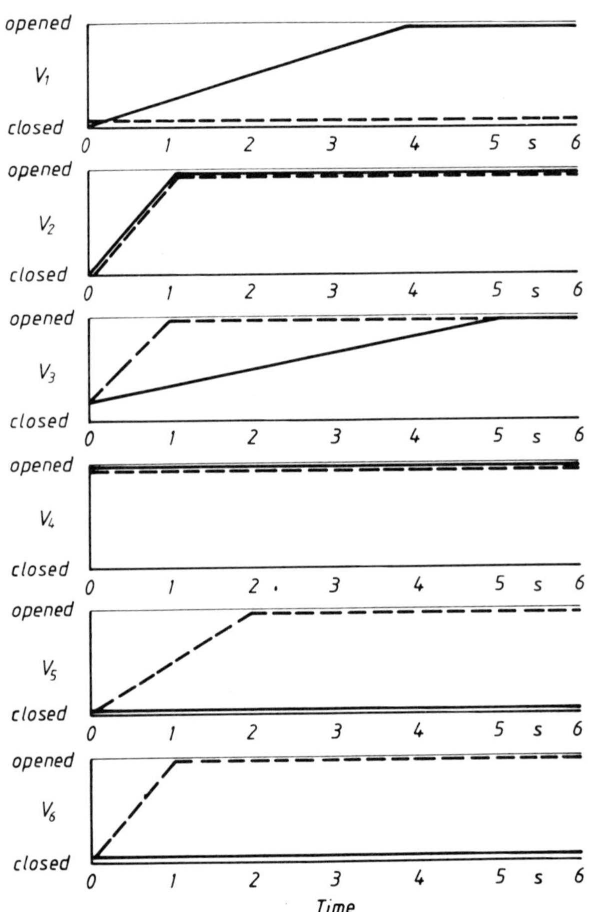

Fig 13    Valve control during simulation

# C41/81

# Aerodynamic performance of centrifugal pipeline compressors

S SAYYED, BS, BSME, Member ASME, FInstPet
Delaval-Stork VOF, Hengelo (O) Netherlands

SYNOPSIS   Site performance of centrifugal pipeline compressors is often predicted from aerodynamic tests carried out in the manufacturer's shop. As the majority of these tests are performed on air at atmospheric conditions, test results need to be translated to actual site operating conditions. This paper outlines the evaluation of some of the major factors to allow accurate prediction of performance of pipeline compressors related to site operating conditions.

NOMENCLATURE

$b$  = exit width of impeller, m

$g$  = acceleration due to gravity, $m/s^2$

$k$  = specific heat ratio, cp/cv

$m$  = mass flow, kg/s

$N$  = rotational speed, rpm

$n$  = polytropic exponent

$P$  = pressure, bar

$Q$  = volume capacity, $m^3/s$

$R$  = universal gas constant, Nm/kmol K

$r$  = pressure ratio

$T$  = absolute temperature, K

$U$  = peripheral velocity, m/s

$z$  = compressibility factor

$hr$ - coefficient of heat transfer for casing in still air, $J/s\ m^2 K$

$H$  = head, Nm/kg

$Ma$ = Machine Mach Number

$Psh$ = shaft power, KW

$Qm$ = total mechanical losses, J/s

$Qr$ = heat loss from compressor casing, J/s

$Qsl$ = external seal loss equivalent, J/s

$Re$ = Reynolds Number

$Rc$ = Reynolds number correction

$Sc$ = heat transfer surface of exposed compressor casing, $m^2$

$tc$ = average temperature of casing surface, K

$ta$ = ambient temperature, K

$\eta$ = efficiency

$\nu$ = gas kinematic viscosity, $m^2/s$

Subscripts

1 = value at inlet

2 = value at outlet

d = design conditions

i = isentropic

t = test conditions

INTRODUCTION

In recent years there has been a significant increase in the use of large capacity centrifugal pipeline compressors to transport natural gas. With the continuously increasing cost of energy more emphasis is being placed on system efficiency and the advantages offered by high performance machinery for new as well as existing gas transmission systems. In a system with a given flow, higher efficiency can mean a reduction in capital outlay for power, an increase in delivered quantity of gas, or a reduction in fuel and operating costs. In a new system it could further mean an increase in station spacing for the same flow and power with the resultant reductions in both operating and capital costs. In view of this trend accurate prediction of performance for pipeline compressors has become increasingly important.

CENTRIFUGAL PIPELINE COMPRESSORS FOR GAS TRANSMISSION

Pipeline compressors for gas transmission can be grouped into single and multistage designs.

Single Stage Design

Almost all single stage pipeline compressors have a cantilever rotor construction supported by two bearings located at the driving end of the impeller. A typical cross-section of a single stage pipeline compressor is shown in figure 1.

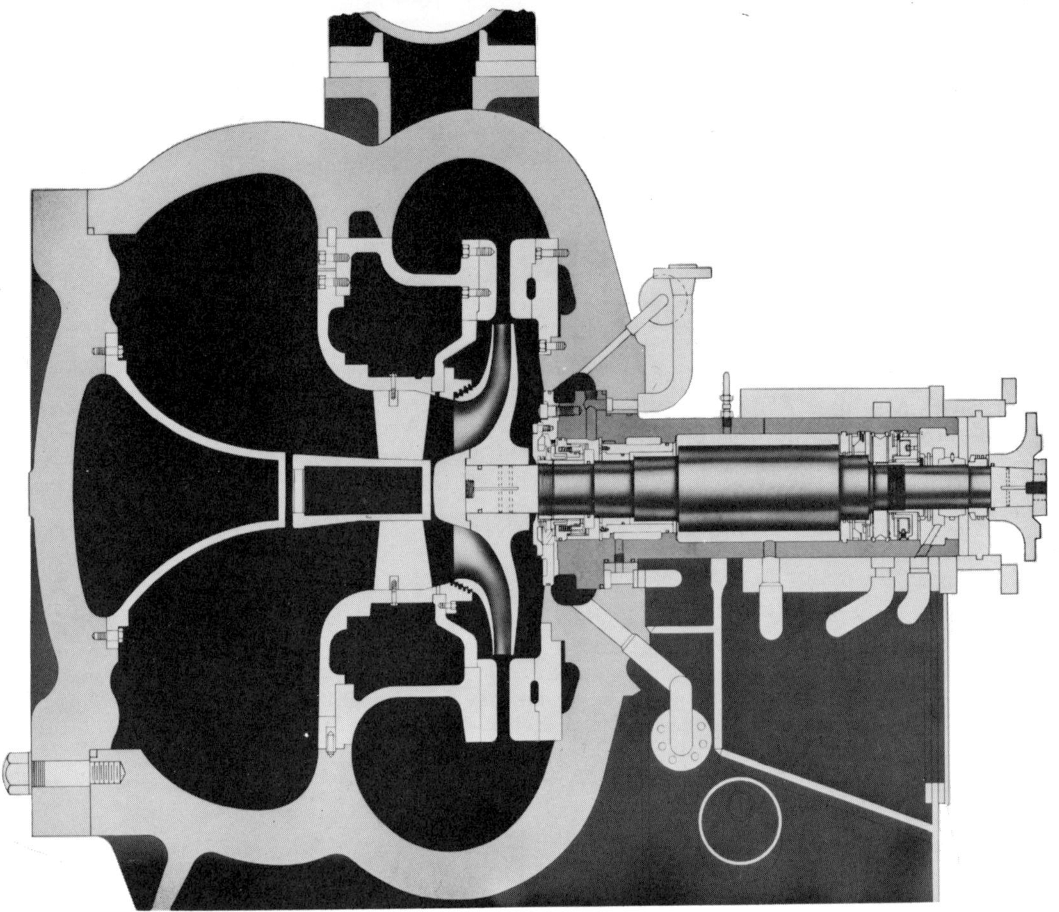

Fig 1    Cross-section of a typical single stage pipeline compressor

This arrangement provides simplicity in design
without compromising maintenance features. In
some applications an axial inlet is used to
maximize performance. Efficiency improvement
using an axial inlet may be of the order of two
or more percentage points.

Multistage Stage Compressor Design

In multistage compressors, the rotor is a
beam design supported by two journal bearings
mounted at each end of the rotor. A typical
cross-section of a two stage fabricated compressor is shown in figure 2.

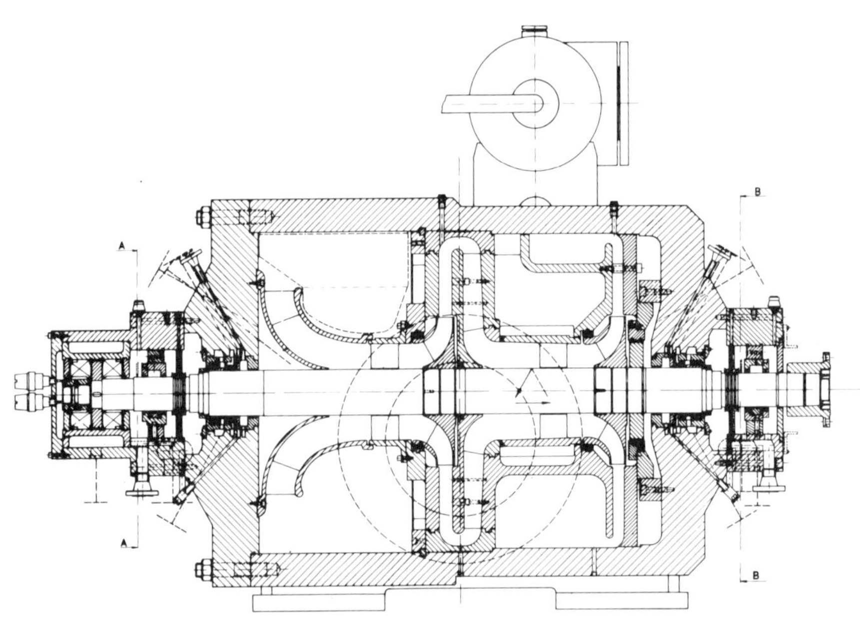

Fig 2    Cross-section of a multistage beam type pipeline compressor

Journal and thrust bearings as well as oil seals are easily accessible for inspection and maintenance without opening the compressor casing. Suction and discharge nozzles are in line opposite one another, a typical feature of pipeline compressors. This nozzle alignment compensates for the thrust applied by the gas piping, prevents distortion of the casing and misalignment of the compressor and its driver. The performance of a multistage pipeline compressor is usually somewhat lower than that of single stage designs.

PERFORMANCE TESTING

The purpose of an aerodynamic test is to ascertain that the compressor performs in accordance with the guarantees given by the manufacturer when operating with design gas at design conditions. To conduct a field performance test in accordance with ASME Power Test Codes PTC-10 1965 (ref. 1) is usually difficult or impractical due to the limitations of some compressor installations, accessibility of piping, and often hard to duplicate guarantee conditions. As a consequence, results of field performance tests are suspect and most users are reluctant to pay large sums for 'hypothetical results'.

To ensure that a unit will perform as guaranteed a full or part load performance test of the compressor is conducted at the manufacturer's plant. These tests are carried out in accordance with ASME Power Test Codes for Compressors and Exhausters PTC-10 1965. This provides the evidence that quoted efficiency levels are met, and in cases where the guarantees are not attained allows for the necessary corrections to be made. Even if a full load test in the manufacturer's shop is conducted, it is not always possible to duplicate exact guaranteed conditions of inlet temperature and pressure.

Shop performance Tests

The majority of pipeline compressors are tested in the manufacturer's shop on air at atmospheric conditions. To obtain an acceptable dynamic similarity, tests are performed at speeds other than the design speed. The test speed is called an 'Equivalent Speed' of the design speed (ref. 2). The aerodynamic performance on air is similar to site performance when dynamic similar flow is obtained everywhere in the flow channels. This similarity is achieved when flow angles, Mach numbers, Reynolds numbers and volume ratios are the same.

Shop air tests for pipeline compressors generally fall under class II of the Power Test Codes and the test results must be converted to expected performance on design gas at site conditions. The limitations for shop air tests and the allowable deviations from design parameters of volume ratio, capacity-speed ratio, Machine Mach number, and Reynolds number are given in table 1.

Calculation of Equivalent Speed for Air Tests

The equivalent test speed for air when inlet Mach numbers are equal is determined from the ratio of sonic velocities from equation 1.

$$N_t = N_d \sqrt{(kgRT_1)_t / (kgZRT_1)_d} \qquad (1)$$

The equivalent test speed when density ratios are equal is determined from the relationship of polytropic compression and is obtained by equation 2.

$$N_t = N_d \sqrt{\frac{\left[T_1 R \, (n/n-1) \, (r^{n-1/n} - 1)\right]_t}{\left[T_1 Z_1 R \, (n/n-1) \, (r^{n-1/n} - 1)\right]_d}} \qquad (2)$$

Table 1    Allowable departure from specified design parameters for Class II and Class III tests

| Variable | Symbol | Range of Test Values Limits - % of Design Value | |
|---|---|---|---|
| | | Min | Max |
| Volume ratio | $Q_1/Q_2$ | 95 | 105 |
| Capacity-speed ratio | $Q_1/N$ | 96 | 104 |
| Machine Mach Number | $M_m$ | | |
|   0 to 0.8 | | 50 | 105 |
|   Above 0.8 | | 95 | 105 |
| Machine Reynolds Number where the design value is | Re | | |
|   Below 200 000 Centrifugal | | 90 | 105 |
|   Above 200 000 Centrifugal | | $10^x$ | 200 |
| Mechanical losses shall not exceed 10 per cent of the total shaft power input at test conditions. | | | |

x Minimum allowable test Machine Reynolds Number is 180 000

For perfect dynamic similarity the test speeds calculated using equations 1 and 2 should be identical. In practice this is seldom possible, indicating that the test simulation is not absolutely perfect. This can be illustrated by the following example :

Example 1

It is desired to find the equivalent test speed of a three stage natural gas pipeline compressor. The compressor's design speed is 6190 rpm and it is to be performance tested on air in the shop according to ASME Power Test Codes PTC-10, 1965. Table 2 provides the data for design and test conditions.

For pipeline compressors where the number of compression stages are few and the differences in molecular weight, ratio of specific heat, and compressibility factor, are relatively small, an average of these two equivalent speeds is taken as the test speed and is considered adequate enough to provide reliable results.

All parameters from table 1 are checked and satisfied as shown in table 3.

Generally the only exception to the above rules is the Reynolds number which often fails to comply with the allowable limits. This is also the case in the above example, although the Reynolds number is greater than minimum allowable value of 180 000. More often Reynolds number for a pipeline compressor tested on air is of the order of 1.6 to 5.5 x $10^5$, whilst operating the same compressor under site conditions with natural gas the Reynolds number may be of the order of 1.2 to 4.5 x $10^7$.

Table 2    Equivalent speed calculation for a three stage pipeline compressor

|  | Design Gas<br>Natural Gas | Test Gas<br>Air |
|---|---|---|
| Absolute suction pressure, bar | 45.97 | 1.013 |
| Absolute discharge pressure, bar | 79.48 | 1.844 |
| Suction temperature, K | 283.15 | 293.33 |
| Discharge temperature, K | 329.24 | 359.11 |
| Suction flow, $m^3$/s | 7.63 | 6.53 |
| Polytropic head, Nm/kg | 74715 | 56025 |
| Polytropic efficiency | 0.846 | 0.846 |
| Polytropic exponent, n | 1.380 | 1.509 |
| Ratio of Specific Heat, k | 1.304 | 1.400 |
| Compressibility factor, z | 0.903 | 1.000 |
| Molecular weight | 16.810 | 28.966 |
| Density ratio | 1.487 | 1.487 |
| Equivalent speed by Density ratio | 6190 | 5360 |
| Equivalent speed by Mach. Number | 6190 | 5240 |
| Average equivalent speed | 6190 | 5300 |

Table 3    Confirmation of allowable departure for shop air test

|  | Natural Gas<br>at 6190 RPM | Air at<br>5300 RPM | Variance | Variance<br>Allowed |
|---|---|---|---|---|
| Volume Ratio $Q_1/Q_2$ | 1.4614 | 1.4799 | 101.3 | 95-105 |
| Capacity Speed Ratio $Q_1/N$ | 2.6125 | 2.6123 | 99.99 | 96-104 |
| Machine Mach number Mm | 0.553 | 0.559 | 101.12 | 50-105 |
| Machine Reynolds number Re | 4.15 x $10^7$ | 5.49 x $10^5$ | 1.32 | 10-200 |

Air Performance Test Arrangement

A typical test arrangement for a shop air test is shown in figure 3, using a shop variable speed electric motor and a gearbox. The type of instrumentation, number of measuring stations and their locations as indicated in figure 3 are specified in the Power Test Codes. The nozzle for measuring flow should be sized so that whole operating envelope of the compressor can be obtained. During the testing, the operating condition of the compressor must be maintained stabilized before recording data. The data recorded during the test consists of readings of pressure, temperature, flow and speed. Measurements of power consumption are normally not available through direct readings and are calculated from other data. All instruments used in measuring performance data are calibrated before and after the tests. Allowances for errors in measurements are subject to agreement between manufacturer and customer prior to testing and are included in the proposed test report.

TEST DATA AND CORRECTIONS

In order to translate test data to design conditions the following corrections are applied to the test results.

Reynolds Number Correction

From theoretical and experimental studies it has been well established that Reynolds number affects the performance of compressors. The magnitude of the corrections which should be applied to performance parameters when translating shop test data to actual site conditions has not yet found universal acceptance. Reynolds number can be defined by the following equation.

$$Re = U_2 \, b_2 / \nu \tag{3}$$

As the air test Reynolds number is smaller in magnitude to that at site, a variation in performance results. The magnitude of change increases considerably as the Reynolds number decreases.

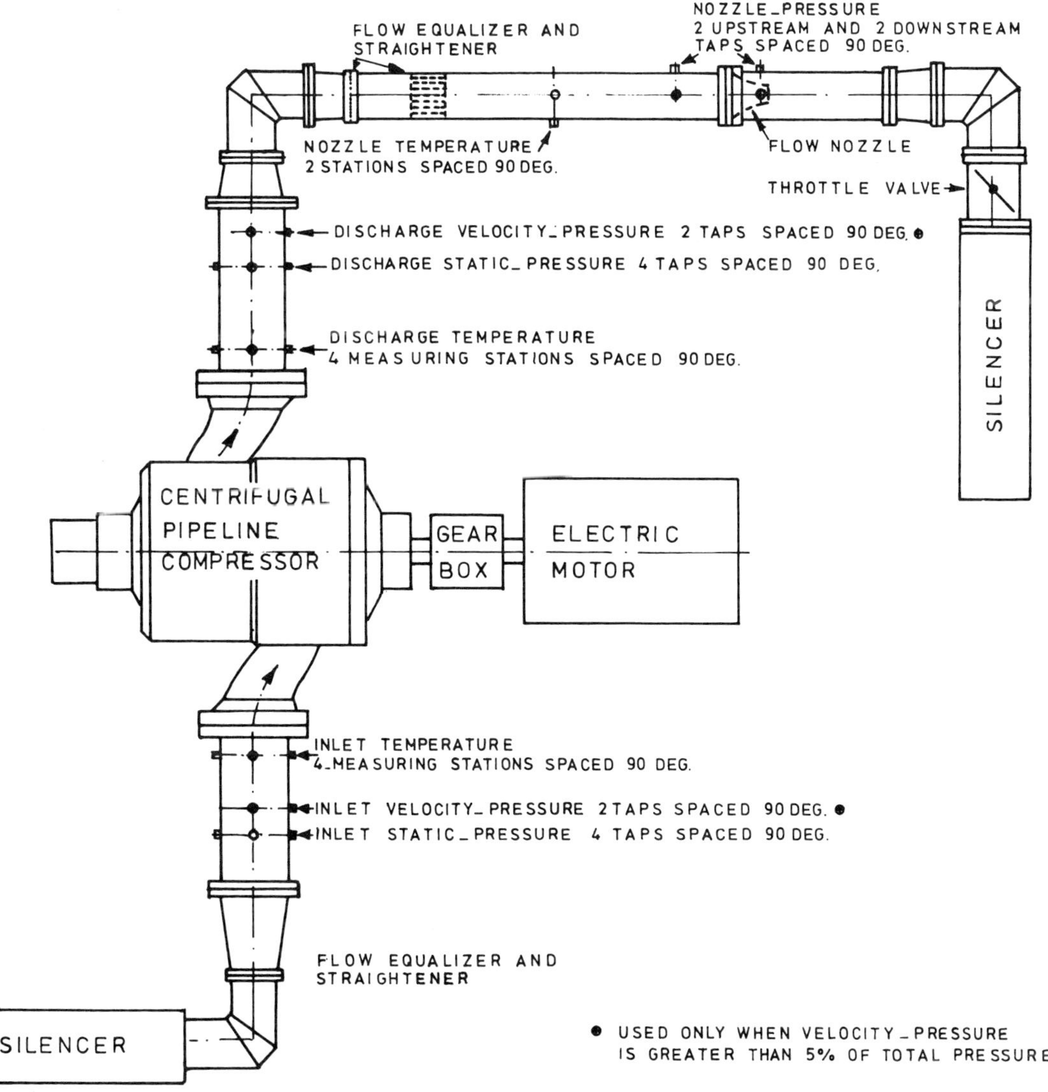


Fig 3   Typical air performance test arrangement

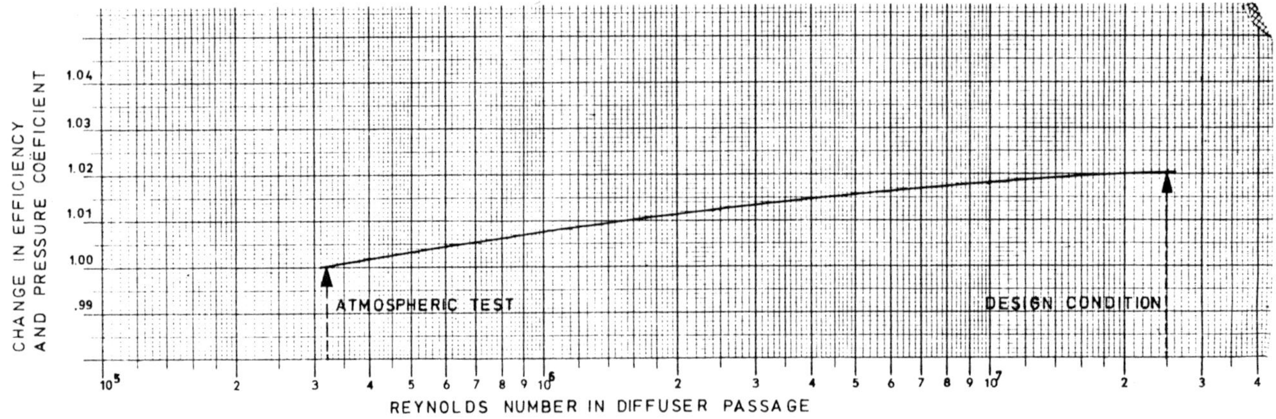


Fig 4   Effect of Reynolds number on performance

In case of pipeline compressors where the variation in Reynolds number between shop air test and actual site operating conditions is significant, the necessity to correct air test data becomes rather important (ref. 3).

Currently corrections applied to compressor performance are principally confined to efficiency and head. A correction for Reynolds number based on accumulated experience is presented graphically in figure 4 (ref. 3).

### Heat Loss Correction

To determine the isentropic efficiency, heat-balance method based on temperature rise is used according to the following equation.

$$\eta_i = T_1 \left[ (P_2-P_1)^{k-1/k} - 1 \right] / (T_2-T_1) \qquad (4)$$

This equation applies only if there is no external heat loss due to convection or radiation from the compressor casing. The external heat loss from a compressor casing can be computed from equation (5) by measuring the exposed surface area, the average temperature of the surface and the ambient temperature.

$$Q_r = Sc \ (T_c - ta) \ hr \qquad (5)$$

When calculating the coefficient of heat transfer assume still air around the compressor casing or moving air if such is the case (refer to heat transfer hand books for methods).
Normally the loss is insignificant and in many instances can be neglected. Insulation of the compressor case and piping further helps to minimize this loss.

### Mechanical Losses and Guaranteed Power

The heat equivalent of mechanical losses from bearings and seals is determined from temperature rise of the lubricating oil and the quantity of oil flow which is determined by calibrated flow meters.
Adding the mechanical losses to isentropic power provides the total power guaranteed.

$$Psh = \frac{mH_i}{\eta_i} + Qr + Qm + Qsl \qquad (6)$$

### Translation of Test Data and Graphical Presentation of Results

In order to translate the corrected test data to design conditions, the following equations are applied.

$$Q_d = Q_t \ (Nd/Nt) \qquad (7)$$

$$Hd = Ht.Rc \ (Nd/Nt)^2 \qquad (8)$$

$$\eta_d = \frac{Nt.Rc}{\eta_t} \qquad (9)$$

The summarized test results are presented in the form of a performance curve of discharge pressure or pressure ratio, power and efficiency versus inlet capacity.

### CONCLUSIONS

With the high and increasing cost of energy, the use of more efficient designs even at higher initial costs can be justified. In order to ensure that guaranteed efficiencies are met performance testing is often specified. When the use of actual gas in shop testing is impractical due to safety considerations or local restrictions, the tests are performed on air and the results are translated to actual site conditions. To achieve reliable results accuracy and consistency in data acquisition is critical. Moreover when test pressure levels are low, effects of Reynolds number and heat transfer must be taken into account to determine precisely the efficiency levels achieved. As efficiency is of prime importance in the evaluation of machinery, the magnitude of the corrections to be applied to test data must be agreed upon between user and manufacturer to obtain meaningful results.


### REFERENCES

1. ASME-Power Test Code PTC-10, 1965. American Society of Mechanical Engineers, New York.

2. Sayyed, S. 'Aerodynamic Shop Testing Multistage Centrifugal Compressors and Predicting Gas Performance' ASME Paper 78-Pet-28 presented at ETCE Houston, Texas. Nov 5-9, 1978.

3. Wiesner, F.J., 'A New Appraisal of Reynolds Number Effect on Centrifugal Compressor Performance' ASME Paper 78-GT-149 presented at the Gas Turbine Conference, London, England. April 9-13, 1978.

4. Koenig, C.F., 'Testing Centrifugal Gas Compressors' Delaval Turbine Inc.

# Offshore gas compression: an operator's experiences

D S T RAUBENHEIMER, BSc, CEng, MIMechE, FIGasE
Shell Internationale Petroleum Company, The Hague, The Netherlands

SYNOPSIS    Since the earliest exploitation of North Sea hydrocarbons, Shell has been an Operator in this field. More recently, in the far north, the hostile environment has compounded the difficulties of high gas pressures, variable flow rates and molecular weights.

The lessons learned are now being translated into designs for future installations, using both centrifugal and reciprocating compressors. This paper describes some of the problems that have been encountered with lube and seal oil systems, piston and liner tribology and noise. Of particular interest has been the selection of compressors for supercritical fluids.

INTRODUCTION

Shell has been an operator in the North Sea since mid 1968 when natural gas from the Leman Bank gas field first flowed ashore at Bacton, Norfolk. The scene moved steadily northwards, and into crude oil discoveries, amongst the most exciting of which is the giant Brent Field, East of the Shetland Islands. Figure 1 shows the extent of Shell's North Sea operations, also in partnership with other participants.

All of the producing installations have gas compression in some form or another. Our gas field installations in the south have large, low ratio, high volume pipeline type centrifugal compressors, whilst on oil producing platforms, very high pressure reciprocating compressors re-inject the gas into the hydrocarbon bearing strata. The drivers are electric motors or gas turbines, determined in each case by economics, process or space considerations.

A platform is an incredibly compact drilling, production, processing, storage, accomodation and recreation complex. It has to be completely self-contained, and an installation which on land might be spread over several square kilometres is crammed into the area of half a football pitch. An offshore platform is arguably the most expensive real estate in the world, it is therefore frugally applied, with severe weight restrictions into the bargain. Add the climate of the North Sea with its capricious weather, gale force winds and sluicing tides, you have what must be one of the most difficult oil patches in the world. Figure 2 illustrates a gas production and compression complex on the Leman field, and figure 3 is our Brent D oil drilling production accomodation and gas injection platform. Note, by the way, the vast difference in the supporting structure, dictated by water depth and sea conditions.

Since the physical laws of fluid behaviour do not change to suit the environment, some novel approaches have been taken to satisfy them within the constraints of the offshore platforms. Such novelty has been found also in the compression installations where considerable ingenuity has had to be exercised to accommodate the compressors, drivers and auxiliaries into extremely cramped quarters, yet maintaining adequate gas cleanliness and access for maintenance. This does not imply that mistakes have not been made, but we hope to translate past experience into future success.

Of the numerous topics that could be discussed in this paper, I have selected a few which I believe will be of interest to this Congress. They cover some of our experiences regarding design and operation of centrifugal and reciprocating gas compressors, environmental considerations and compressing of supercritical gases.

CENTRIFUGAL COMPRESSOR SEAL OIL SYSTEMS

To prevent gas leakage from the compressor along the rotor shaft, oil film seals are customarily used. The seals and sealing systems are designed to API standards API614 and API617. Oil at a pressure above the gas pressure at the seal chambers is supplied to the seal faces, see figure 4. The bulk of the oil flows to the atmospheric pressure side of the seal and returns to the reservoir, but a small quantity flows towards the gas side. This is drained away with a small gas flow to a contaminated seal oil trap as shown in figure 5. The traps are drained of contaminated oil under level control.

On our 13 500 kW Leman compressors, the seal oil system is combined with the lube oil system. The contaminated seal oil rejoins the oil circuit after passing through an atmospheric

degassing section of the reservoir. Shell Turbo
T 32 oil, with a specified viscosity of 32 cS at
38°C and a flash point of 200°C is used in the
system.

Shortly after commissioning the compressors, it was found that the viscosity had
degraded to around 28 cS at 40°C, and worse, the
flash point had deteriorated to as low as 21°C.
Samples of the lubricant were found to contain
up to 5% ww hydrocarbons in the range, $C_1$ to $C_{15}$.

The source was clearly the seal oil
passing towards the gas stream, and intermingled
with it in the contaminated seal oil traps.
The atmospheric degassing system was evidently
not successful, and the only means of controlling the viscosity and flash point was to dis-
card the entire lubricating oil inventory every
fortnight. As each of the four compressors contained 10 000 litres, this was a costly exercise.
An added impetus to find a solution was the
discovery that the atmosphere in the oil tanks
had an oxygen content varying from 2% to 20%,
well within the flammable range. In addition,
laboratory tests indicated that the oil being
returned to the reservoir could be electrostati-
cally charged. There had already been a serious
fire due to lightning flash igniting the vent
stack and lube oil tanks vents during a severe
storm in December 1978; avoidance of any risk
of fires within the lube oil tanks became a high
priority.

Blanketing of the oil reservoir with
gas is impracticable. Considerable quantities
of air are entrained in the gravity drain lines
from the compressor and power turbine bearings.
Over-pressuring the reservoir to prevent this en-
trainment would inhibit the free flow of the
gravity drains, and gas could migrate from the
bearing chamber into the hot parts of the power
turbine. The line of attack had to be to remove
the contaminants from the recirculating seal oil.
The contaminated seal oil flow is too large to
contemplate discarding it, as is done on some
smaller centrifugal compressors on process
duties. There was also no convenient way of disposing of the contaminated seal oil.

At our Stanlow, U.K. refinery, we have
been successfully reclaiming seal oil contaminated with hydrocarbons and hydrogen sulphide from
a platformer recycle compressor. The reclaimer
consists principally of a packed column, in which
the heated oil is stripped of its contaminants
by a counter current of nitrogen. Whilst successful, it was inapplicable offshore as a ready
supply of nitrogen was not available in the
quantity required.

Several manufacturers of proprietary
lubricating oil clean-up equipment were approached and eventually a vacuum degassing system was
selected. Contaminated seal oil is collected
in a heated tank, and is then dispersed over a
large surface area of inert material in a vessel
evacuated to 270 Pa ( 2 torr) pressure. Light
hydrocarbons distil off, which can be assisted by
the introduction of a carrier gas, and are boos-
ted to the vent system by two exhausters in
series.

At the time of writing, the degassing systems are being commissioned. Initial
results are encouraging  in that during the runs
achieved so far, the degassing units have significantly improved the quality of the lubricating
and seal oil. Amongst the teething troubles has
been the seizure of the primary exhauster, due to
dilution of its lubricant by the light hydrocarbons distilled off the seal oil.

We are also testing the effective-
ness of anti-static additives, as used in avia-
tion fuels, in reducing the static electricity
charge in the lubricating oil reservoir. No
results are available at present, but the aim is
to eliminate this possible source of ignition of
the vapours in the reservoir air space.

Degradation of viscosity and flash
point of the lubricating oil has not been unique
to the Leman compressors. We have found that
wherever mixed hydrocarbon gases are being compressed, and the contaminated seal oil is re-
circulated, the problem is the same.
We are thus specifying separate lube and seal oil
systems on future applications, and the installation of contaminated seal oil reclaimers where
the discarding of contaminated oil is unacceptable. Inert gas blanketing of the seal oil tanks
apparently does not positively prevent ignition
of the vapours, as was recently demonstrated by
an ignition in one of our refineries. Perhaps a
strong air draught through the reservoirs to keep
the atmosphere below the lower flammable limit
will be the answer. Further research is being
undertaken to determine efficient, economical
methods of reclaiming contaminated seal oils.

COMPRESSORS FOR SUPERCRITICAL FLUIDS

Oil fields are not as the name describes, producing only liquid crude oil, but are
accumulations of mixed hydrocarbons, some of which
are gaseous at atmospheric pressure. When the oil
is produced, the gas is flashed off, and it can
be in prodigious quantities. Far offshore, it
may be uneconomical or impractical to exploit the
produced gas, and it is therefore frequently flared.
Such waste is unacceptable today, and the only
alternative may be to re-inject the gas into a
suitable formation.

The gas can be recovered at a later date, in the
meantime the re-injected gas can assist in maintaining the pressure of the producing formation.

There are a variety of techniques for
calculating vapour/liquid equilibria of the produced hydrocarbons, and for predicting the behaviour of the vapours released during the oil/
gas separation process. The behaviour of most of
these vapours can be predicted with reasonable
accuracy up to around 150 bar pressure, which
covers most of the compression applications for
gas exploitation. For re-injection however, much
higher pressures are encountered, up to 700 bar,
and the behaviour of the vapour or gas becomes
more difficult to predict, particularly near the
critical point.

An analysis of a gas we are to inject on a North Sea platform is shown in figure 6. The molecular weight is 22, and had to be compressed from 138 bar and 27°C after dehydration, to 415 bar. The phase envelope is given in figure 7. However, the critical point is markedly influenced by the properties attributed to the $C_7$ and heavier fraction. These constituents are in extremely small quantities and thus difficult to identify in the analysis of a discovery well-stream, upon which the entire process design may be based. The $C_7$ and heavier fraction above was given a molecular weight of 104, and the mixture critical conditions of 117 bar and 260 K. With slight variations in composition it is conceivable that during the compression process the 'gas' will be near its critical point.

What is the harm done by this condition? Compressors are designed to handle compressible fluids; if an incompressible fluid is aspirated, particularly by a positive displacement compressor, mechanical damage is the inevitable result. It is therefore important to have an accurate prediction of fluid behaviour, also to maintain the suction temperatures at a sufficient level to be safely in the vapour zone.

A further problem has emerged. Under supercritical conditions, that is, in the dense fluid phase, we have found that compressor cylinder lubrication has been inadequate. In spite of injecting enormous quantities of lubricant into the cylinders, up to ten times the compressor manufacturer's recommended rate, we have experienced severe cylinder and piston scuffing and very high wear rates. There are indications that under these conditions mineral lubricants are rapidly and totally absorbed by the compressed fluid and no longer perform their intended function. Lubricants heavily compounded with animal fats have been more successful, but it is feared that they will ultimately damage the structure of the reservoir rock into which the gas is injected.

The practical manifestation of the foregoing problems have been legion. The injection fluid volumes we have had to compress have not been sufficient for application of centrifugal compressors, and we have had to resort to reciprocating compressors. Severe difficulties have arisen on several gas injection installations. I have selected two for discussion here, which have notable tribological and mechanical problems.

The first group run at 1000 rpm driven by 2.5 MW gas turbines, 120 mm stroke x 114 mm diameter tail rod cylinders. Suction pressure is 131 bar and 32°C, discharging at 310 bar. The aluminium bronze piston coating has scuffed rapidly in contact with the cast iron cylinder liners, and piston ring gaps have increased from 1 to 5 mm in 40 hours operation. We have been using a mineral lubricant, 10% compounded with tallow which has reduced the scuffing, but gives other problems. It is unsuitable for the lubricator pumps. Cams and followers wear at a considerable rate, leaving shavings which ultimately choke the pumps and check valves.

Although we are injecting 7 litres of lubricant per hour into the cylinders compressing 30 700 $m^3$/hr gas, none of this lubricant separates from the fluid after cooling, and there is evidence that it is plugging the reservoir rock, reducing its injectivity.

On another group of injection compressors, each of the six cylinders has the first stage at the head end, and the second stage at the crank end, see fig. 8. The cylinder bore and stroke are 187 mm and 483 mm respectively, and the piston rod is 114 mm diameter. The piston and piston rod are a single steel forging. Each of these compressors handles $2.23 \times 10^5 m^3$/h from 138 bar to 415 bar, and runs at 296 rpm, driven by an 8500 kW electric motor. With this staging arrangement, the pressure differential across the piston ranges from 27.6 $MN/m^2$ (4000 psi) on the inward stroke to 0.35 $MN/m^2$ (50 psi) on the outward stroke. The lubricant used is similar to the first group of compressors mentioned, as is the injection rate. They have not experienced the scuffing problem, but loss of reservoir injectivity is a problem. The initially fitted compounded PTFE rider bands were of insufficient cross section and were rapidly destroyed, partly by being blown out through the discharge valves. The pistons had to be remachined to accept thicker, segmental bronze rider rings. There have been failures of the pistons themselves in 900 to 1500 hours service. The actual mechanism of failure is under investigation, but it appears to be fatigue initiating in the corner of the piston ring groove nearest the crank end. This area is highly stressed during the inward stroke.

All our injection compressors have an unsatisfactory valve life, breakages are not limited to any style of valve, or to suction or discharge. Some breakages are akin to those resulting from liquid ingestion, suggesting that the fluid compressibility is less than predicted by calculation.

Our experiences have led us to specifying the following design criteria for compressors handling hydrocarbon mixtures under supercritical conditions:

1    Ensure that the suction temperature is always above the predicted dewpoint, after allowing for expansion through the suction valve.

2    Use slow running speeds, 450 rpm maximum and 5 m/s maximum piston speed.

3    The piston should be short with respect to the stroke. All of the cylinder wall should be exposed to the fluid during one crank rotation. Frictional heat removal is thus enhanced.

4    Provide some means of relieving pressures generated by low compressibility fluid inadvertently aspirated. Use either spring loaded discharge valve cages or 'stretching' cylinder head bolts, such as applied to hyper type compressors used for polyethylene manufacture.

5        Not more than 5 pressure rings
are needed, preferably with the rider band in the
centre of the piston. Double acting single
stage cylinders are preferred, but often impracticable if crosshead thrust reversals are to
be achieved. Tailrod cylinders are avoided.

The above criteria were applied
to the selection of two injection compressors
for a new platform in the North Sea. These
compressors have passed their works test and are
currently being installed in their modules.
They are scheduled to enter service in March 1981.

The problem of cylinder lubricant
remains unsolved; intensive research is being
undertaken on two fronts. The first is to establish the effect of non-mineral lubricants on
the reservoir rock, the second is to develop
lubricants and piston trim materials that will
give satisfactory performance. The conventional
lubricators also require attention, the cylinder
lubricants are unsuitable for the slow speed
cams and followers. Diesel engine fuel pump
types are under consideration, trials have been
encouraging. Further attention needs to be
given to valve and piston rod packing design
to improve service life.

There is considerable urgency in the
resolution of the problems associated with supercritical fluid compression. With worldwide
emphasis on energy conservation, we foresee an
increasing demand for gas injection compressors
in the future.

NOISE LEVELS

Noise pollution is becoming an increasingly environmental topic, and legislation
such as the U.K. Health and Safety at Work Act,
sets out maxima for noise exposure at workplaces.
On offshore platforms the situation is rather
different, as the place of work is also the
place of residence. Personnel are subjected to
the same type of noise whether at work or at
rest. True, the accommodation modules are
acoustically treated to provide quiet during
rest periods, but some tones are remarkably
resistant to treatment.

This is exemplified on our Leman AK
compressor station. A noise survey revealed a
noise peak at 1100 Hz with second and higher
harmonics which exceed the desired noise level
ranking curves at several locations on the platform. Fig. 10 shows the measured noise level
compared with ISO curve 85 (90 dBA), and fig. 11.
shows the noise levels at distances from the
compressors. The noise appears to be aerodynamically generated in the centrifugal gas compressor,
and is radiated by the casing, suction and discharge piping. It occurs mainly during peak power
output of one compressor, being slightly atte-
nuated when at reduced power output, or with
more than one compressor on line.

The principal annoyance of a distinct-
ly tonal noise is in its apparent inescapability.
Even in areas where noise levels are well within
required criteria, these tones are still detect-
able. Unlike a conventional place of work, personnel have to endure this noise at all times the
compressors are running in the mode which generates the noise. During a current expansion
plan, steps are being taken to attenuate the
noise at source.

The foregoing illustrates the need
to pay careful attention to the elimination of
noise at the design stage of an offshore installation. More attention requires to be paid
to attenuating noise at source. Wearing ear
defenders is commonly accepted practice, but
leads to discomfort and fatigue if worn for
long periods at a time. The environment in the
accommodation modules will also be improved.
We have learned that it is not an easy exercise
to add acoustic attenuation equipment once a
platform is in operation.

TESTING

Many of our offshore projects have
been delayed from the target commissioning date.
The delays fall broadly into two categories,
namely, those due to work being incomplete
onshore on the load-out date, and those due to
untested equipment encountering commissioning
problems. I include design defects in the
latter category.

On a land based project, during the
last phases of construction, it is possible to
flood the site with personnel to accelerate the
pace of work when earlier delays have been
encountered. Similarly, during commissioning
it is a relatively easy matter to bring the
appropriate specialist to site, or to send
equipment away for repair when problems arise.
Not so offshore. The pace of construction
and commissioning work is determined primarily
by the number of beds available on or adjacent
to the platform. If there is an equipment
failure during commissioning, it may be weeks
before it can be replaced or repaired due to
weather conditions restricting offshore transport operations. Space restrictions also make
access difficult for equipment removal for
repair.

A further restriction is that once a
platform is 'live' and has oil or gas in production, simple tasks such as welding and heating
or using open electrical tools become a major
undertaking. The ever present risk of flammable
vapours in the atmosphere dictates that stringent
safety rules must be observed, and such jobs
cannot always be done when desired.

We have also experienced a very high
incidence of plant damage due to construction
debris remaining in vessels, pipework, etc.
Several of our reciprocating gas injection compressors have suffered severe damage to bearings,
crankshafts, cylinders and valves due to debris
in suction and lubrication systems.

In an attempt to reduce commissioning
delays, we are now specifying more rigorous works
testing of our compressors and drivers. The
compressors and drivers are first individually
works tested to the appropriate codes, at least
one centrifugal compressor of each type will be
subjected to a full aerodynamic performance
test. Having done this, we then perform a complet

string test of driver and driven machines, using all contract auxiliary equipment. In the case of a gas turbine driven centrifugal compressor this would include the lube and seal oil systems, control and monitoring systems. We will permit omission of the gas turbine air intake filter and exhaust silencer, these are not readily assembled and dismantled without damage. The entire train has to undergo several fully automatic starts and stops in the test, including four hours continuous run at design speed.

The vexed question is whether or not to call for a full load string test. Large volumes of high pressure flammable gases on the test site are not desirable. Thus for low pressures and light gases an inert gas cocktail is relatively easy to prepare to provide a satisfactory load on the compressor train. Not so for high pressures and heavy gases, the inert gas mixture often does not match the properties of the design gas, or requires a reduced speed run to avoid driver overload. Such a test is of no interest, indeed impossible with an electric motor driver. We are therefore usually content to have a no-load string test. The principal merit of the string test is to demonstrate that all components will run satisfactorily together, that auxiliaries have been assembled correctly, that the control and monitoring systems function correctly. A satisfactory no load string test therefore gives more confidence that offshore work load and commissioning time will be within expectations.

Those string tests that we have conducted to date have been of immense value in debugging the compressor trains before leaving the vendors works. Oil leaks, faulty auxiliaries, vibrations, incorrectly wired controls and component inaccessibility have been amongst the many offences which have been found. Whilst they may be relatively easily located and rectified when in the works, the increased complexity, time and logistics effort for offshore rectification fully justifies the time and expense taken for the works string test.

ACKNOWLEDGEMENTS

The author wishes to acknowledge the considerable assistance and advice received from his colleagues in the preparation of this paper. He also wishes to thank Shell Internationale Petroleum Maatschappij for permission for its publication.

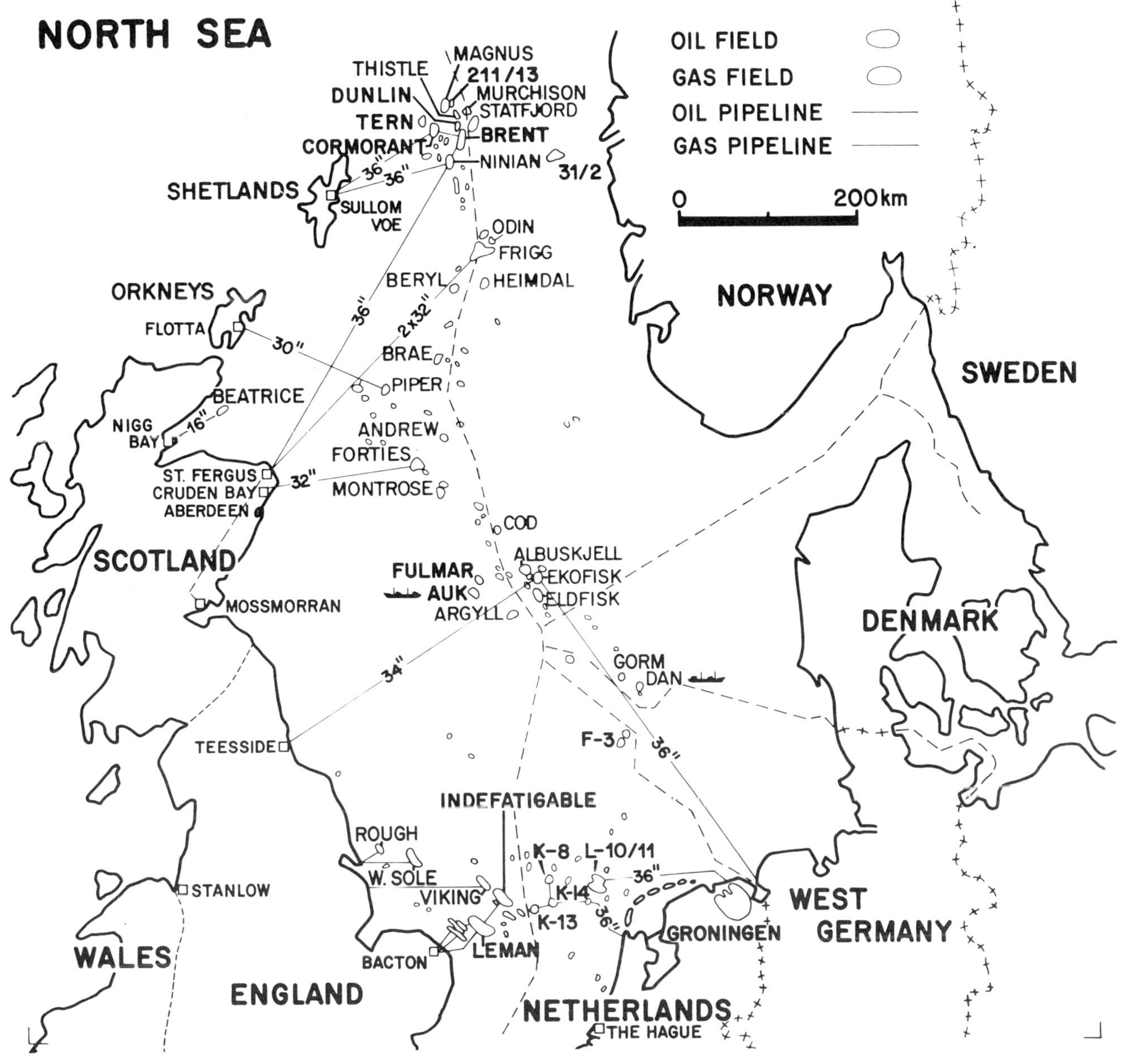


Fig 1    Map of North Sea petroleum and gas production

Fig 2   Leman Field — gas production and compression complex

Fig 3   Brent 'D' production platform

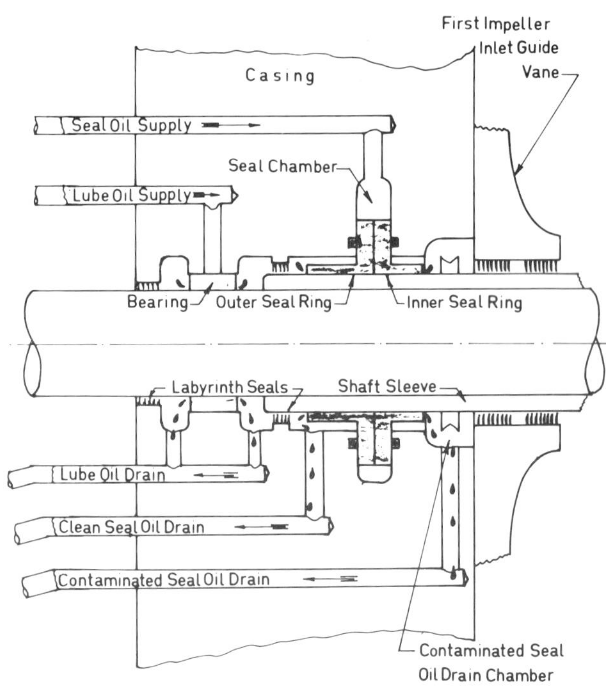


Fig 4   Diagrammatic arrangement of centrifugal compressor oil film shaft seal

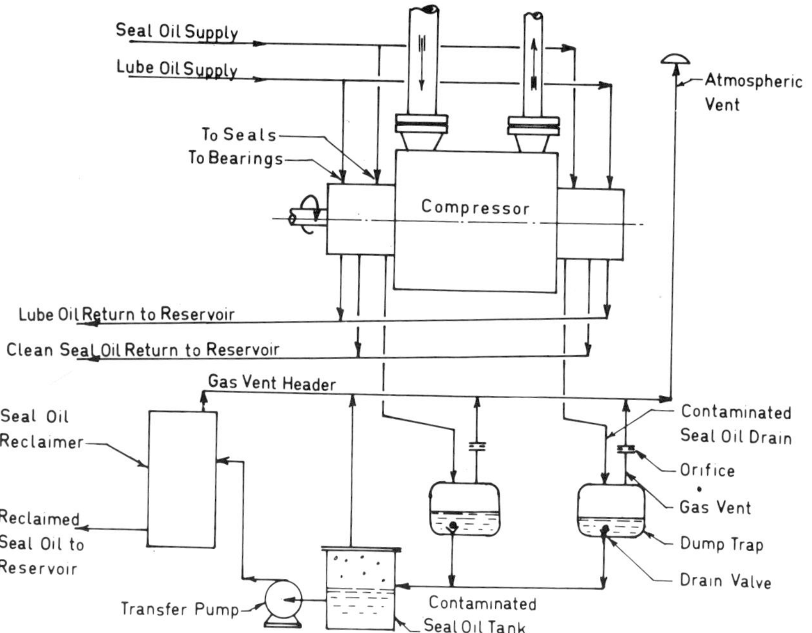



Fig 5   Contaminated seal oil drain system

| Component | Mol % |
|---|---|
| $N_2$ | 0.579 |
| $CO_2$ | 0.705 |
| $CH_4$ | 79.058 |
| $C_2H_6$ | 8.637 |
| $C_3H_8$ | 6.033 |
| $iC_4H_{10}$ | 0.953 |
| $nC_4H_{10}$ | 2.158 |
| $iC_5H_{12}$ | 0.614 |
| $nC_5H_{12}$ | 0.615 |
| $C_6H_{14}$ | 0.347 |
| $C_7H_{16}$ and heavier | 0.301 |

Molecular weight       21.72

Molecular weight of $C_7H_{16}$ & heavier component  104.37

Critical pressure       117.24 bar

Critical temperature   260.56°K

Fig 6   Denmark — Corm Field — analysis of injected gas

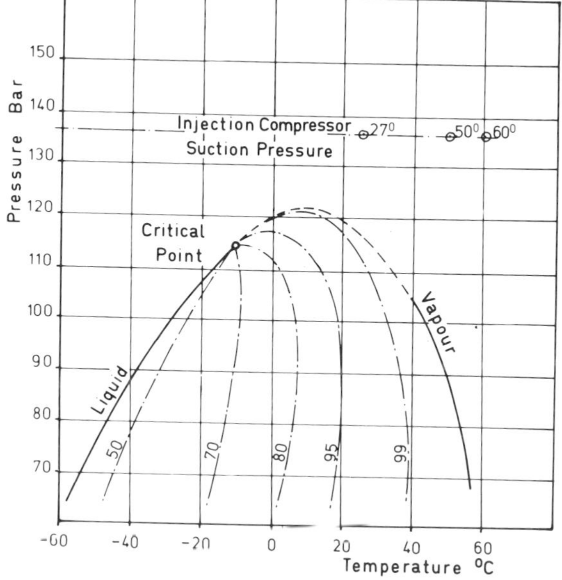



Fig 7   Denmark — Gorm Field — pressure *vs* temperature diagram of injected gas

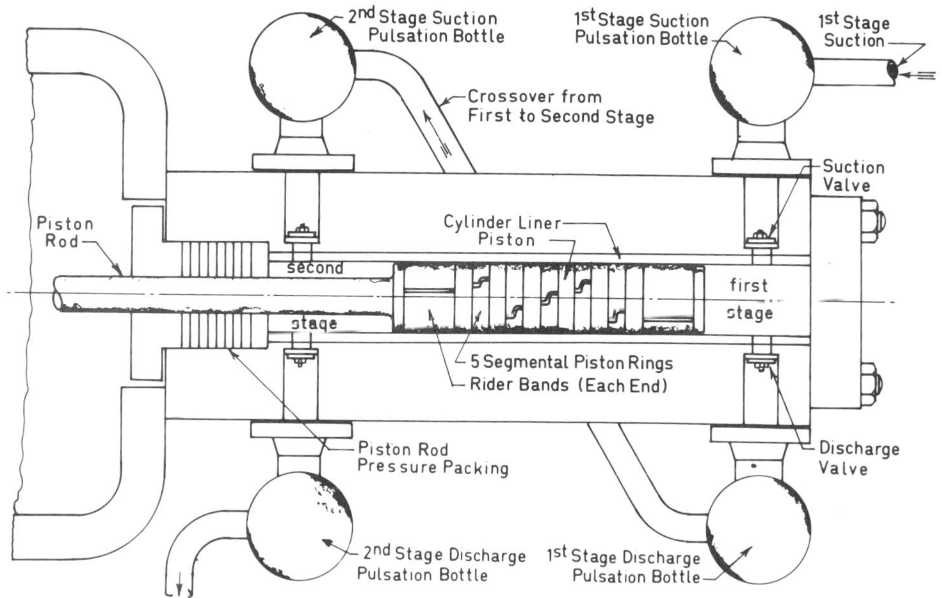



Fig 8   Diagrammatic arrangement of two stage compound cylinder

Fig 9    Reciprocating compressor for high pressure gas injection

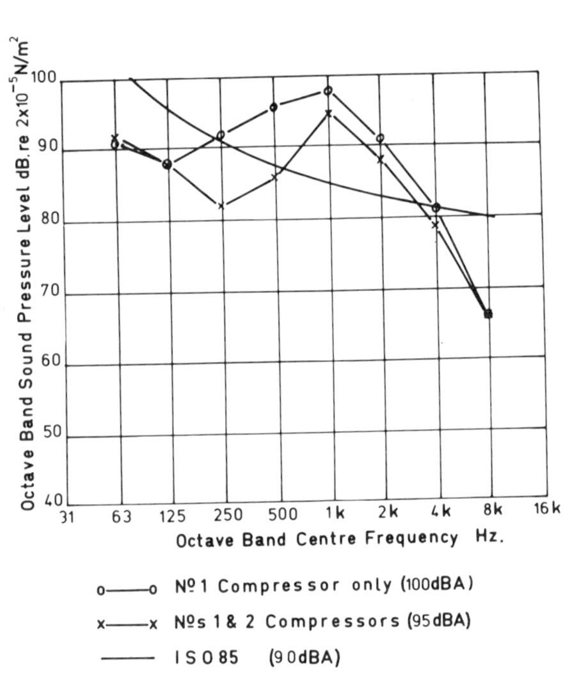



Fig 10    Leman AK — noise level two metres from compressors

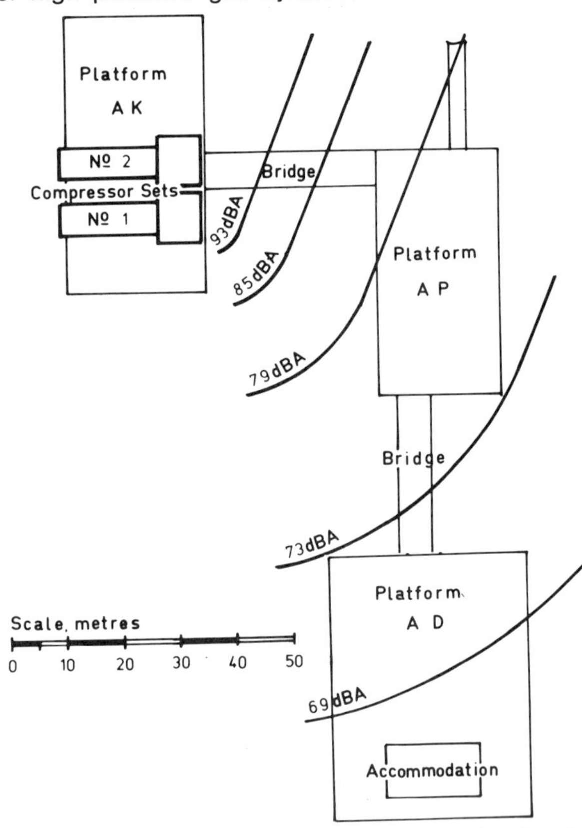



Fig 11    Leman 'A' production group — noise levels radiated from compressors

# C50/81

# The design and operation of a LNG boil off vapour compressor

W WONG, CEng, MIMechE
Bechtel Great Britain Limited

SYNOPSIS   This paper describes the application of a centrifugal compressor required for compressing cold methane at -145°C.   Considerations in its design, selection and problems encountered in its manufacture and operation are explained together with the results of site operating experience.

## 1.   The application

The production of LNG (liquid natural gas) is a continuous process.  The user is invariably in another continent and transportation is usually by LNG tankers and is, therefore, intermittent. Loading takes place about every four days and the production from the plant has to be stored in large insulated tanks.  Each tanker load is about 125 000 cubic metres of LNG.

Between each loading operation the tanks are gradually filled.  Just prior to each loading operation some LNG is taken from the tanks and used to cool the pumping installations and pipe lines.  The tanker arrives empty and the storage space must also be cooled to about -140°C before loading can commence.  All these operations and the action of heat influx into the system result in large volumes of boil off vapour being produce   The quantity of this vapour also varies with the day and night reaching a peak when vapour is produced from the cooling operations described.

The operation of the LNG plant requires a large amount of energy, and the boil off vapour can be used as a source of fuel if a boil off vapour compressor is provided to transfer the gas into the plant fuel system.  Most plants in the world are powered by steam turbines and the fuel mains is at low pressure.  In the application described the plant electric power is supplied by 100 MW of gas turbine driven generators and the main refrigeration loops are provided by 180 MW of gas turbine driven refrigeration compressors.  Gas turbines require relatively high pressure fuel gas.  This application is unique because of this last reason and has entailed the solution of quite a number of design problems not hitherto encountered.

Apart from the major compression units all other services were to be motor driven.  This meant that the boil off vapour compressor had also to be electric motor driven.  The resulting need for rapid start contributed to the design problems to be solved.

## 2.   The design choice

The head requirements of the duty meant that a centrifugal compressor having about 10 stages would be needed.  Preference was to use compressors of proven design but for such low inlet temperatures no machines of more than five to six stages existed, and a train of two cases in series would, therefore, have been necessary. Ethylene compressors with nine stages are quite numerous but inlet temperatures for ethylene compressors are in the region -100°C not -145°C. The disadvantage of using two casings would be the increased dynamic mass of two rotors and the increased thermal mass of two machines in series with the interconnecting pipework.  The capital cost of such an installation was also more expensive than using a single case machine. It was attractive, therefore, to consider the feasibility of a single case machine.  The design risk was in having a large number of impellers on one shaft.  There was a fear of thermal distortion due to the rapid changes of temperature involved from starting and stopping. This was especially worrying as it was only possible to use an electric motor drive.  There could be no gradual increase in speed that would have been possible with a turbine drive.

A comparison between the two designs is given in table 1.

Demag, the manufacturer chosen, had considerable experience in oxygen compressors.  Oxygen compressors have a fire hazard due to possible ignition of metal caused by localized energy release from particle impact.  A high conductivity impeller is used on such machines in order to minimize this danger.  This same feature was considered attractive in avoiding danger of impeller or rotor distortion from rapid temperature changes.

For the same reasons it was also attractive to have a shaft of symmetrical design.  The Demag standard use of shrunk-on impellers helped to achieve this objective because it avoided keyways in the shaft.  The use of circumferential grooved and split rings for location of the impellers was also considered beneficial as rotors constructed with spacer sleeves between the impellers had been known to give rise to shaft deflection. The method of impeller location is indicated in Fig. 1

Before finally proceeding with the single casing it was also necessary to establish if the basic rotor dynamics would be satisfactory. It was considered that when the rotor is rapidly cooled during start up some thermal distortion could take place which would lead to rotor imbalance. Under these circumstances it was necessary to demonstrate that the rotor design was conservative in concept and stable in characteristic.

In order to compare the proposed design with machines in successful operation it was decided that a survey of other machines of similar bearing span and rotor mass should be carried out.

## 3. Establishing rotor dynamic stability

Rotor stability depends on the relation of its first critical speed to the design speed. It is also dependent on the exciting and damping forces. The two main forces are:

a. Exciting forces from rotor imbalance: these are dependent on the mass and speed of the rotor. A light high speed rotor is inherently harder to balance than a heavy low speed rotor and hence exciting forces resulting from out of balance masses are relatively larger for a light high speed rotor than those of heavy low speed rotor.

b. Exciting forces from gas flow past the rotor: the magnitude of these forces depends on the gas density. The higher the density the larger the forces. Their effect on the rotor is largely a function of the absolute stiffness of the rotor. A lightweight high speed rotor is far more sensitive to such exciting forces than a heavy low speed rotor.

Figure 2 shows on the y-axis the allowable ratio:

$$N_D/N_{C1}$$ i.e. Design speed/first critical speed

as a function of

$$M_L/N_D$$ i.e. Rotor mass kg/design speed

for different gas densities at discharge conditions expressed in $kg/m^3$. It should be stated that the gas exciting forces are in no way dependent on the number of stages, whether a rotor has 5 or 9 stages, these forces act along the whole length of the rotor.

It was also necessary to survey the Vendor's experience in manufacturing such a rotor. It can be shown that for a given design the rotor shaft diameter is a function of the following main factors:

a. The required first critical speed.
b. The rotor mass.
c. The bearing span.

thus:

$$D \sim \left\{ W_{ki} \quad (m_L \ L^3)^{\frac{1}{2}} \right\}^{\frac{1}{2}}$$

where D is shaft diameter
  $W_{ki}$ is first critical frequency
  $m_L$ is rotor mass
  L is bearing span

The above relationship can be derived from the equation for bending of a simply supported beam:

$$y = m_L \ L^3/EI$$

where y is the static deflection due to the shafts weight.
  E is Young's modulus
  I is 2nd area moment

and noting that by definition $W_{ki} = (g/y)^{\frac{1}{2}}$

where g is gravitational acceleration.

Thus if two different rotors possessed the same value of the above factor then their shaft diameters should be identical and hence their rotors dynamically similar. The statistics from a number of selected machines in operation were plotted using such a factor and the proposed rotor also shown. This is given in Fig. 3.

Based on all the foregoing considerations a single casing design was chosen.

Table 1 Design Comparison

| Design | A | | B |
|---|---|---|---|
| Casing No | 1 | 2 | 1 |
| No of impellers | 6 | 5 | 9 |
| Diameter of impellers | 560 mm | 420 mm | 560 mm |
| RPM | 8700 | | 8740 |
| **Material of Construction** | | | |
| Impellers | 9% Nickel | | Cu Ni |
| Shaft | 9% Nickel | | 9% Nickel |
| Casing | Ni Resist | | 10½%Nickel |
| **Method of Construction** | | | |
| Impeller drive | Key | | Shrink Fit |
| Impeller Location | Shaft Sleeve | | Recessed shaft |

Operating Specifications

| | | |
|---|---|---|
| Inlet Pressure | bara | 1.013 |
| Inlet Temperature | °C | -145 |
| Gas Mol. weight | | 16.24 |
| Discharge Pressure | bara | 16.21 |
| Discharge Temperature | °C | 61 |
| Polytropic Head | kJ/kg | 300 |
| Shaft Power | kW | 2620 |
| Volume Flow (at inlet conditions) | M³/h | 13000 |

## 4. Manufacturing problems

The shaft was forged from 9% nickel steel and
the first batch of material was scrapped due
to hydrogen induced internal cracking.  This
metallurgical problem was resolved after some
investigation and the shaft was sucessfully
made.

It should be noted that there was originally
some suggestion that the shaft could be made
of a lesser alloy steel.  The machine is
driven from the discharge end which would be at
a more normal temperature.  The coldest end of
the shaft would then be at the lowest stress
and would, therefore, be less affected by loss
of ductility and strength.  This was an
attractive idea and was considered when the
metallurgical problems were encountered with
the 9% nickel shaft.  Fortunately the idea was
rejected otherwise it would not have been
possible to solve the next problem encountered
which was the cracking of impellers.

In order to do away with keys and pins the
impellers were shrunk on to the shaft.  Due to
the differential expansion between the impeller
material and the shaft material it was necessary
to have a shrink fit which was sufficient to
take account of the suction thermal transients
as well as the loss of shrink fit due to centrifugal force.  Some $350^{\circ}$C preheat was required
for each impeller prior to assembly on to the
shaft, but subsequent metallurgical tests showed
cracks to have developed in the impeller
materials.

Finally after intense merallurgical investigation
it was found that Cu Ni has a tendency to crack
with high temperature but by control of composition and limiting the maximum temperature to
$250^{\circ}$C cracking of the impellers could be avoided.

However, it was still necessary to adhere to the
required interference fit and this was achieved
by cooling of the compressor shaft prior to
assembly with the impeller.  Cooling of the
shaft was done by rifle drilling of a hole down
its centre and pumping liquid nitrogen through
it.  This solution was only possible due to the
adoption of a 9% Ni shaft and was in a way a
vindication of not using the lesser material
originally proposed.

## 5. Design problems

It was ironic that the very ideas which were
chosen to avoid problems of distortion and rotor
instability also gave rise to contraints on how
the machine could be operated.

The ambient temperature on the site was $35^{\circ}$C and
operation of the machine with warm gas had to be
avoided.  For example, start-up with gas inlet
temperatures above $-70^{\circ}$C would lead to surge.
If very much higher inlet temperatures were
allowed then the discharge temperature could
rise to above $250^{\circ}$C which could lead to impeller
cracks.  It was imperative, therefore, to ensure
an adequate cool down of the machine and to
ensure that the gas into the suction would not
exceed $-80^{\circ}$C.

On the other hand due to the design of the
electrical systems there could be considerable
voltage dip on start-up and if the temperature
of the gas was allowed to fall much below $-80^{\circ}$C
then the increased mass flow would cause the
motor to be overloaded in spite of being 20%
oversized.

During normal operation the compressor suction
was expected to be $-140^{\circ}$C and the discharge to
be $61^{\circ}$C.  During shut-down there would be a
tendency to equalize out the rotor temperature.
For gas turbines and steam turbines with long
rotors a period of barring over would be
recommended to ensure that the rotor would not
sag.  For an electric motor drive the machine
would stop rotating within minutes.

In order to solve the problems of starting and
stopping and in an attempt to control temperature change it was decided to make use of methane gas which was available at $10^{\circ}$C.  This gas
was arranged to be injected during start up to
dilute the expected $-140^{\circ}$C gas supply in order
to obtain an inlet temperature of $-80^{\circ}$C.  Once
the machine started hot gas was then available
to recycle back to suction for the same purpose.
The injection of hot gas was to be timed over
a period of 30 minutes and gradually reduced to
control the rate of cooling.

In a similar way, on shut-down of the machine
the $10^{\circ}$C methane gas was injected for a period
of 30 minutes to ensure that the rotor would
slowly rise in temperature.

It can be imagined that the control system of
the machine became quite complicated and the
flow sheet for the machine is shown in Fig 4.
The flow control of the machine was also
complicated.  For flows less than surge the
machine had to work with recycle back to the
storage tanks and the hot discharge gas cooled.
This was done by injection of LNG spray into a
section of the recycle pipework back to the
tank.

It was necessary to control the capacity of the
machine in order to maintain the storage tank
pressure between 400 and 1200 mm $H_2O$.  This was
done by discharge throttling instead of suction
throttle in order to avoid vacuum conditions at
the compressor suction pipe and hence possible
ingress of air.  Many safety precautions had to
be taken to prevent either over pressure or
under pressure of the tank due to failure of
the compressor capacity control.

6.  A performance test to ASME code PTC was
required to check the machine's performance at
the works before shipment.  Calculation, however,
soon revealed a problem.  It was not possible
to maintain equivalence of volume ratio throughout the machine because of the differing ways in
which the properties of methane and the test
gas, Refrigerant 22, varied during compression.
Table 2 illustrates this point.

Table 2

| Stage | Methane | | Refrigerant 22 | |
|---|---|---|---|---|
| 1 | Z = 0.975 | K = 1.378 | Z = 0.988 | K = 1.163 |
| 4 | 0.976 | 1.35 | 0.983 | 1.1436 |
| 9 | 0.98 | 1.31 | 0.974 | 1.122 |

where Z is compressibility factor and K is the
ratio of specific heats.

This problem was solved by running two test speeds, one comparable with high flow and the other with low flow and interpolating (1).

## 7. Start-up experience

One great concern considered during the design stage was rapid temperature change. It was imagined that the machine would be warm due to the high ambient with cold methane gas present in the suction pipework. Starting of the electric motor would cause cold gas to be drawn into the machine resulting in a temperature change down to -140°C in about twenty seconds. In reality there was no problem. The boil off gas piping interconnecting the tanks was so long and the heat inleak from the sun so large that the problem was getting the inlet temperature down to design in reasonable time.

The first start of the machine was abortive. Because the inlet temperature would not come down sufficiently, the discharge temperature rapidly increased and the machine surged due to lack of pressure rise available to get the gas through the recycle valve. The effect of inlet temperature on compressor performance is shown in Fig. 5.

The second attempt to start was by using the recycle valve together with the manual bypass. This was successful and a typical start-up profile is shown in Fig. 6.

This experience demonstrates that such machines should be started up in the overload or choke mode, that is with maximum flow and minimum head. At the design stage it was planned to start the machine at near the rated point, however because the inlet temperature cannot be accurately predicted it is correspondingly difficult to adequately size the recycle valve. Use of a manual bypass and operating in overload overcomes this problem.

## 8. Part load operation

As mentioned previously the compressor installation was equipped for low throughput operation by recycling gas back to the storage tank after being cooled by liquid LNG. However, in spite of this provision it was not possible to return the gas back to the tank without some superheat.

The warm gas returning to storage was found to stratify in the vapour space and in fact a gas temperature gradient was established between the liquid surface and the tank top. This is very much in line with what can happen with the liquid itself. Because of this tendency prolonged operation with gas recycle soon led to an increase in the gas inlet temperature to the machine. The gradual increase in inlet temperature caused inturn a gradual decrease in compression ratio until eventually the machine was unable to deliver at the required discharge pressure and went on full recycle.

In point of fact it can be shown that operating the machine on recycle reduces LNG production because of the quantity of liquid required to cool the recycle gas. This liquid is evaporated and has to be pumped into the fuel system. Generally speaking the value of LNG loss soon became more than the value of vapour saved.

## 9. Operating problems

The machines have now been in operation since 1978 and there have been no mechanical problems to date and the mechanical design concept can be considered proven.

In performance however, a problem has been encountered due to the compressor internals fouling with oil after a period of time. This requires the machine to be shut down and allowed to warm up. Any concealed oil liquefies and is drained out and the machine performance is restored. The cause of this oil fouling is still being investigated.

It is suspected that the drain oil from the process side of the seal is cooled so that the drain passages slowly become choked due to a layer of congealed oil. Eventually the gas packing flow is cut off and oil passed into the process stream. This theory is at present unproven but should be considered for future designs.

## 10. Acknowledgements

I wish to express my thanks to Bechtel Great Britain Limited, my employer, and to Demag A.G., the manufacturers of the compressor described for permission to write this paper.

I am also indebted to Demag for the concept of rotor dynamic similarity and the data shown in Fig. 2 and Fig 3.

## Reference

(1) Acceptance testing of centrifugal compressors and the application of ASME Power Test Code PTC10-1965. I. Mech E paper C25/78.

Author W. Wong



 C50/81

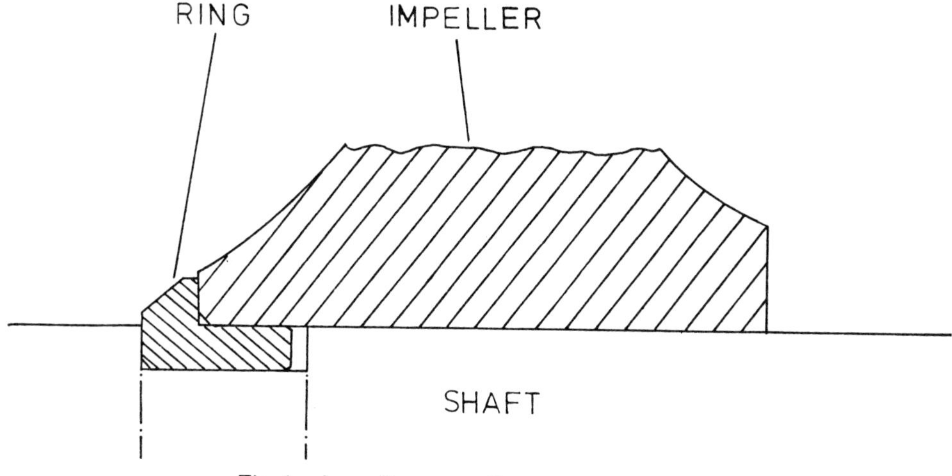


Fig 1    Impeller mounting arrangement

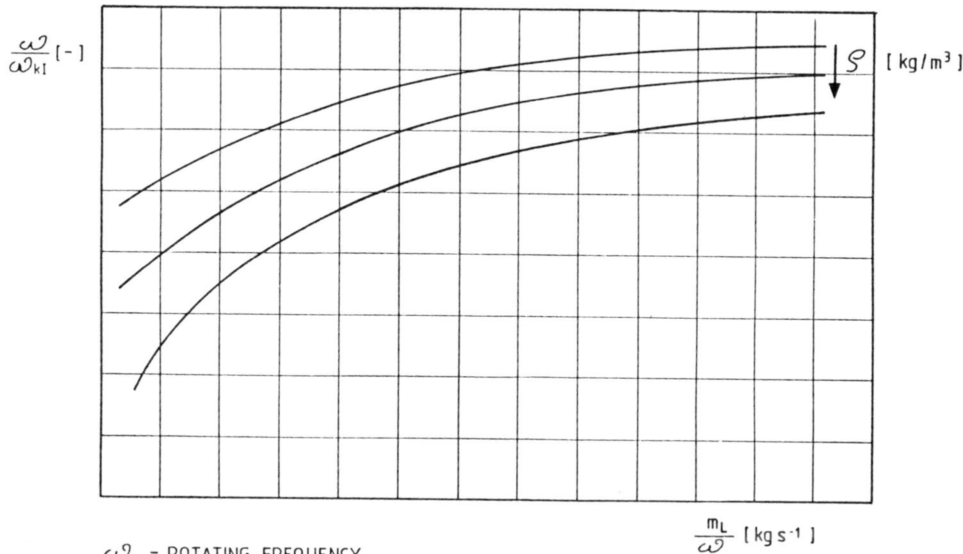


Fig 2    Ratio of rotating frequency over first lateral critical frequency as a function
of ratio of rotor mass over rotating frequency and spec. density

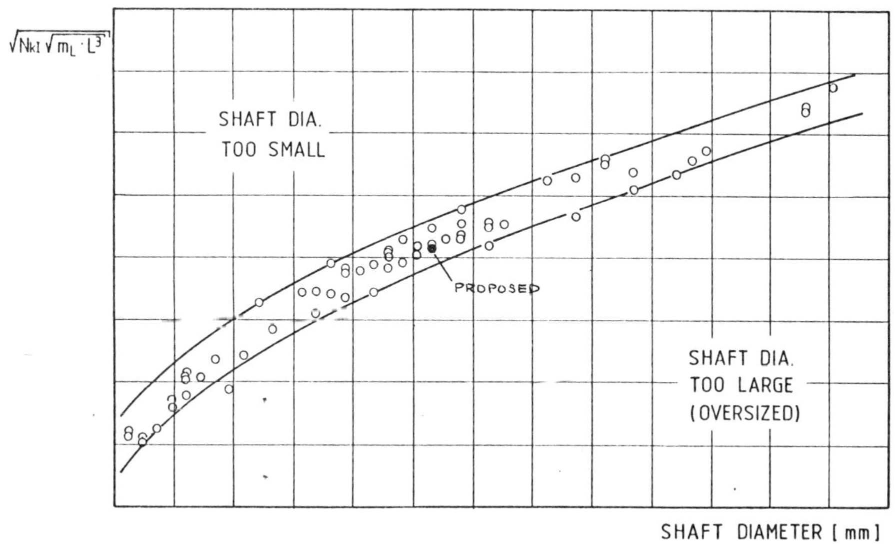


Fig 3    Rotor similarity

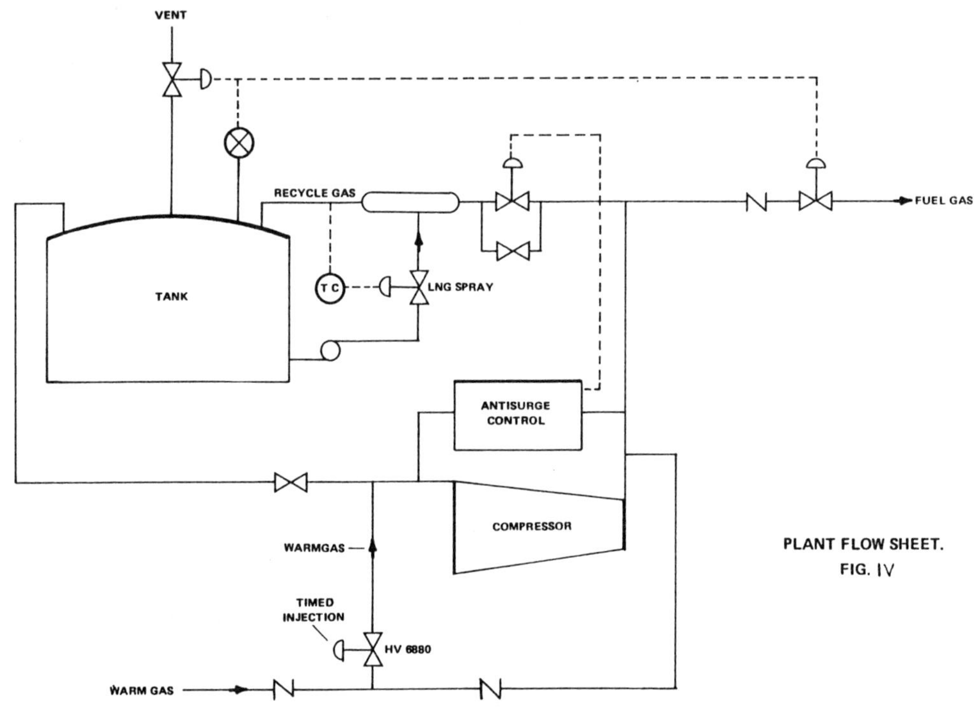


Fig 4    Plant flow sheet

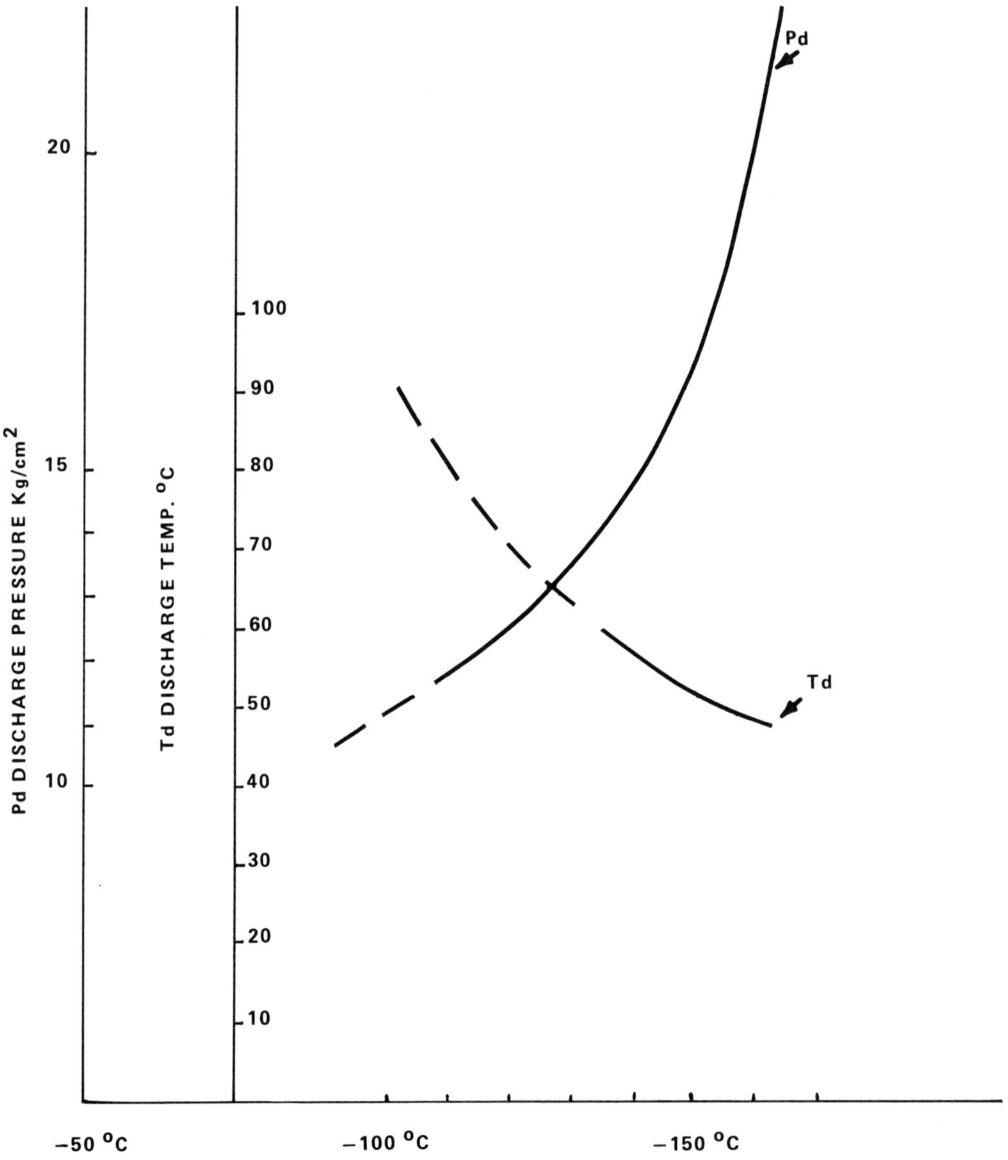

Fig 5    Variations of discharge conditions with inlet temperature

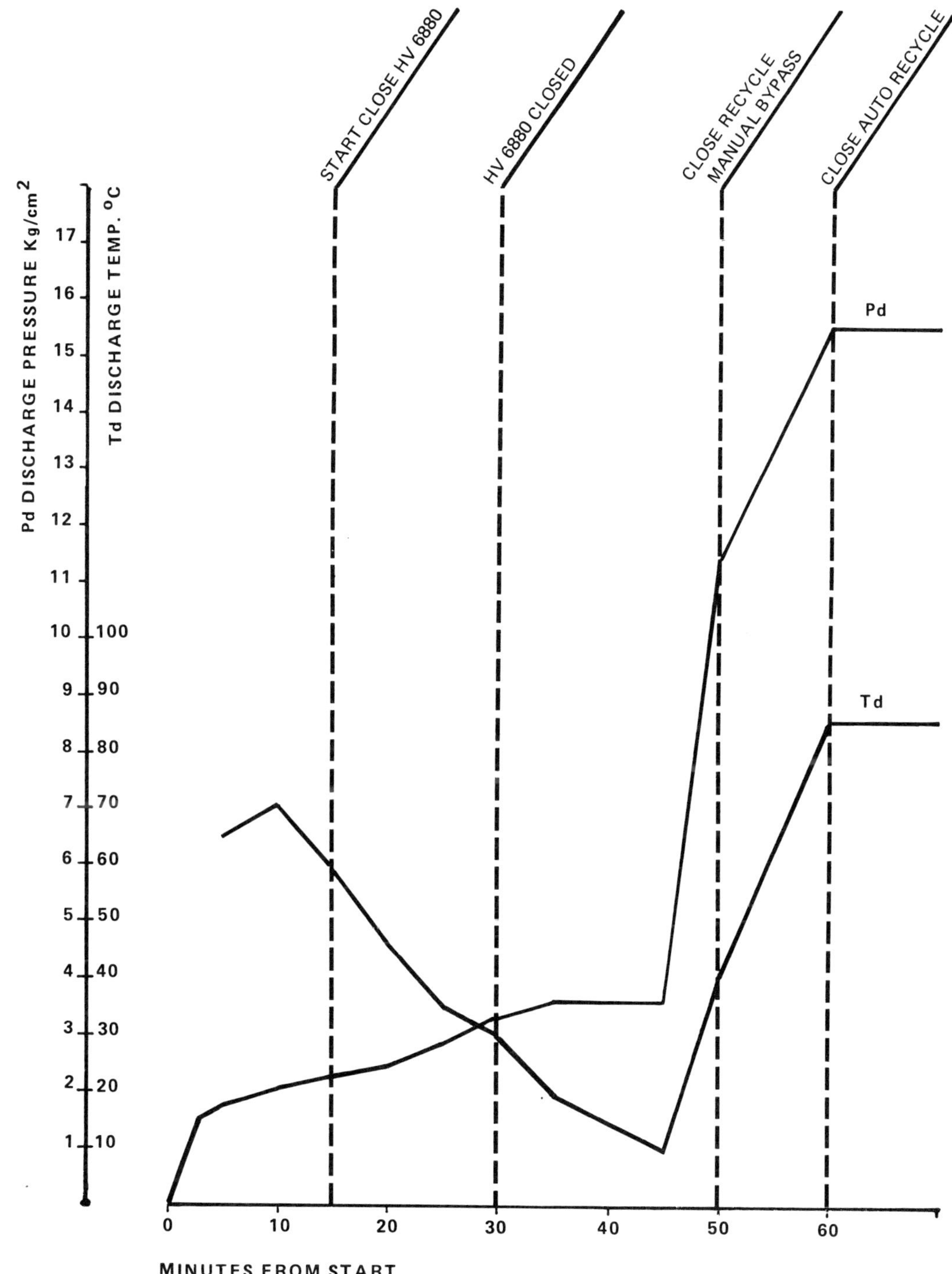


Fig 6   Starting characteristics

# Centrifugal compression on the Arun high pressure injection project

N J COLETTI, BSME,
Dresser Clark Division, Dresser Industries, Inc, Olean, New York, USA
M E CRANE, Jr, Mobil Research and Development Corp, Princeton, New Jersey, USA

SYNOPSIS  This paper briefly discusses the design, testing, operation, and maintenance considerations experienced on the Mobil/Indonesia Arun High Pressure Gas Injection Project.  Although the hydraulic performance of these compressors in such applications follows normal and accepted design parameters, many other unexpected problems have been encountered.  Avoiding such problems in the design stage or correcting them in the field requires careful application of new philosophy and criteria in the mechanical design of such compression systems.  Since the subject is extremely broad, the intent of this discussion is to only highlight the main problem areas and present the corrective actions which have been utilized to provide reliable centrifugal compression facilities in this pressure range.

GENERAL COMPRESSOR DESIGN

The Mobil Indonesia Arun Project consisted of three compressor trains, each train consisted of a 272B4/4 low stage compressor and a 181B3/3 high stage unit.  These were driven by a General Electric Frame Five Gas Turbine through a Maag Gearbox.  The first train was started in May of 1977.  Following is a summary of the design operating parameters.

Table 1

272B4/4

| | |
|---|---|
| Pressure inlet | 1100 PSIA |
| Temperature inlet | 115°F |
| Pressure discharge | 4200 PSIA |
| Temperature discharge | 242°F |

181B3/3

| | |
|---|---|
| Pressure inlet | 4177 PSIA |
| Temperature inlet | 125°F |
| Pressure discharge | 7115 PSIA |
| Temperature discharge | 168°F |
| Compressor speed | 10080 RPM |
| Design Gas Flow | 170 MMSCFD |
| Design Horsepower | 25200 BHP |

Gas Composition:
$CH_4$ - 75%, $CO_2$ - 15%, $C_2H_6$ - 5.5%, $+C_2$, $C_3$, etc.
$H_2S$ - Trace.

Both compressors have a one-piece barrel like outer case (see Fig. 1).  Open ends of the barrel case are closed by removable covers, 'heads.'  The heads are held by segmented retaining rings of the Clark patented shear ring construction.  Conventional o-rings are used as the gas seal between the heads and case.  The inner case, consisting of diaphragms and guide vanes, has horizontally split components for accessibility.

Rotor construction consists of a forged one-piece shaft with interference fit impellers.  Impeller spacers and sleeves are used under all labyrinth seals and oil film seals.  The impellers are arranged in two sections back-to-back and discharge from the centre of the case.  This arrangement has advantages with regards to thrust loading and also shorter bearing span when intercooling is a requirement, as is the situation here.  Impellers, spacers, and sleeves are shrink fit by design and assembly is obtained by heating.  Both torque and axial loading are transmitted by the interference fit without the use of keys.

Couplings between units are continuously lubricated gear type and have performed well.

MATERIALS

Materials of construction for natural gas reinjection compressors really are not much different than other process gas compressors at the same pressure levels.  The major concern is o-ring material and protection against particulate erosion.

Basic Materials

Cases and heads are forged carbon steel. Components such as diaphragms and inlet guide vanes are standard ductile iron.  Impellers are AISI-4330 welded construction with the added requirement of 90000 maximum yield and RC22 maximum hardness for stress corrosion protection. This has been standard even for trace amounts of $H_2S$.

Other reinjection compressors use standard cast aluminum labyrinths.  This project also uses them but differs from other units in that they are DYKOR coated.  This is a Teflon coating designed to protect the aluminium from attack by mercury present in the gas.  Actually, laboratory tests proved that mercury does not attack the aluminium directly.  Instead it reacts with aluminium oxide.

Since the gas contains no $O_2$, coating the aluminum probably is unnecessary.

Oil film seal rings, journal bearings and thrust bearings are all steel backed, SAE-11 babbitt faced.

## O-rings

O-rings are used throughout the centrifugal compressor for internal sealing of process gas, seal oil and lube oil, as well as the main process gas to atmosphere seal at the head to case junction.

The standard o-ring material for natural gas has been Viton. A problem developed here that is unique to high pressure applications. Basically, the gas permeates the o-ring at high pressure, then fractures the o-ring on depressurization. The more the unit is up and down, the shorter the o-ring life. Most o-rings never completely fail to seal. Although the o-ring itself is not a major problem to replace the down time encountered in a remote location is costly.

To use an elastomer type seal, it is necessary to use as impermeable a material as possible while still being compatible with the gas. Among o-ring manufacturers and users, the consensus of opinion is that fluorocarbon rubber, such as Viton and Fluorel with a high durometer results in the most practical solution.

Although fluorocarbon can be made as hard as 95 durometer, 90 durometer is the highest readily available. The 90 durometer has resulted in increased life over the 75 durometer originally furnished. Actual life expectancy is difficult to predict because of the many variables involved especially the number of pressurizations and depressurizations. More research is needed on this problem.

## Shaft sleeve erosion

An erosion problem occurred under the shaft labyrinths resulting in excessive clearances. This in turn caused higher than expected gas leakages and, therefore, lower compressor efficiencies.

The problem was thought to be drilling mud particles in the gas causing an abrasive erosion of the rotating shaft sleeves. The aluminium labyrinths, although much softer than the shaft, are not affected. Their diameter remains constant. The shaft sleeves, however, are severely affected as shown in Fig. 2. The appearance of the grooves are polished with rounded edges, as opposed to a rub which would be sharper and usually show evidence of the labyrinth material.

It was assumed this would alleviate itself as the gas became cleaner with time. However, the erosion persisted resulting in several damaged rotors. Expensive filters were considered but deemed impractical for the volume of gas being processed.

A simple but effective cure was incorporated. The oil film seal sleeves had always been hardfaced with Colmonoy #6 giving good dependable service. It was decided to employ this same design to the labyrinth sleeves.

This was installed in the field units. Performance tests were conducted immediately at start-up and once a month thereafter. To date (eleven months after installation) there has been no deterioration in performance. Previously, deterioration was noticeable within weeks and actual performance was unsatisfactory within one month. Typically, compressor train throughput would be reduced 10 − 20%.

THRUST LOADING

In any centrifugal compressor, each impeller transmits an axial thrust force to the shaft towards its inlet. This thrust force on any high pressure compressor is more than a thrust bearing can withstand. A thrust bearing large enough would have too high a surface speed to be practical. Therefore, this high thrust is balanced in basically two different schemes. The still widely used design, especially at low pressure, is the thrust balance piston. This is a disc at the high pressure end of the compressor with one side at discharge pressure and the other side piped to the compressor inlet. These two pressures are separated by a labyrinth seal at the outside diameter of the balance piston. The balance piston thrust opposes the impeller thrust. With the proper diameter selection of the balance piston, the thrust can be balanced. The problem with this scheme is the reliability of the balance piston labyrinth at high pressure differentials. Any failure of the labyrinth results in uncompensated impeller thrust and castastrophic failure. As an example of the magnitude of the thrust, the 181B unit on this project has a gross impeller thrust of 55200 pounds with a thrust bearing area of 31 inches.

This brings us to the second scheme for thrust balance which is the back-to-back impeller arrangement as employed on this project. With this design there is no balance piston. Instead the rotor is arranged in two sections with the impellers in one section back-to-back with impellers of the second section. Thus, with proper impeller selection and, if necessary, adjustment of the division wall labyrinth diameter, the thrust can be completely balanced.

There are two distinct advantages to this arrangement. First, there is no balance piston labyrinth with a high pressure differential to fail causing rotor failure; and second, a convenient means is provided to take the flow out of the case for intercooling and return it for further compression.

Thrust loading then is designed to be balanced in the back-to-back arrangement. But what about off design conditions? Many users have expressed concern with this. To aid in operation and in control design a simple thrust nomograph is made as shown in Fig. 3. It can be seen that with two variables, namely the pressure differential in each section, the thrust bearing loading can easily be determined. In this case, even though the 181B discharges at 7100 psi, the graph shows that it is not particularly thrust sensitive to operation. This is due to the large thrust bearing and low flow impellers with their small unbalanced areas.

The thrust disc to shaft mounting was a light interference fit from line-to-line to .001 inch. Relative movement between the two resulted in fretting corrosion, after some months of operation, making it extremely difficult to disassemble the bearing. Consequently, the rotors are being retrofitted during overhauls to hydraulic fit thrust assemblies. Fig. 4 shows the basic design. Note that a tapered bushing is used between the shaft and disc. Its purpose is to facilitate rotor axial positioning with respect to the stationary components. The inside diameter of the bushing is a straight bore with a slight slip fit so its axial position on the shaft can be adjusted by grinding the flanged end face. The heavy interference fit exists between the tapered outside diameter of the bushing and the inside diameter of the thrust disc. Because the bushing is slender with respect to the disc it will conform to the disc bore and transmit the high interference fit to the shaft.

The application of a high interference fit prevents relative movement of disc and shaft end, therefore, prevents fretting corrosion.


SEALING SYSTEM

Normal sealing pressure in low and medium pressure centrifugal compressors usually is balanced to the machine's suction. However, in this application, the high stage unit has an inlet pressure of 4180 PSI. Although seal oil pumps and necessary hardware exist for this pressure, it was felt that sealing at a lower pressure would provide the most dependable operation.

Advantages of lower seal pressure include:

A)  Lower seal system cost as a result of: 1) lower pressure rated piping and vessels, gages, etc., and 2) lower compressor seal oil flow. This makes the overhead tank capacities smaller as well as having a lower pressure rating. The pumps are also small with smaller drivers.

B)  Less seal system pump problems. High pressures (above 3000-4000 psi) usually result in piston pumps being selected. These have more wearing parts than rotary type and therefore more problems.

C)  Rotor dynamics - lower seal pressures simplify seal design with respect to rotor dynamics. Seal pressure break-down to atmosphere ultimately results in axial load on the seal. This axial load is important to seal effect on rotor dynamics and is much easier to control at lower pressures.

Lower seal pressures are accomplished quite simply when there is a lower pressure compressor in the train. Specifically, this involves pressure break-down across labyrinths with the low pressure side piped to a convenient connection in the low pressure compressor system. In this particular case the train suction was chosen as the seal pressure for both compressors. This is the preferred arrangement because of the constant pressure seal system, independent of speed or process conditions.

A few problems did develop in the field that required modification. First, hydrates formed as a result of the high pressure breakdown. This will be discussed in detail in a later part of this paper. Second, the high velocity under the labyrinths eroded the shaft causing increased leakage flow through the blowdown pipe. This caused a high pressure differential in the pipe increasing the seal pressure, but more important resulted in erratic seal pressure and loss of the steady 5 PSI differential at each end of the compressor. This occurred even though each end has its own seal oil overhead tank. This is associated with a gas flow turbulence in the labyrinth porting and compressor head drilling rather than seal system design. This problem was solved by installing a shaft sleeve utilizing hardfacing as discussed in the section under "shaft sleeve erosion."

Typically, natural gas compressors utilize separate lube oil and seal oil systems. The two main functions of these are; one, to prevent any gas contaminated oil from contacting the hot gas turbine, and secondly, to minimize the quantity of oil requiring de-gasifying. Actually, in the compressor there is no need for a separate system. The bearings and seals are usually made of the same materials, in this case steel backed babbitt. Oil that is satisfactory for sealing is also satisfactory for bearing lubrication. A new approach to the problem of separation was used on this project. A labyrinth seal with air buffer was installed on the compressor to gear coupling to separate the compressor lube system from the gear/turbine system. The compressor then used a common lube and seal oil system without any contamination of the gear/turbine lube oil. This eliminates the requirement for a separation labyrinth between bearing and seal of the compressor.


FULL LOAD TESTING WITH ROTOR DYNAMIC CONSIDERATIONS

Testing of the compressors in the manufacturer's facility in Olean, New York on this project consisted of the standard API mechanical and aerodynamic performance plus a full pressure, full power string test. The mechanical and aero test proved to be quite routine, but the full load test yielded some interesting results.

Although both the low and high pressure compressors proved to be quite stable at low operating pressures the excitation at high pressures was a bit more than the original rotor system could dampen. Subsynchronous vibration appeared and as a precaution the maximum pressure was limited. A basically stable rotor system is not self-exciting, meaning that the bearings and seals do not cause the instability. The machine will run to maximum continuous speed free of subsynchronous vibration at low pressure, such as the standard mechanical spin test. However, there may be some pressure level or operating condition that causes aerodynamic excitation which exceeds the dampening ability of the rotor system. This point is important because it seems popular to assume the bearings and seals are causing the instability. In some cases this may be true, such as machines with sleeve bearings and bushing seals. But sophisticated machinery such as this with tilting-pad journal bearings and tilting-pad seals are not usually subject to half frequency whirl or other self-excited vibration instabilities.

The tilting-pad seal is a patented sealing device that has influence on the dynamics of a centrifugal compressor rotor. It utilizes tilting pads similar to those commonly used in journal bearings, but with a floating cage that also functions as a bushing type seal. Fig. 5 shows a photograph of a tilting-pad seal. By controlling the geometry, the designer can add considerable dampening and stiffness to the rotor system.

Large excitations are required to cause instability in a well designed rotor system. These units were originally designed with double balance pistons such that each back-to-back section had its own balance piston. During testing at pressure near the design of 7100 psi, subsynchronous vibration appeared. The dominant frequency was the first critical speed of the machine. The subsequent test program included installation of pressure transducers in all piping including the balance piston connection lines. Large pressure pulsations were observed as a result of the high density gas flow across the balance piston labyrinths.

The double balance piston concept was replaced with a division wall and labyrinth including the following modification to the division wall. Holes were drilled in the division wall as shown on Fig. 6 to alter the flow dynamics associated with the last stage impeller and division wall labyrinth seals.

In addition to these changes aimed at decreasing excitation forces, a squeeze-film damper was installed in the tilt-pad journal bearings. This bearing utilizes an oil film around the bearing cage which holds the pads. The cage is supported only by this oil film and the two o-rings which also contain the damper oil (see Fig. 7).

The cage and pads themselves are identical to the standard bearing. Five off-set pivoted pads are used with end seals on the cage for oil containment and flow control.

With the modifications the units were operated at full pressure including a four hour acceptance test with no subsynchronous vibration problems.

As a result of the full pressure shop test, no field design changes were necessary with respect to rotor dynamics. This proved to be a considerable savings of time, and, therefore, money to both the end user and manufacturer.

OPERATIONAL CONSIDERATIONS

Surge control

After the compressors, gears, panels, and oil systems had been completely and successfully tested in the factory, they were shipped to the field and installed as part of a matching system of piping, coolers and controls. The simplified system diagram in Fig. 8 illustrates how these components are integrated with the compression train.

Each of the two compressor bodies are back-to-back re-entry barrels and form two compression sections which are cooled, controlled, and surge-protected individually. This provides a total of four compression sections in the train. Each section recycles through an intercooler and suction scrubber, as required by the anti-surge controller for that section (to maintain adequate flow to avoid surging). In addition to the individual anti-surge controls for each section, the system is provided with a hand-indicator-controller (HIC) which, through a low signal selector, will operate all four anti-surge control valves simultaneously. This HIC is used for starting the train and shutting it down. Normal start-ups and shutdowns are performed with all anti-surge valves in the full open position.

Before final sizing of the piping, valves, vessels, and coolers in this system, a detailed computer dynamic simulation was performed to predict the actions and reactions of all components during normal operating conditions and during system upsets. This dynamic simulation was a very valuable tool and provided the means for eliminating many problems during the system design stages. The correlation between the simulation and actual operation was so accurate that the valve stroke timing and instrument response settings now in use are essentially the same as those developed in the simulation.

The commissioning problems encountered with the anti-surge systems were limited to instrument reliability and piping vibration during recycle operation. The instrument reliability problem was the result of a frustrating series of component failures in the anti-surge controllers. The causes of these failures have now been rectified. The valve and piping vibration problem during recycle operation was due to high gas velocities and turbulence in the area of the recycle valves. This has been corrected by changing valve trim size and by piping revisions which provide a smoother expansion path for gas immediately downstream of the valve. The recycle valves utilized a plug and cage design. The small slots in the cages acted as strainers in the piping system, collecting residual contamination and making the valves inoperative. This was corrected by operating the train on recycle at full rate with the trim removed from these valves until the piping system was clean.

The anti-surge control systems are now operating well and provide adequate surge protection. It should be noted that any system such as this requires regular preventive maintenance. Routing calibration of inputs and set points is necessary to assure that all components function properly.

## Hydrate Formation

As noted in the design presentation, these compression trains utilize only one seal oil pressure level. This is accomplished by the use of end seal labyrinths, which are all referenced through compensating lines to first section suction pressure. In theory, this is a satisfactory arrangement and caused no problems in the first two compression sections. In practice, however, the high differential pressure across the labyrinths in sections three and four resulted in sufficient cooling of the gas to cause hydrating in the seal chamber. The hydrates plugged the ports to the compensating lines (see ports labeled "S" in Fig. 9), and the seal pressures increased beyond the capabilities of the seal oil system. The seal oil console was designed for maximum pressure of 2700 psi.

By design, the pressure differential on the third section labyrinth is 3100 psi (4200 psi to 1100 psi), and on the fourth section labyrinth is 4500 psi (5600 PSI to 1100 psi). Examination of the Katz hydrate chart reproduced in Fig. 10 indicates that both of these pressure reductions are in the hydrate forming range at the design suction temperature of 125°F.

The initial corrective action for this problem was to increase the inlet temperature to sections three and four to 160°F. Seal oil pressure recorders were installed on sections three and four to monitor the results. Careful manipulation of inlet temperatures to sections three and four proved that normal operation could be maintained at suction temperatures above 146°F. Since the initial temperature lines on the hydrate chart are close to vertical in both ranges, the initial temperature requirements for both sections were essentially the same. Also for that reason, the initial temperature was the most logical point to take corrective action. For the sake of expedience, the compressors were operated for a period of time using inlet temperature control to sections three and four to avoid hydrating. This was not the best solution for the following reasons:

The temperature was controlled by louvres on the air fan coolers and was not as accurate as desired. In addition, sudden heavy rainstorms had a drastic effect on controllability.

The increased inlet temperature reduced compressor throughput capability.

Increased discharge temperatures from section four caused expansion problems in the long, underground injection lines.

The ultimate correction was the installation of short two inch diameter bypass lines from section two discharge line to the 'D' port on the section three labyrinth, and from section three discharge line to the 'D' port on the section four labyrinth (see Fig. 9). This provided uncooled discharge gas at 190° - 200°F from the previous compression section to the initial temperature port on each of the section three

and four labyrinths. The hydrating problem was eliminated and the trains could be operated at design inlet temperature.

## Field Instrumentation

Because of the critical nature of the compressor train operation and the lack of industry experience, these units were instrumented rather heavily. It was considered very important during the commissioning stages to know as much as possible regarding the general operating conditions and any transient conditions encountered during start-ups, shut-downs, and rate changes. The machines were all equipped with X-Y radial probes at each bearing, thrust position probes on each shaft, keyphasors on the compressor train and main driver, and a complete array of temperature, pressure and flow indications for all lube oil, seal oil, and process functions.

In addition to the permanent on-stream monitoring equipment, a fairly complete set of instrumentation was provided for analysis of vibration. This included an 8-track FM recorder, complete real time analyser, various oscilloscopes, filters, and plotting equipment.

All of the monitoring and analysis equipment has been used extensively on these compressor trains with no signs of rotor instability indicated.

## CONCLUSIONS

The compressors have performed according to aerodynamic design, with the exception of the rapid loss of efficiency experienced due to the 'O' ring failures and the heavy grooving of the shaft sleeves under the end seal and centre wall labyrinths. In early December, 1979, an overhaul was performed on one train which incorporated all of the revisions developed during the first two years of operation. The revisions included the following:

1.    Colmonoy hardfaced sleeves under all labyrinths.

2.    Hydraulically fitted thrust discs.

3.    Careful removal of all sharp edges in the bore of the barrel (to avoid 'O' ring damage during bundle insertion).

4.    Viton o-ring hardness increased from 75 to 90 durometer.

5.    Hot gas bypasses to third and fourth section end seal labyrinths.

6.    Vertical bundle assembly to improve bore concentricity.

7.    The overhaul was completed and the train was put on-stream December 10, 1979. It has operated since that time with minimal mechanical difficulties and no apparent loss of efficiency.

8.    In addition, it should be noted that in order to obtain satisfactory performance from a

sophisticated centrifugal compression train such
as this, it is essential to develop operation
and maintenance training programs that will
provide qualified field personnel. No matter
how well the machines are designed, built, and
commissioned, they must be carefully operated
and maintained, if they are to continue to perform the job for which they are intended.

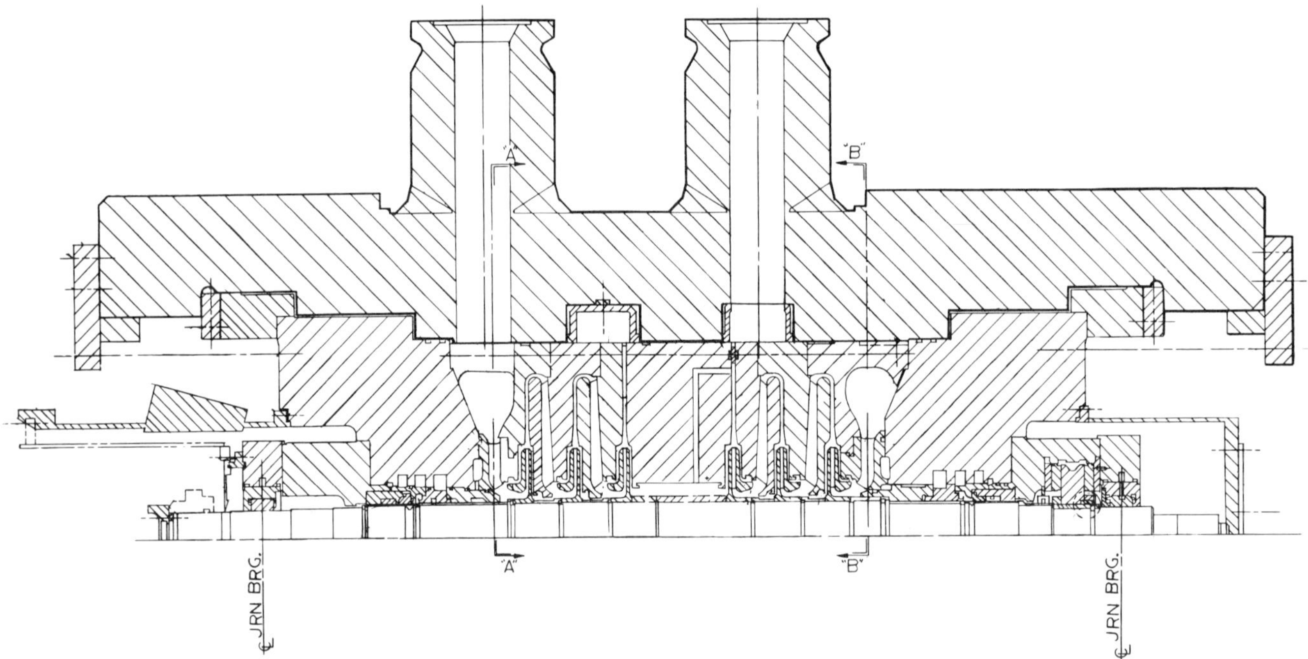

Fig 1   Unit cross-sectional drawing

Fig 2   Eroded shaft sleeve

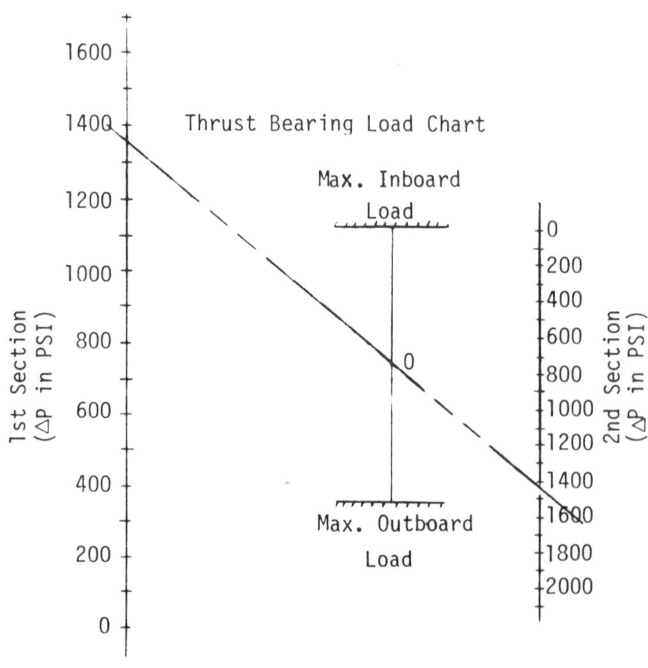


Fig 3   Thrust nomograph

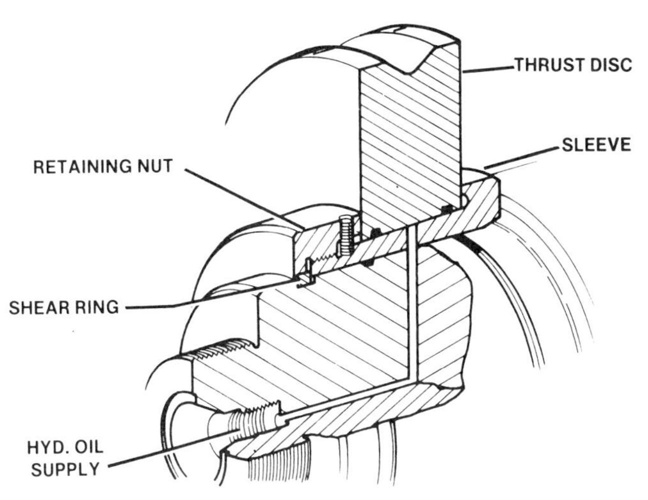



Fig 4    Hydraulic fit thrust assembly

Fig 5    Tilt-pad seal ring assembly

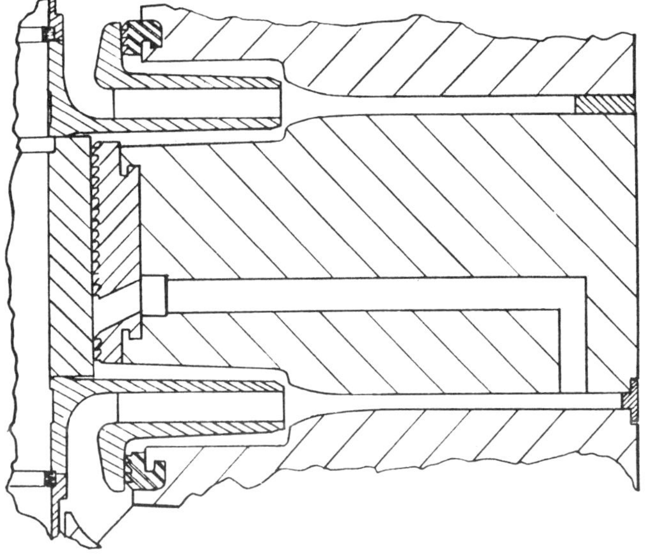

Fig 6    Modified division wall assembly

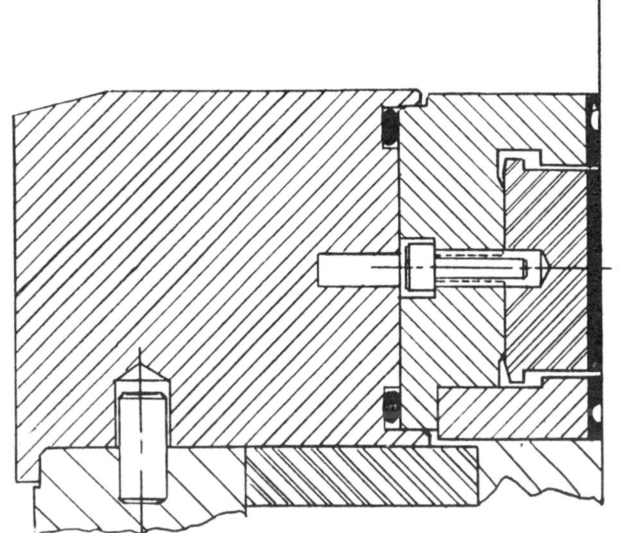

Fig 7    Tilt-pad journal bearings

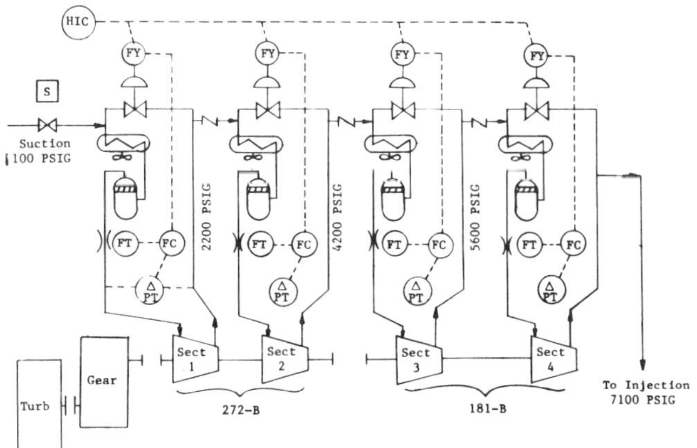



Fig 8    System diagram

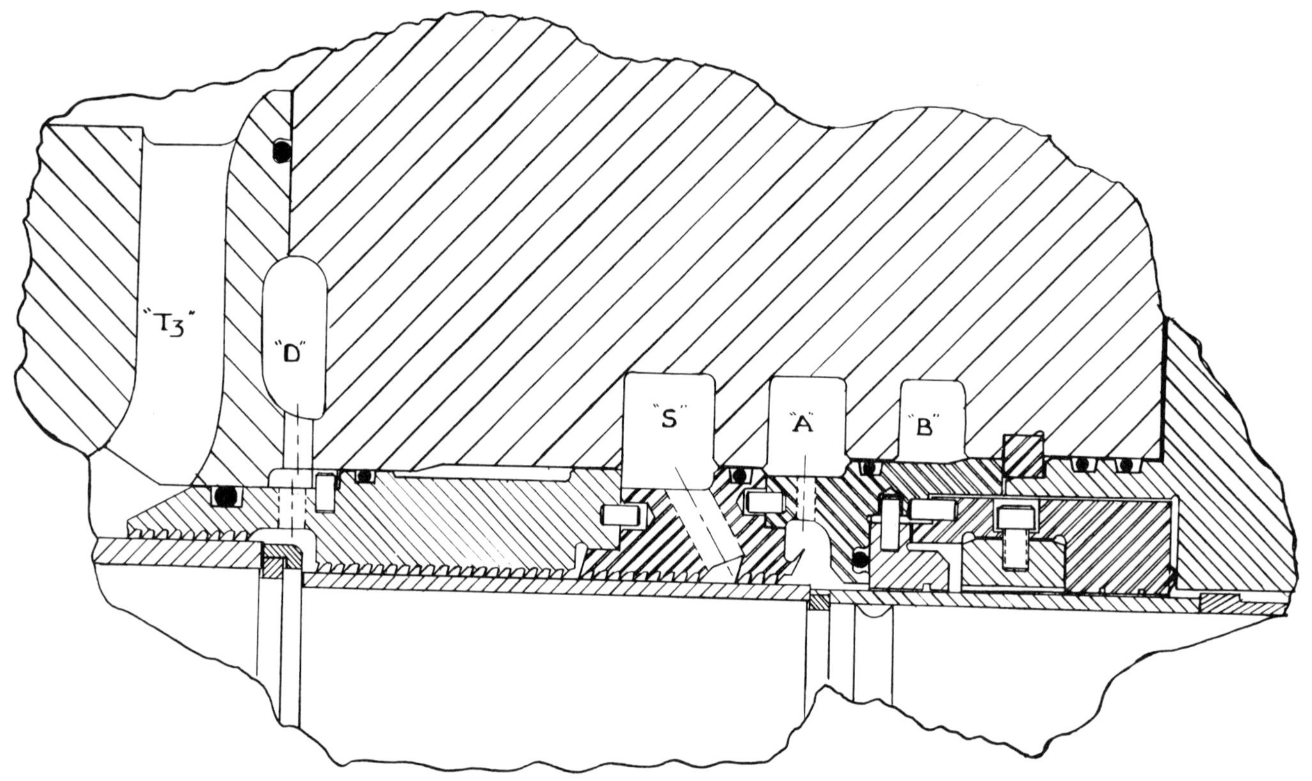


Fig 9   Labyrinth and seal assembly

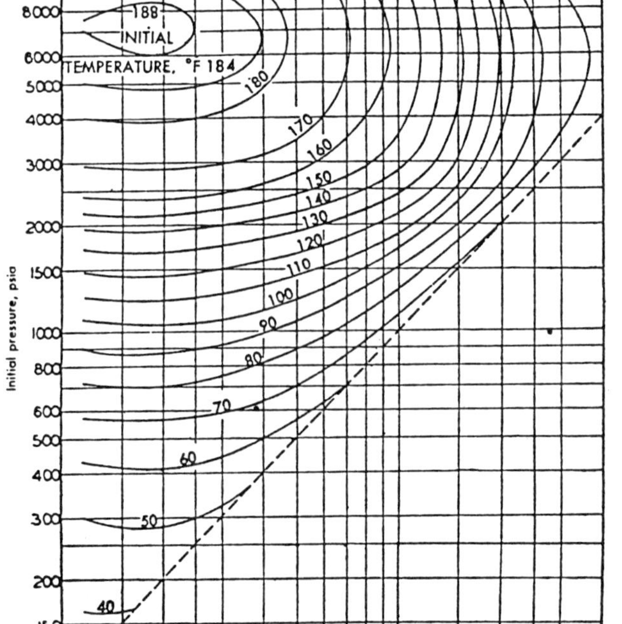


Fig 10   Katz hydrate chart

# C51/81

# Full load testing of centrifugal natural gas injection compressors

D SABELLA, L TERRINONI and A TIMORI
Nuovo Pignone, Florence, Italy

SYNOPSIS    This paper shows the results of full load tests of high pressure natural gas reinjection centrifugal compressors.
These tests are performed in order to verify the mechanical reliability of the compressor when high density gas is handled and to confirm experimentally thermodynamic performance (efficiency, pressure ratio, choking to surge range).
Tests results of standard centrifugal stages utilized for these compressors  are compared with full load tests results.

## LIST OF SYMBOLS

$D_2$ = Tip Diameter

$H_p$ = Polytropic Head

$Q$ = Suction Flow

$Reu_2 = \rho_2 U_2 D_2 / \mu_2$ : Reynolds Number

$U_2$ = Impeller Tip Speed

$\mu_2$ = Dynamic Viscosity at Outlet of Impeller

$\rho_2$ = Gas Density at Outlet of Impeller

## INTRODUCTION

The use of centrifugal compressors operating at pressure over 15 M Pa has been rapidly increasing over the last few years, especially for natural gas services.

The high degree of reliability achieved in the field of high pressure centrifugal compressors is shown by the results of full load tests performed and confirmed by the large number of machines installed to date. Testing under actual operating conditions (ASME PTC 10 Class 1) makes it possible to acquire experience otherwise unobtainable  by mechanical performance (API 617) and similitude thermodynamic tests (ASME PTC 10 Class 2,3). The main objectives of full load testing with natural gas are:

(1) To confirm that machine performance fully complies with all design data and to verify mechanical characteristics under actual operating conditions;

(2) To arrive at more advanced thermodynamic and mechanical design criteria.

From the standpoint of the reliability of a centrifugal compressor for high pressure natural gas reinjection, or any other high density service, the main factors to be considered are:

(a) Lateral behaviour of the rotor: here it is necessary to verify that there are acceptable vibration levels over the whole compressor operating range, as excitations may originate from the oil seals or may be aerodynamically induced. Due to the high density of the gas, these can be large enough to provoke asynchronous vibrations of the rotor.

(b) Axial thrust which must be accurately predicted so that it can be suitably minimized.

(c) Predictions of the compressor's thermodynamic performance: one feature of these machines is, in fact, that they run with high Reynolds numbers and gases whose characteristics differ greatly from ideal. It is thus necessary to have an in-depth experimental knowledge of the performance of the single stages used.

## STANDARD STAGES

Aerodynamic design of stages for process compressors must necessarily take into account the great number of variables and requirements that have to be satisfied. Stages capable of meeting a wide range of operating requirements have been developed by Nuovo Pignone through extensive analytical calculations and single stage testing techniques. Systematic individual stage testing represents a basic and substantial part of the development effort.

The stages used in Nuovo Pignone centrifugal compressors come from various families of optimized standard stages. These are made up of impellers having different design parameters capable of matching most of the process applications required. The use of stages previously subjected to systematic rig testing allows reducing the uncertainties of compressor performance predictions, since not even the most sophisticated calculation models are able to fully describe the structure of the flow field inside a centrifugal compressor stage. The stages are tested in the same configurations they have on the compressor, as this is the only way to obtain reliable indications to be adopted in compressor design.

Specifically, the machines treating high density natural gas especially require the use of standard stages with intrinsically stable characteristics, since it has been experimentally proven that non-stationary fluid dynamic phenomena (e.g. rotating stall) can excite asynchronous vibrations of the rotor (1), (2), (3), (4), (5), (6), (7), (8).

The performance of the stages tested for various values of impeller tip speed Mach numbers (see Fig. 1 in which the stage's standard surge and choke limits are indicated) get memorized in the computer and are entered in a calculation programme that allows selection of the standard stages best suited for use in the compressor designated to provide the desired performance. Naturally, operating conditions of the standard stages can differ from those on the rig and thus the computer programme allows for corrections pertaining to gas properties.

In fact, tip speed Mach number and inlet flow coefficient being equal, the ratio of the impeller's inlet and outlet gas densities, and thus head coefficient, are dependent upon the value of the gas specific heat ratio $C_p/C_v$,

It is therefore necessary to correct performance in order to return to similitude conditions. Furthermore, as a precaution, it is assumed that the Reynolds number ($Reu_2$) affects stage efficiency and head solely through windage losses (disc friction). The increase in efficiency expected for stages working at high pressures is, as a result, smaller than that calculated by other procedures such as the one suggested in PTC 10.

In addition, it is assumed that no further variations in the friction coefficient exist for $Reu_2 \geqslant 2 \times 10^7$.

## CHARACTERISTICS OF THE FOUR COMPRESSORS TESTED

The compressors tested here indicated below as A, B, C and D, are designed for natural gas reinjection services (Figs. 2 and 3 show the longitudinal cross-section. of compressor C and D). Compressors C and D are directly coupled to the driver by means of a diaphragm coupling, whereas A and B are coupled by means of a step-up gearbox using the same kind of connection. The design features of the compressors are summarized in Tables 1 through 4.

## TEST INSTALLATION

Upon successful completion of the mechanical tests, the compressors were installed on the outdoor facilities for full load testing. Two shaft gas turbines served as drivers. The gas was circulated in a loop provided by an air cooler. The gas system was completed by a throttling valve on the final delivery line, a blow-off valve and pressurizing station in which reciprocating compressors were used to boost the inlet pressure to the desired value. All mechanical and thermodynamic data were processed through a data acquisition system and gas composition was constantly checked throughout testing. Shaft vibrations were detected by two no-contact probes at 90° on each bearing. Two axial probes were also provided, one to monitor the shaft axial movement, the other to monitor the phase angle. Vibrations could thus be measured by means of oscilloscopes and a real-time frequency analyser. The results were recorded on magnetic tape, allowing the data to be studied in-depth at a later date.

## EXPERIMENTAL RESULTS OF THE THERMODYNAMIC TESTS

Fig. 4 shows the expected and experimental curves of the polytropic head and efficiency vs. suction flow as percentages of the reference value at design speed of compressor A.

The compressor was tested in accordance with ASME PTC 10 Class 1 using natural gas. The expected curves were obtained from the results of rig testing of the standard stages according to the foregoing description. Actual performance is extremely close to the predicted curves, at both design and off design conditions.

Compressor surge at constant speed occurred at a flow of 67 % of reference value. The polytropic head at surge was equal to 115 % of reference value. The slight differences between actual vs. expected performance at surge may be due to Nuovo Pignone's precautionary method for setting up the compressor's stable operating range (see Standard Stages). In fact, even if one of the stages is operating for a flow where $\dfrac{dHp}{dQ} > 0$ for the stage itself,



   C51/81

thus potentially unstable, a value of $\frac{dH_P}{dQ} < 0$ which corresponds to stable operating conditions, can result for the compressor as a whole. Fig. 5 shows compressor B's expected and experimental curves as percentages of the reference value at design speed. The compressor was tested according to ASME PTC 10 Class 1 with natural gas. The same considerations made about compressor A hold true for compressor B. Surge at constant speed occurred at a flow of 73 % of reference and at a polytropic head of 114% of reference.

Fig. 6 shows the expected and experimental curves of compressor C tested according to ASME PTC 10 Class 1. The same remarks made about compressor A are also true in this case.

Fig. 7 shows the expected and experimental curves, in percentages, of compressor D. The compressor is characterized by the highest delivery pressure and lowest suction flows of those tested. Testing, according to ASME PCT 10, was conducted with natural gas at speeds ranging from 74 to 112% of design speed and at a maximum delivery pressure of 49 M Pa. The compressor presented an extremely wide operating range, with a 20% increase in polytropic head between the reference flow value and the minimum value at constant speed. The fact that the operating range at higher flows was greater than expected demonstrates better behaviour in relation to choking for impellers having a very low flow coefficient such as the ones used on compressor D. A possible explanation could be the high Reynolds number which, in fact, entails a decrease of friction losses that are more pronounced in larger-than-design flows and, as a result, improvement in efficiency under such conditions.

LATERAL BEHAVIOUR OF THE ROTOR

The relatively low specific speed of reinjection compressor stages has no negative effects on the rotor's flexural stiffness since, due to the reduced axial length of each individual stage, the bearing span can be kept to a minimum. Nevertheless, high powers and great pressure rises require the use of couplings and thrust bearings which, due to their sizes and locations on the rotor, tend to reduce its stiffness and affect its lateral vibration modes. Theoretical calculations for analysing the lateral behaviour of the rotor during the design stage are based upon synchronous and asynchronous responses and analyses of the system's damping effectiveness (9).

Figs. 8 and 10 show for compressors B and D the rotors' responses to synchronous excitations (i.e. unbalancing) which were located in the middle and at both ends of the rotor to amplify the first and second modes. Amplitudes were calculated at the vibration probe locations.

Figs. 9 and 11 show the rotors' naturally damped frequencies with logarithmic decrements in the operating ranges of the compressors. The same figures also indicate the rotors' frequency responses to an asynchronous excitation. Amplitudes were again calculated at the vibration probe locations.

This latter type of diagram has been devised to analyse rotor sensitivity to the effects of aerodynamic asynchronous excitations which are of especially great importance due to the high density of the gas treated.

The calculations are performed by varying the frequency of the excitation force between 0 c p m and a frequency over the compressors' maximum continuous speed, assuming that the excitation be applied on the final impeller, which, from an aerodynamic analysis carried out for the various compressors, resulted as most critical.

Figs. 12 through 18 show the orbits and spectral analyses corresponding to the vibrations detected on the shaft near the journal bearings, both at design conditions and approaching surge.

The experimental results demonstrate that only when approaching surge do asynchronous vibrations occur, whereas they fail to appear over the whole operating range of all four compressors.

The following remarks refer to all of the compressors tested:

- The frequency of the asynchronous vibrations are in an approximately constant ratio to rotation speed which accords with most theories and experimental data regarding rotating stall;

- analyses of the asynchronous vibration components show that the amplitude on the delivery side is always higher than on the suction side as theoretically predicted in view of the location of excitation on the last stage of each rotor;

- the asynchronous vibration amplitudes decreased as the impeller tip speed decreased and a gas density decreased, indicating that the aerodynamic excitation force is a function of tip speed and density.

## Conclusions

The new information and experience acquired allowed the building of a very reliable high pressure centrifugal compressor family.

Full load testing bore out the validity of the analytical-experimental prediction methods used for forecasting thermodynamic and mechanical performance. The use of standard stages made it possible to achieve adequate compressor performance at both design and off-design conditions. The experimental results confirmed that the use of stages individually tested on the rig is instrumental in preventing the appearance of considerable aerodynamic asyn-

chronous excitations over the compressors' whole
operating ranges. Disturbances were observed
in the zone nearing surge, which could be explained by aerodynamic excitations on the last
stage.


REFERENCES


(1) BONCIANI L., FERRARA P.L. and TIMORI A.
    "Aero-Induced Vibrations in Centrifugal
    Compressors", Workshop on Rotordynamic
    Instability,Texas A & M University, May
    12 - 14, 1980.

(2) FERRARA P.L. "Vibrations in Very High Pres-
    sure Centrifugal Compressors", ASME paper
    no. 77 DET  15.

(3) ACTON O. "Studio sperimentale o teorico
    dello stallo in un compressore centrifugo",
    Acts of the Turin Academy of Science, vol.
    95, 1960  1961.

(4) GREITZER E.M. "Surge and Rotating Stall
    in Axial Compressors" Part. III JOURNAL
    OF ENGINEERING FOR POWER, TRANS. ASME, Se-
    ries A, Vol. 98, Apr. 1976, pp. 199 217.

(5) ABDELHAMID A.N., COLWILL N.H. and BOLLOWS
    J.F. "Experimental Investigation of Unstea-
    dy Phenomena in Vaneless Radial Diffusers",
    ASME Journal of Engineering for Power,
    vol. 101, January 1979, p. 52 - 60.

(6) DAY I.J., GREITZER E.N. and CUMPSTY N.A.,
    "Prediction of Compressor Performance in
    Rotating Stall" ASME paper No. 77-GT-10.

(7) TOYAMA K., RUNSTADLER P.W. and DEAN R.C.,
    "An Experimental Study of Surge Centrifugal
    Compressors" JOURNAL OF FLUIDS ENGINEERING
    TRANS. ASME, March 1977 pp. 115-131.

(8) FABBRI J. "Rotating Stall in Axial Compres-
    sors, NASA TT F-11, 765 December 1968.

(9) LUND J.W., "Stability and Damped Critical
    Speeds of a Flexible Rotor in Fluid-Film
    Bearing", ASME paper no. 73-DET-103.


TABLE 1:

Design features of compressor A

| | |
|---|---|
| Suction pressure (MPa) | 9.85 |
| Suction temperature (K) | 310. |
| Discharge pressure (MPa) | 16.6 |
| Design speed (rad/s) | 628. |
| Max continuous speed (rad/s) | 681. |
| Gas handled | natural |
| Molecular weight | 19.64 |
| Bearing type | tilting pad |
| Driver | gas turbine |
| Coupling | diaphragm type |


TABLE 3:

Design features of compressor C

| | |
|---|---|
| Suction pressure (MPa) | 6.95 |
| Suction temperature ( K) | 361. |
| Discharge pressure (MPa) | 22. |
| Design speed (rad/s) | 965.5 |
| Gas handled | ethane |
| Molecular weight | 30.38 |
| Bearing type | tilting pad |
| Driver | electric motor |
| Coupling | diaphragm type |


TABLE 2:

Design features of compressor B

| | |
|---|---|
| Suction pressure (MPa) | 16.05 |
| Suction temperature ( K) | 333. |
| Discharge pressure (MPa) | 25.2 |
| Design speed (rad/s) | 628. |
| Max continuous speed (rad/s) | 681. |
| Gas handled | natural |
| Molecular weight | 19.64 |
| Bearing type | tilting |
| Driver | gas turbine |
| Coupling | diaphragm type |


TABLE 4:

Design features of compressor D

| | |
|---|---|
| Suction pressure (MPa) | 17.17 |
| Suction temperature ( K) | 323. |
| Discharge pressure | 39.4 |
| Design speed (rad/s) | 1079. |
| Max continuous speed (rad/s) | 1199. |
| Gas handled | natural |
| Molecular weight | 20.76 |
| Bearing type | tilting pad |
| Driver | gas turbine |
| Coupling | diaphragm type |

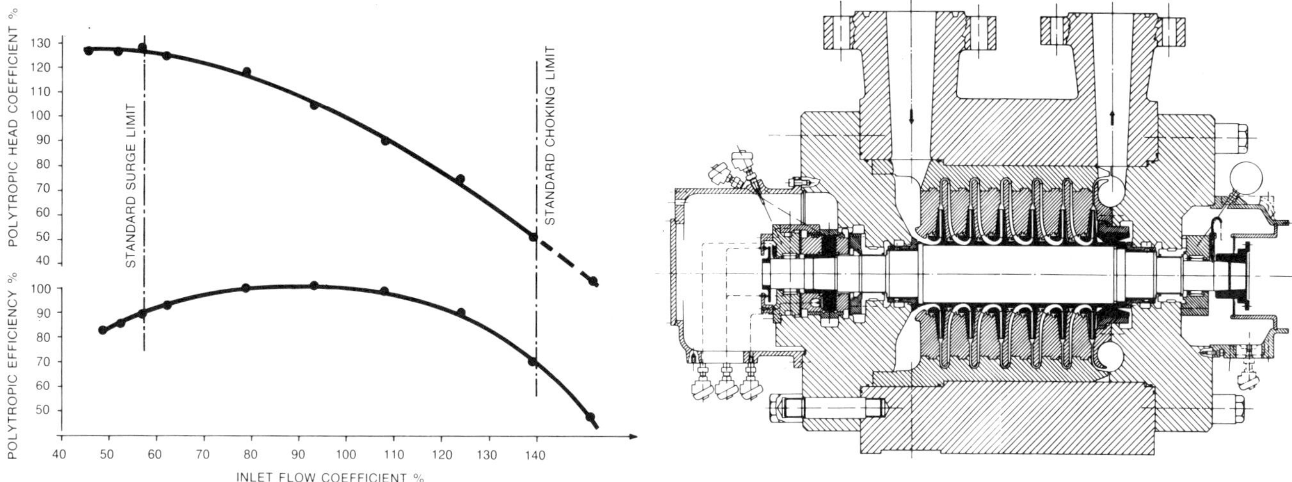

Fig 1    Typical standard stage performance curves

Fig 2    Compressor C (type BCL 400/B)

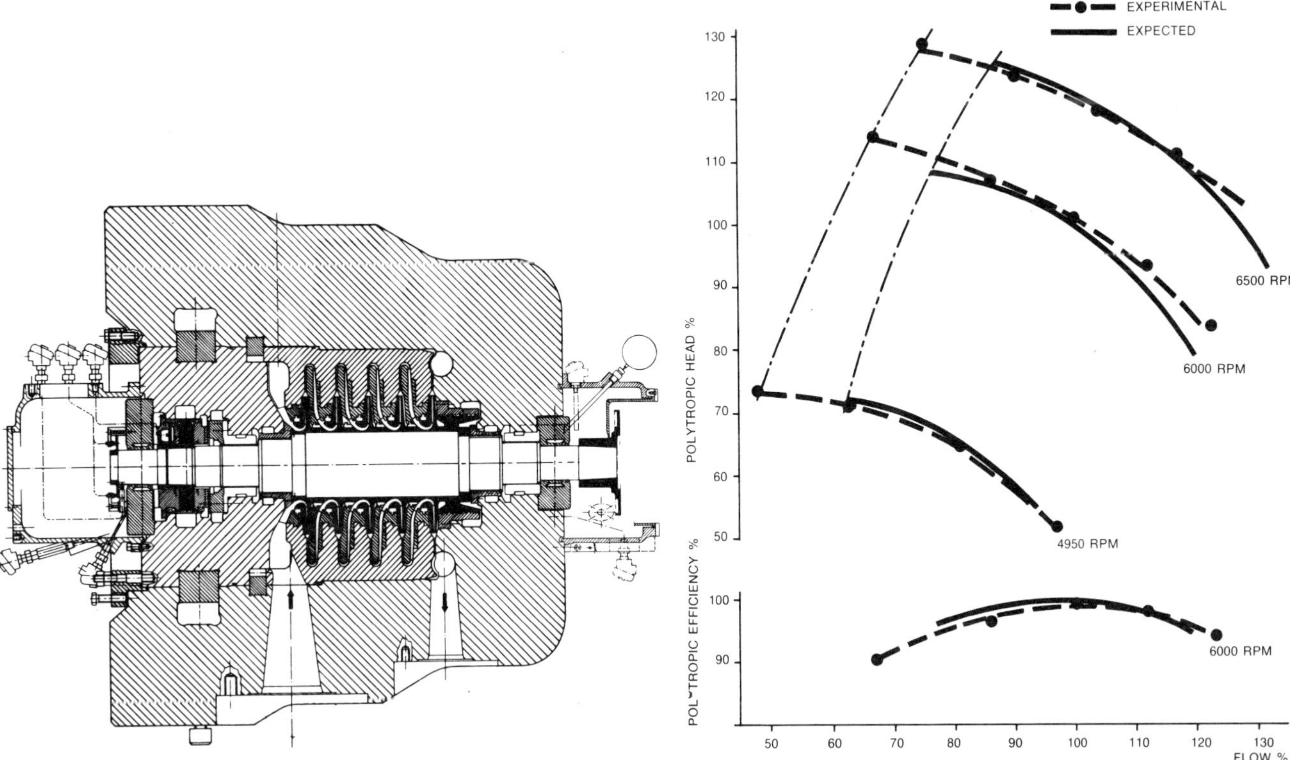

Fig 3    Compressor D (type BCL 400/C)

Fig 4    Expected and experimental performance curves for compressor A

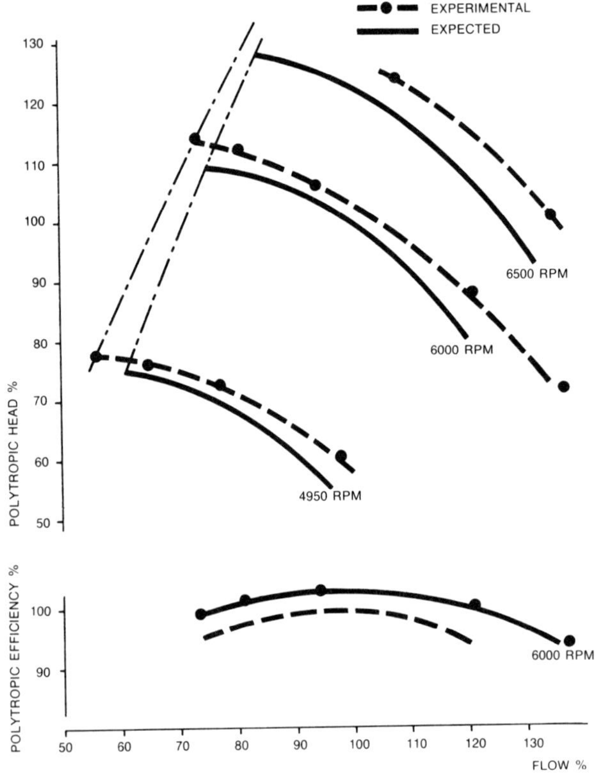



Fig 5 Expected and experimental performance curves for compressor B

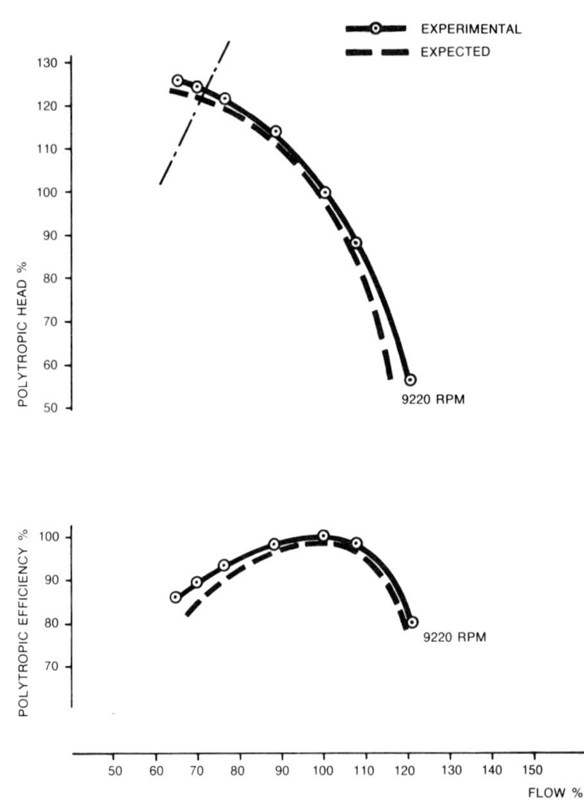



Fig 6 Expected and experimental performance curves for compressor C

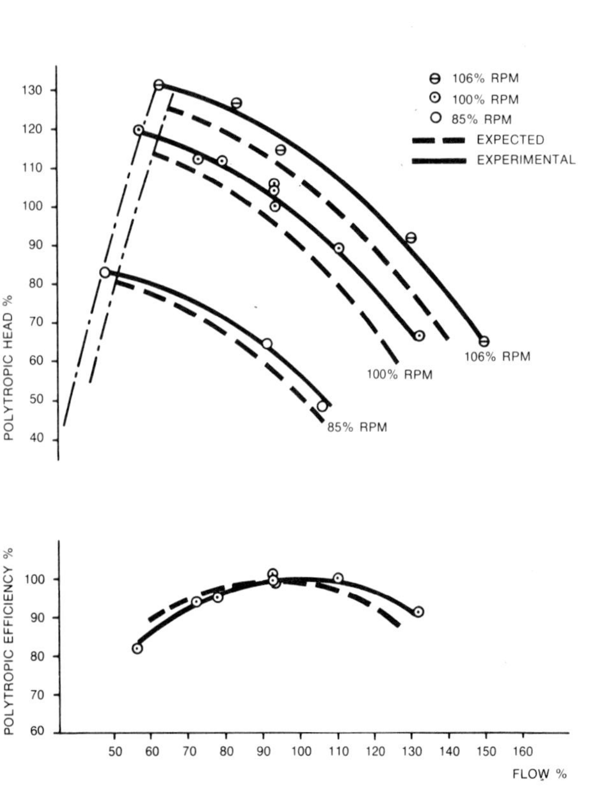



Fig 7 Expected and experimental performance curves for compressor D

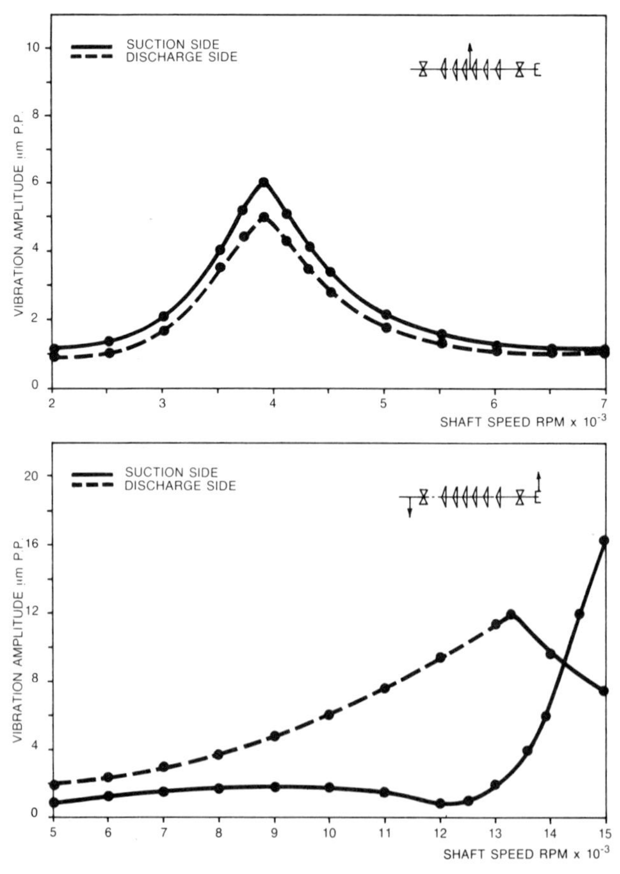



Fig 8 Compressor B synchronous response

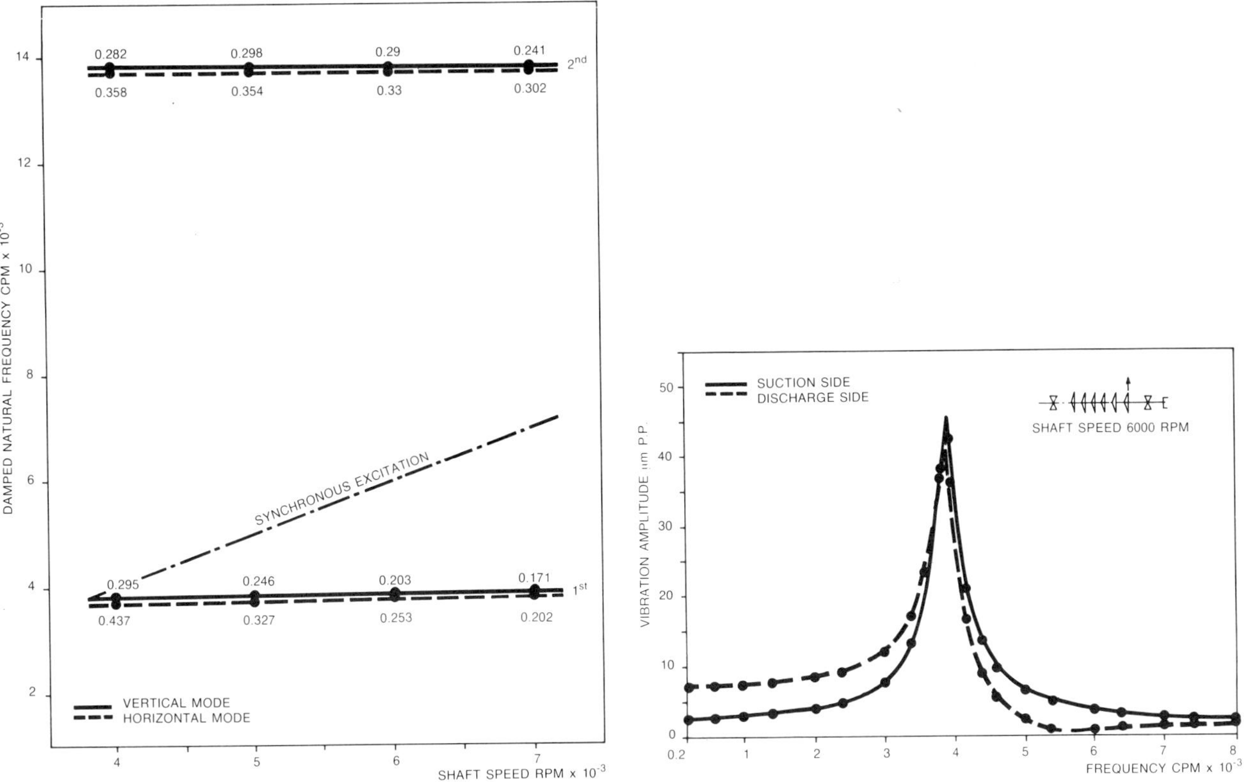

Fig 9    Compressor B damped natural frequency and asynchronous response

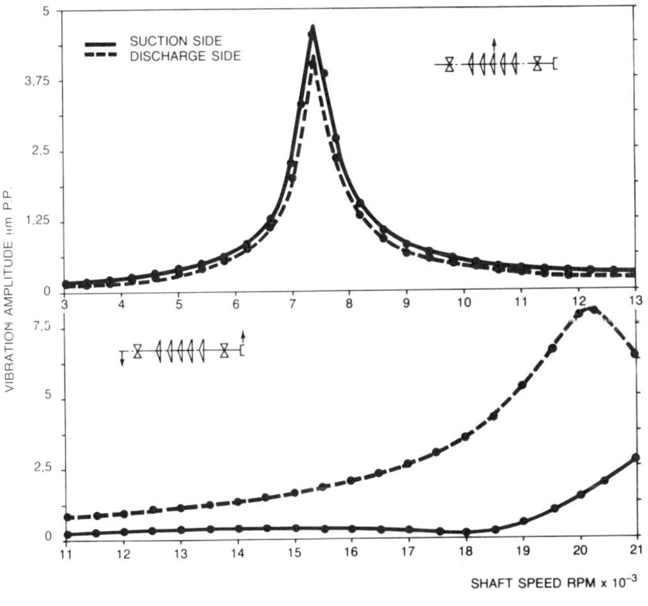

Fig 10    Compressor D synchronous response

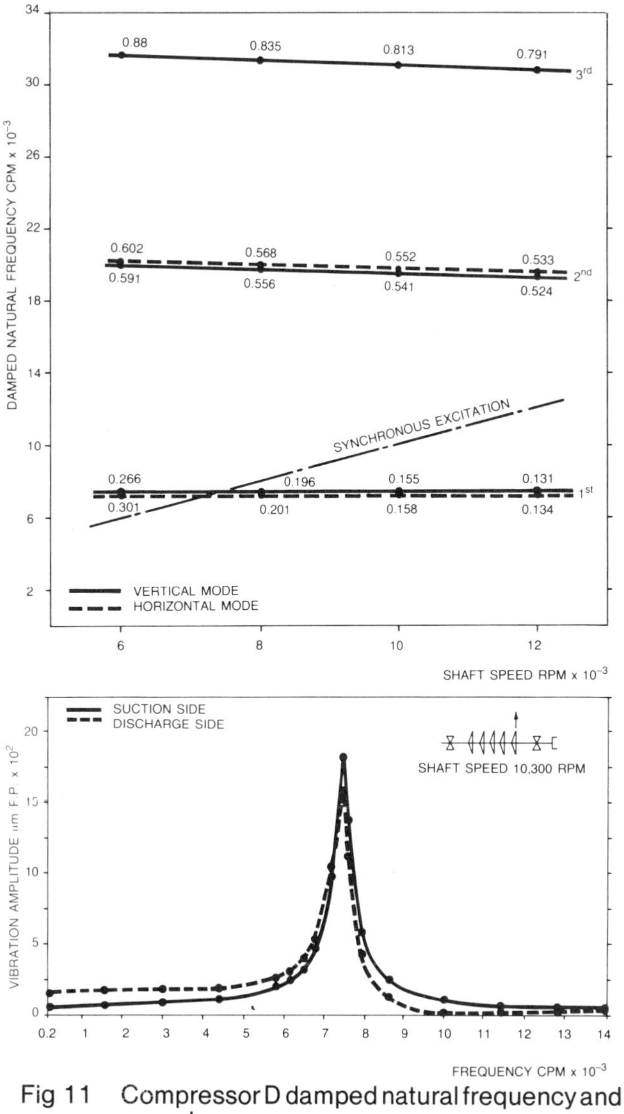

Fig 11    Compressor D damped natural frequency and asynchronous response

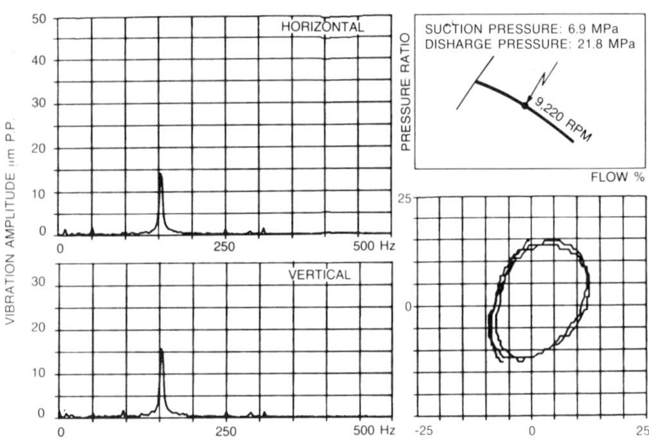

Fig 12 Compressor C discharge side vibration spectrum and shaft orbit at 9220 rev/min

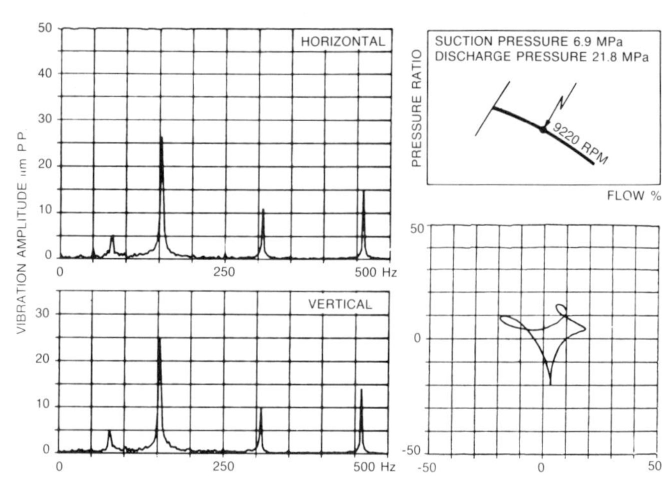

Fig 13 Compressor C suction side vibration spectrum and shaft orbit at 9220 rev/min

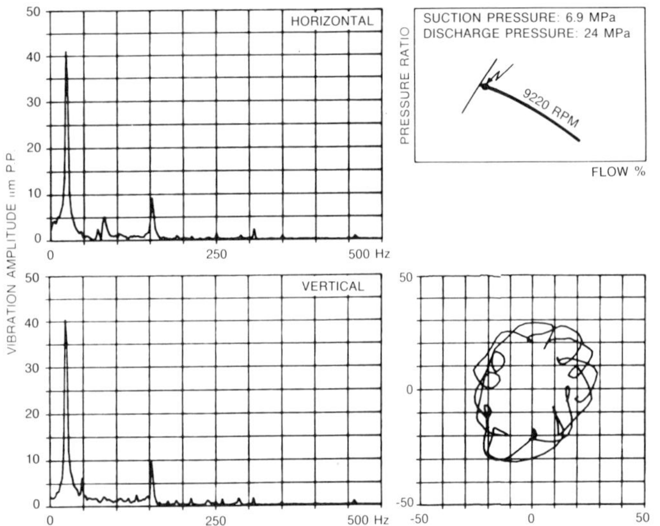

Fig 14 Compressor C discharge side vibration spectrum and shaft orbit at 9220 rev/min near surge

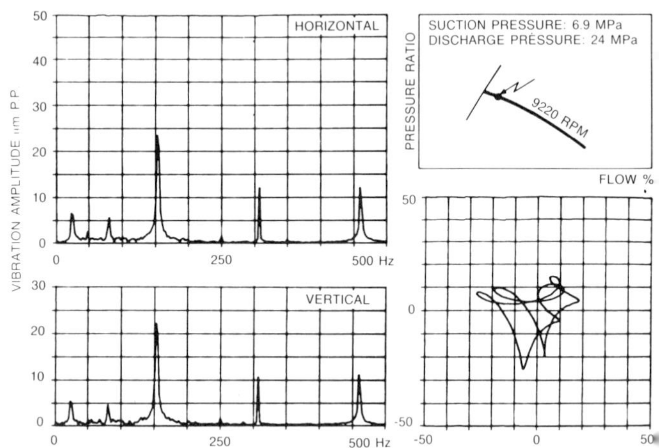

Fig 15 Compressor C suction side vibration spectrum and shaft orbit at 9220 rev/min near surge

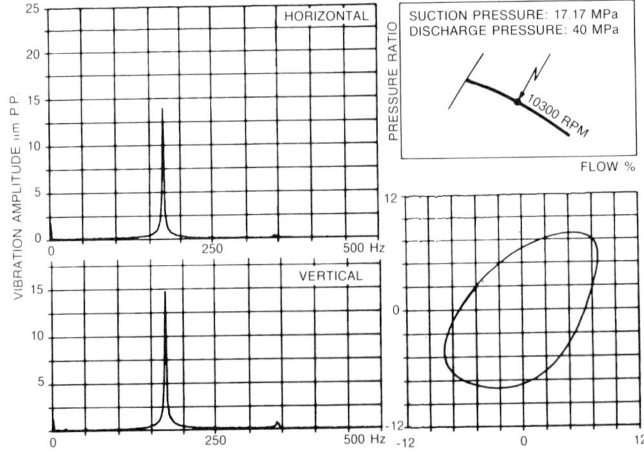

Fig 16 Compressor D discharge side vibration spectrum and shaft orbit at 10 300 rev/min

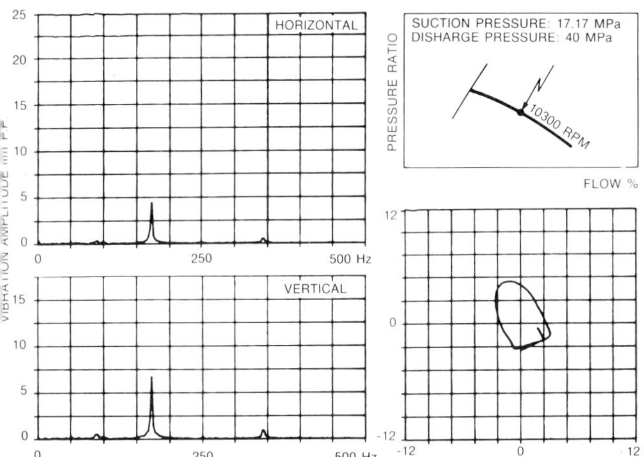


Fig 17   Compressor D suction side vibration spectrum and shaft orbit at 10 300 rev/min

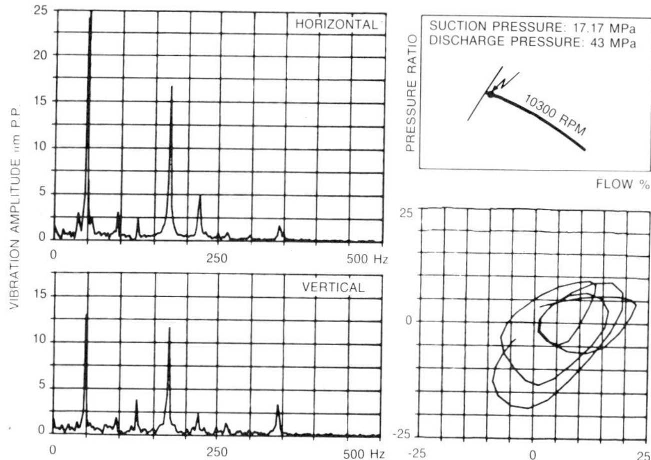


Fig 18   Compressor D discharge side vibration spectrum and shaft orbit at 10 300 rev/min near surge

# C43/81

# Design of packaged reciprocating compressors for high pressure reinjection of associated gas for off-shore oil and gas processing platforms

A TRAVERSARI, S PECCHI AND A PARTISANI
Nuovo Pignone SpA, Florence, Italy

SYNOPSIS   This paper deals with high pressure reinjection of natural gas which is increasingly
applied to recover such a valuable energy source.
Reciprocating compressors are an adequate solution for this heavy service provided that a thorough
thermodynamic study of the hydrocarbon mixture to be compressed is performed in order to achieve
proper control of the dew point along the compression path,  thereby  limiting the harmful diluting
effect of the gas handled on the cylinder lubricant. Particular pains must be taken with lubrication,
friction heat removal and fatigue stresses in the cylinder design of these compressors. In the
case of off-shore installation, special emphasis must be laid upon compressor balancing and
packaging.

## INTRODUCTION

Soaring fuel costs have brought about the need for more intensive use of the so-called secondary recovery systems that allow oil and gas wells to be exploited to the utmost. Natural gas reinjection, for instance, has the dual aim of maintaining the pressure in the wells' underground formations, thus increasing the production of crude oil and, at the same time, of not losing its associated gases. Although up to a short time ago this gas was burned, nowadays such practice is no longer acceptable, not only due to the ensuing atmospheric pollution, but also because it involves wasting a valuable source of energy.

Recompression of the gas in the wells at injection pressure is achieved by special compressors. First developed for on-shore applications, they are currently being used on off-shore installations where tight space and difficult environmental conditions demand top efficiency and reliability. A specific example of a reciprocating compressor designed for natural gas reinjection service is the one shown in Fig 1 installed on platform C of the Gorm field in the Danish sector of the North Sea for gas reinjection to 370 kg/cm$^2$ (36.28 M Pa) with gas suction at 140 kg/cm$^2$ (13.73 M Pa).

## THERMODYNAMIC STUDY OF THE GAS

It is common knowledge that the products obtainable from the well head, their composition, delivery pressure and temperature can vary from field to field, well to well, and even within the same well over periods of time. In the North Sea field, the associated gas is mainly composed of methane (over 75%), along with varying amounts of $N_2$, $CO_2$, as well as sulphur compounds and heavy paraffins. Therefore, a special gas treatment has to be included so that these undesirable substances can be removed prior to the arrival of the gas in the compressor for reinjection. This consists essentially of dehydrating the gas with glycol to prevent hydrate formation and then cooling it for better control of the dew point and higher NGL recovery prior to reinjection in the well.

Nevertheless, owing to differing operating conditions, either temporary or longterm, gas streams may exhibit rather marked variations in temperature, pressure and composition. These variations may in turn affect - even appreciably - compressibility characteristics, $C_p/C_v$ ratio, molecular weight, dew point conditions and the like, whereby influencing the performance of the machines and their accessories. In the case of reciprocating compressors, if not taken into due consideration, they can jeopardize proper functioning of the cylinder valves, by-pass valves and pulsation suppression devices as well as increase the vibration level over the whole piping circuit. It is thus necessary to conduct a thorough investigation of the thermodynamic behaviour of the gas. The first thing to check is the phase behaviour of the gas mixture in the p,t, diagram in order to find a suction temperature that is a suitable compromise between safe operation, with suction conditions far enough from dew point, and limited energy consumption.

Determining the optimum temperature also allows, in two stage units, compression without resorting to any inter-cooling, maintaining final discharge temperature within safe limits, saving space (always welcome on an off-shore platform) by excluding the inter-cooler and avoiding inter-stage condensation of liquid hydrocarbons, whose presence could jeopardize lubrication in the high pressure stage. In fact, as is well known, gaseous hydrocarbons tend to dissolve in oil and the amount of gas dissolved depends on their partial pressure. Even if what goes on inside the compressor cylinders is not easily predictable, for some time it has been observed that the hydrocarbon mixtures have a diluting effect on the lube oil which becomes more and more marked as operating pressures increase. The first consequence of this is a considerable drop in the lube oil viscosity in the cylinder and packing so that for services as severe as natural gas reinjection the use of high viscosity lube oils is recommended. As we have already mentioned, it is extremely important that the compressor's suction temperature be sufficiently far from the gas mixture's dew point, but it is likewise important to equip the machine with scrubbers since liquid hydrocarbons, even more than gas, dilute oil and can negatively affect lubrication even when compression is performed in a single stage.

Over the last few years, various methods for calculating the thermodynamic properties of hydrocarbon mixtures, mostly based on the use of the Redlich-Kwong equation (1), have been devised. In the event of mixtures at high pressures, it is preferable to adopt the Redlich-Kwong and Soave equation (1) that provides reasonably accurate data up to pressures of 700 kg/cm$^2$ (68.65 M Pa) and also allows for adequate approximation of the convergence region, even in the presence of heavy paraffins and aromatics. Fig 2 shows the p, t, phase graph obtained using the NGPAK*H programme and the compression starting points for the reinjection service performed by the compressor in Fig 1.

In order to calculate the compressibility factor Z, which is of prime importance, especially in proper sizing of the compressor cylinders, it is advisable to use the Benedict, Webb and Rubin equation (2) as modified by K.E.Starling (3). The graph in Fig 3 shows the compressibility factors calculated according to this method and diagrammed by a plotter. Long-term experience has shown that, even at high pressures, it provides Z values amply lying within the accuracy limits required.

COMPRESSOR DESIGN

In selecting compressors for a service as special and severe as that described above, various aspects must be simultaneously considered:
- critical importance of piston and piston rod seals;
- high delivery pressures;
- number of stages required to optimize energy consumption, volumetric efficiency and temperature of the compressed gas both within the compressor and at gas injection points;
- distribution of loads amongst the various compressor cranks and their development during each crank revolution;
- balancing of the reciprocating masses to limit free inertia forces and moments;
- need to reduce the space requirement while at the same time retaining easy access to the machine parts for maintenance.

Selection of the compressor

The basic design involves the selection of the crankgear, with an eye to the stroke-speed ratio. In fact, with the same piston speeds, two options are normally available:
- short stroke high speed;
- long stroke low speed.
The first option has the advantage that the unit is more compact and lightweight and, in addition, less costly. These advantages are all of great interest, especially the first two for platform installation where machine weight and size affect the weight and size of the platform itself. On the other hand though, there is a drawback in the short stroke mechanism, i.e. it is harder to get rid of the friction heat generated at the seal rings. In fact, the shorter the stroke, the less the piston rod and the liner, respectively engaged by the packing and piston rings used as gas seals, are uncovered during stroke. This negative factor tends to progress in step with the increase in the differential pressure to which the seals are subjected, because it brings about a corresponding increase in the sealing section length of the piston rod and liner in addition to a higher contact pressure between rings and sliding surfaces. Should sealing section lengths exceed the stroke, entire zones of this elements are never uncovered by the sealing rings and thus fail to come into contact with fresh gas, which is one of the means for removing friction heat. As a result, the wall temperature in these zones rises, whereby laying the groundwork for the occurrence of inconveniences stemming from improper lubrication, already precarious in gas reinjection services due to the oil dilution problem mentioned before.

The second option offers greater operating reliability since, as evidenced by the foregoing considerations, longer stroke allows better heat removal off the sealing surfaces.

As platform-installed compressors must
necessarily be designed for minimum maintenance,
the second option, despite the greater size,
weight and expense involved, is to be considered
the better of the two.

## Compressor cylinders

Design of the compressor cylinders must be
approached from two main standpoints: functional
and constructional.

From the functional standpoint, cylinder
design must:
- enable seal rings to operate under the best
  possible conditions;
- transmit to the crosshead pin bearing a load
  which reverses during the cycle so that there
  is an opportunity for the oil to be replenish-
  ed in the clearance space between the pin and
  bearing during the repeated application of
  load ('load reversal' condition) (4);
- facilitate compressor connections to the gas
  lines.

In the case of the compressor shown in
Fig 1, a four-cylinder arrangement was chosen
for balancing reasons (these will be discussed
in greater detail later on). Three cylinder
solutions were examined.

The first solution (see Fig 4a) achieves
compression using double-acting cylinders with
tail rods. Compression is possible with either
one or two stages, requiring, in the first case,
four equal cylinders and, in the second, two
first stage and two second stage cylinders.
Both solutions assure good load reversal since
the outboard and inboard piston areas are perfectly equal. On the other hand, the number of
packings is doubled due to the fact that packings
are also required on the tail rod side. So,
piston rods are lengthened, making seal alignment harder and subject to alteration from
thermal deformations of the cylinder body. There
is also an added drawback: solution (a) requires
the most floor space.

A second solution (see Fig 4b) entails
four single-stage cylinders balanced on the inboard end with the gas taken from the delivery
line. Compared to solution (a), (b) has the advantage
of not requiring tail rods and their packings,
while at the same time ensuring proper load
reversal in the crankgear. Its shortcoming is
that the packings are subjected to practically
constant pressure. This situation causes non-
-uniform pressure drops along the packings so
that few of the downstream seal rings have to
bear most of the differential pressure (5), (6).
As a result, these are compelled to work under
such severe conditions that their lubrication -
in itself a difficult matter as we mentioned
before - becomes even more troublesome.

Solution (c) envisages compression by means
of four identical cylinders, each performing a
first compression stage on the outboard end and
a second stage on the inboard end. Using this
kind of cylinder structure, it is possible to
achieve better compressor performance owing to
the pressure fluctuations the packing is subjected to, which uniform the pressure drops
across the packing seal rings, to load reversal
on the crankgear and to limited use of floor
space, without complicating alignment. This
solution is thus preferable to the other two and
for this reason was adopted for the compressor
in Fig 1.

From the constructional standpoint, two
configurations can be used in a double-acting
cylinder for high pressures. The simpler entails
a monoblock of forged steel that performs the
functions of the cylinder proper and that of the
valve housing as well (Fig 5 a). Most commonly
used since it is the simplest and most economical, this solution has several drawbacks.
Amongst these are the difficulties in properly
cooling the cylinder's sliding zone and the
complications involved in adequately radiusing
the cross-bored areas such as valve ports and
lube oil passages which are the points of major
stress concentration. In addition, especially
when dealing with large-size cylinders, it
should not be forgotten that it is rather difficult to produce large-size forgings whose inner
parts - those subjected to the greatest stress -
have mechanical properties suitable for the
stress levels of the cylinders of high pressure
compressors (7).

The second cylinder structure is illustrated in Fig 5 b. This solution separates the
functions of the cylinder proper from that of
the valve housing. In fact, the cylinder assembly is composed of three separate reduced-size
forgings whose manufacture, heat treatment and
machining are greatly facilitated, especially
in the critical points cited above. Cooling of
the sliding zone results is much better since the
cylinder in that zone is not as thick as in the
foregoing solution, and also because it is performed by a specially-finned external jacket
that allows high speeds of the cooling fluid,
and thus increased heat exchange. Assembly of
the three pieces is performed using extra long
bolts whose elasticity can be exploited in the
case of overpressures due, for example, to
exceptional liquid entrainments. The extra long
bolts may, in fact, stretch, thereby affording
enough space between head and cylinder so as to
allow the opening of special discharge passage-
ways which serve to relieve pressure. The
structure of the cylinder in Fig 5 b is especially suitable for reinjection compressors
from the standpoint of the fatigue resistance
exhibited by components subjected to pressure
fluctuations, as well as from the standpoint of

piston seal lubrication. This solution has the
drawback of being costlier than solution (a) and,
as a result, is not as frequently adopted (7).
Solution (b) has an extra added advantage though.
In fact, within certain limits, it allows variations in compressor capacity that can be
matched to the variation in well conditions
just by changing the centre section of the
cylinder(s).

As far as the seal rings are concerned,
the material that has proved most reliable in
this kind of service is bronze (8). Furthermore, the piston should be as short as possible
to reduce the ratio between its length and
stroke for the foregoing reasons, and also to
allow fresh gas to come into contact with the
part of the cylinder actually swept by the
piston rings. This would not occur should the
distance between the first ring and the front
of the piston be too great.

## Crankgear

Considering the aforementioned functional
characteristics of the compressor cylinder, one
may conclude that the crankgear should be of
the long stroke, moderate speed type. In addition, the environment in which off-shore machines
operate calls for extremely high reliability of
the crankgear. Thus, to achieve maximum compressor operating safety, all available means,
including the following, are used:
- bearing made of an anti-friction metal with
  high embeddability to avoid damage to the
  shaft in the event of exceptional contamination
  of the lube oil;
- cross-heads with replaceable shoes to facilitate maintenance;
- lock-tightening the piston rod to the crosshead by means of a hydraulic system, which is
  the best way to ensure proper execution of this
  very important operation;
- adequate instrumentation for operation control.

Another important factor in off-shore
applications is the dynamic action of both reciprocating and rotating masses. This must be
reduced to a minimum or, better still, entirely
eliminated. High inertia forces and moments
could cause excessive fatigue stresses in the
module's structure. Accordingly, the design
criterion used on off-shore applications is
different from that of conventional onshore ones.
For instance, whereas the conventional four
crank compressor has pairs of cranks phased at
90° so that the irregularity of torque transmitted is minimized, the compressor in Fig 1
(designed for platform installation) has been
equipped with the following features serving
to minimize dynamic forces and moments:
- reciprocating masses connected to adjacent
  opposed cranks of equal value;

- crankshaft with two pairs of cranks phased
  at 180°.
This solution allows elimination of the forces
and moments of the first order (running at
compressor speed) both vertical and horizontal.
Only a second order moment whose value is
negligible remains on the horizontal plane. A
situation of inertia forces and moments is
shown in Fig 6.

As far as absorbed torque is concerned,
the presence of a flywheel with adequate
moment of inertia serves to achieve a satisfactory irregularity degree to assure proper
operation.

## Piping

Pulsations and vibrations are of great
significance in reciprocating compressors,
especially in those used for off-shore applications. High pressure pulsations can have a
negative influence on the compression cycle,
causing reductions in volumetric efficiency,
increases in absorbed power, abnormal stresses
on the compressor's mechanical parts, mal-
functioning or breakage of the cylinder valves,
as well as high vibration levels in the piping.
These vibrations can in turn provoke excessive
fluctuating stresses that might result in
breakage due to fatigue (9).

Special features had to be incorporated
into the design of the unit shown in Fig 1.
In fact, analogue calculations have to be
performed to pinpoint and eliminate acoustic
resonances and high pressure pulsations that
hinder efficient operation (10), (11). In
addition, the natural mechanical frequencies
over all the piping section had to be determined so that resonances with excitating forces
could be detected and eliminated by modifying
the position and/or kind of supports (9).

Also, during the analogue calculations,
it is necessary to check that the acoustical
pulsations do not alter the compression cycle,
thus reducing throughput, and that they do not
produce valve fluttering.

In order to curb pulsations and reduce
them to acceptable levels, it is usually
necessary to install pulsation suppression
devices on each stage's suction and delivery,
as can be seen in the compressor in Fig 1.

The piping network must also be analysed
from the flexibility standpoint to make sure
that, even during the highest differential
temperatures, the ensuing stresses will not
exceed the maximum allowable values for the
material used. The lower the ambient temperature, the greater the importance of this
calculation.

THE COMPRESSOR PACKAGE

Electric motor

When reciprocating compressors are used in reinjection compressor units, an electric motor as a driver may be a suitable choice, due to its compactness and light weight (the compressor in Fig 1 is an example).

The electric power plant is thus sized to feed the electric motor driver(s) as well. Since the generators on platforms normally produce limited amounts of power, start-up current, when absorbed power is considerable, must be reduced. In the case of the compressor in Fig 1, to get around this, start-up was achieved by use of a special double-cage motor to provide the required starting and maximum torques and minimum start-up current. The coupling to the compressor is direct-flanged which means that the motor-compressor unit forms a very rigid group from the torsional and flexural standpoint. Consequently, the unit's natural torsional and flexural frequencies resulted very high and the ensuing stresses quite low.

Assembly

Units comprising reciprocating compressors for platform installation are generally package-assembled on metal frames. Depending on the circumstances, various criteria may be adopted in designing such a structure. Evidently, the skid building problems to be faced are different when the motor-compressor unit is assembled on the same site where the module is being built as opposed to when the motor-compressor unit is assembled elsewhere. In the first case, a module structure able to receive the motor-compressor unit can be designed whereas, in the second, the normal procedure is to use a skid(s) mounting the already-aligned motor-compressor unit and ancillaries, then transferred onto the module.

The unit in Fig 1 shows the solution of all-metal skids to which the motor-compressor assembly is anchored. As far as skid design is concerned, the main objectives are to obtain a structure providing the following features:
(a) easy, accurate compressor assembly as well as reliable reference points for proper levelling and alignment of all the unit components;
(b) sufficient stiffness so that excessive deformations are avoided when the complete assembled machine is hoisted.

The second objective may be achieved without resorting to oversizing of the skid and thus having to deal with ensuing weight problems, using computer programmes such as those developed by MIT (ICES-STRUDL II) or the University of Southern California (SAPV2) to analyse the elastic deformations. The same programmes may also be used to study the effects of the various static and dynamic loads acting on the structure during operation.

CONCLUSIONS

In designing reciprocating compressors for natural gas reinjection, preference is to be given to long-stroke, low-speed compressors, in order to achieve better dispersion of the friction heat generated by the gas seal rings which is due to the precarious lubricating conditions typical of this service. Naturally, it is necessary to take all the precautions which, oriented towards both a thermodynamic study and a proper treatment of the gas, serve to prevent liquid hydrocarbons from entering the compressor cylinders, thereby causing oil dilution or, even worse, liquid hammer. However, as a safety measure, the cylinders must be equipped with suitable devices so that, should liquid hammer happen to occur, great over-pressures may be avoided. In addition, the cylinder design should also entail limiting stress concentrations to prevent fatigue failures.
Lastly, dangerous vibrations of the compressor and its nearby accessories must be avoided on platform installations. This may be achieved by a special crankshaft arrangement and conducting an acoustical and vibration analysis over all the gas lines.

ACKNOWLEDGEMENTS

The authors wish to thank Nuovo Pignone S.p.A. for permission to publish information regarding company reciprocating compressors, as well as Dansk Boreselskab Als, Solar Construction Services and Odense Steel Shipyard Ltd. for their kind cooperation.

REFERENCES

(1)    SHERWOOD, T.K. and REID, R.C. "The Properties of Gases and Liquids", McGraw Hill, 2nd edition 1966, Appendix E, 626.

(2)    BENEDICT, M., WEBB, G.B. and RUBIN, L.C. "An Empirical Equation for Thermodynamic Properties of Light Hydrocarbons and Their Mixtures", Chemical Engineering Progress, August 1951, 419.

(3)    STARLING, K.E. "Fluid Thermodynamic Properties for Light Petroleum Systems", Gulf Publishing Company, 1973.

(4)    WILCOCK, D.F. and BOOSER, E.R. "Bearing Design and Application", McGraw Hill, New York 1957.

(5)    TRUTNOVSKY, K. "Berüthrungs Dichtungen an Ruhenden und Bewegten Machinenteilen", Springer Verlag, Berlin, 2nd edition 1975, 266.

(6)    TRAVERSARI, A. and GIACOMELLI, E. "Some Investigation on the Behaviour of High Pressure Packings Used in Secondary Compressors for Low Density Polyethylene Production", 2nd International Conference on High Pressure Engineering, University of Sussex, Brighton, July 8-10, 1975.

(7)    MANNING, W.R.D. and LABROW, S. "High Pressure Engineering", Leonard Hill, London, 1974, 188-189.

(8)    ENGLISH, C. "Kolbenringe", Springer Verlag, Vienna, 1958, vol. 2, 25.

(9)    MICHELINI, M. "Pulsation and Vibration Analysis of High Pressure Piping of Ethylene Compressors", 83rd International AIChE Congress, Houston, Texas, March 20-24, 1977.

(10)   CARLI, R. and CASALE, P. "Simulation of Pressure Pulsations in Gas Compression Plants", Quaderni Pignone 3, Florence, 1965.

(11)   CARLI, G., SPALLANZANI, G., LACITIGNOLA, P. and MICHELINI, M. "Description of the SNAM Simulator of Pressure Pulsation and Its Application  to the Study of the Acoustical Behaviour of a Typical Installation", Quaderni Pignone 4, Florence, 1966.

Fig 1    Packaged reciprocating compressor for high pressure natural gas reinjection, ready on the trailer for transportation to the yard for assembly inside the module

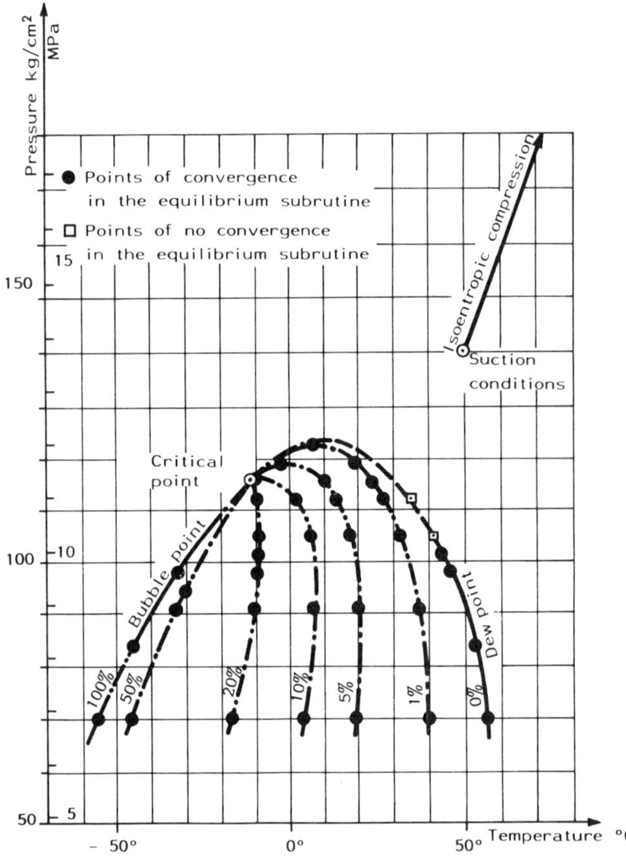


Fig 2    P, T, phase graph of the reinjection gas

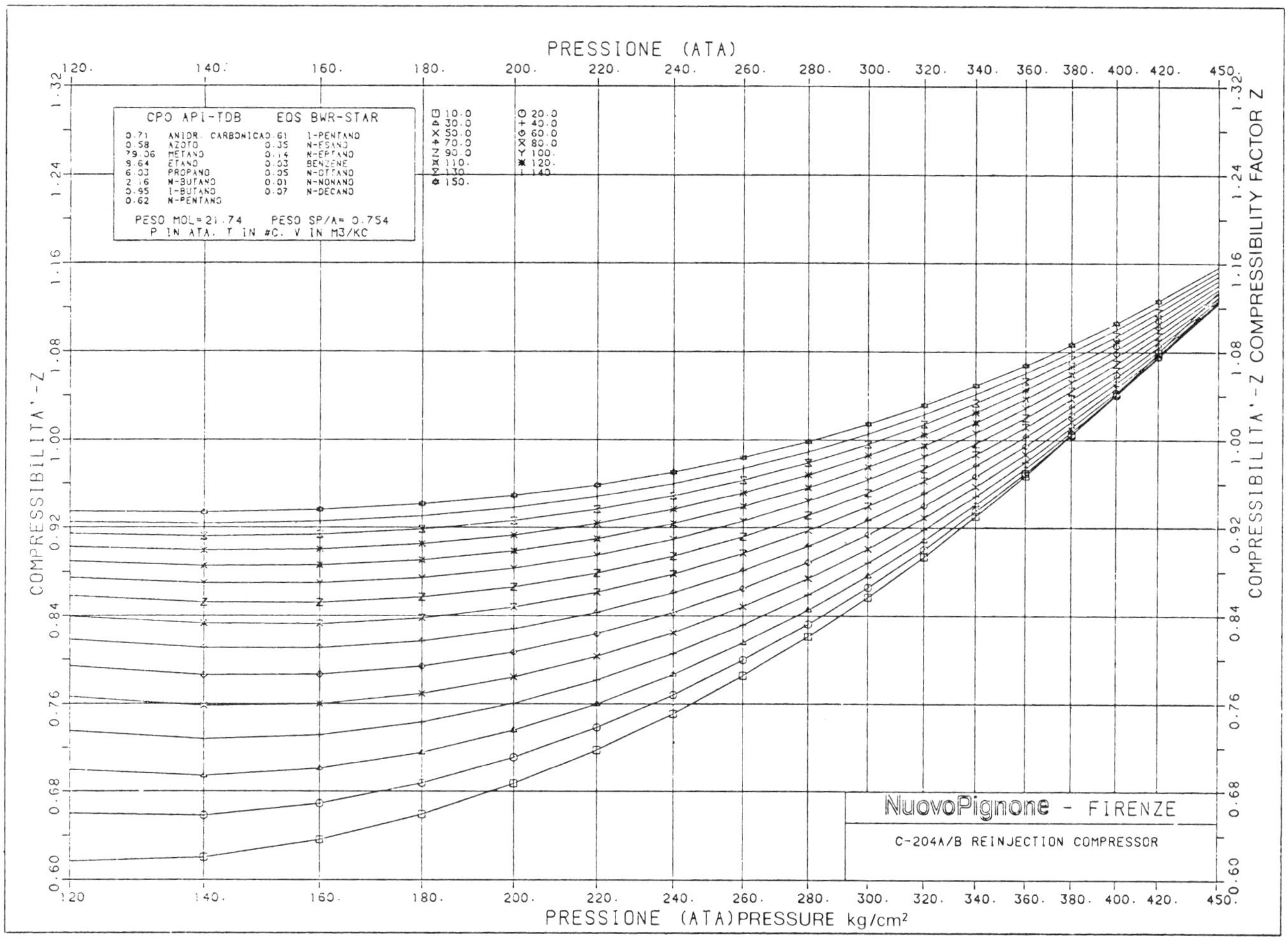

Fig 3    Compressibility factor Z of the reinjection gas

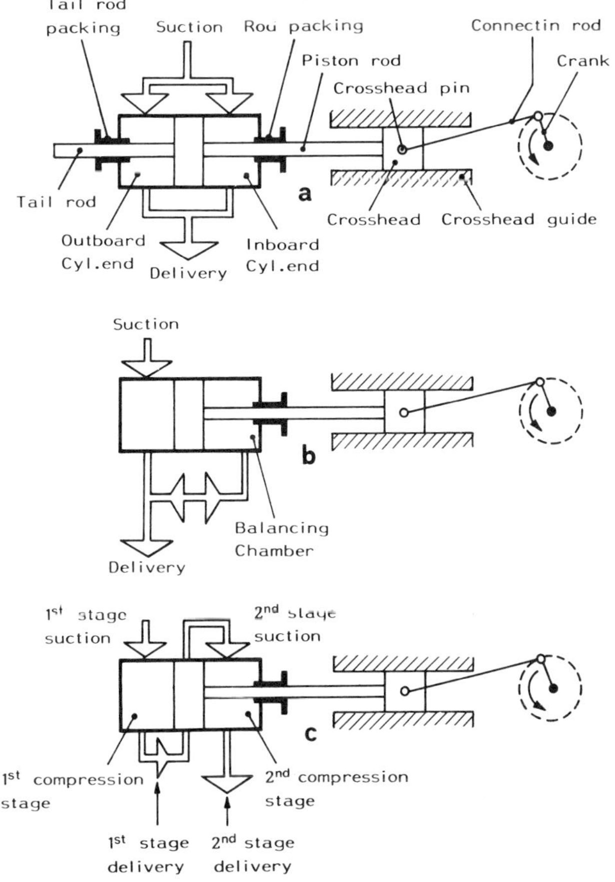


Fig 4    Different compressor cylinder arrangements for high pressure natural gas reinjection

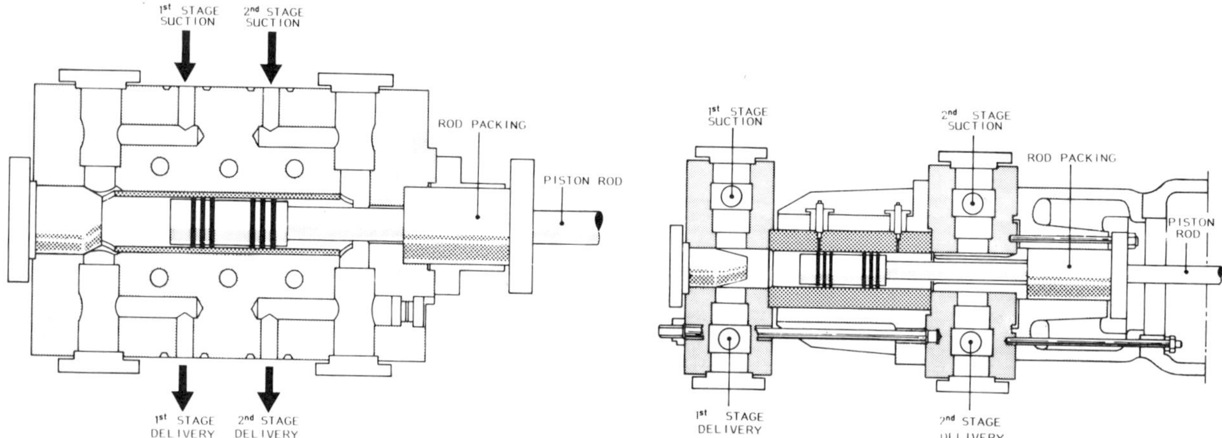

Fig 5a    High pressure monoblock cylinder in forged steel

Fig 5b    Three-piece high pressure cylinder

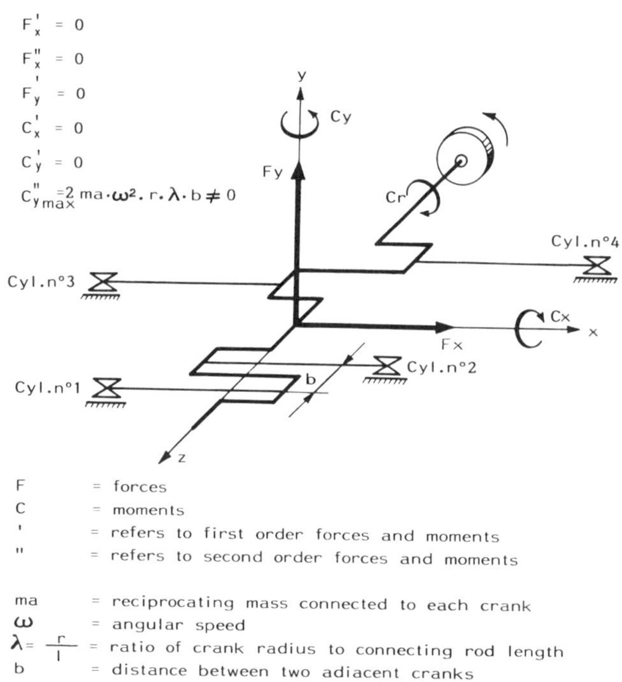


$$F'_x = 0$$
$$F''_x = 0$$
$$F'_y = 0$$
$$C'_x = 0$$
$$C'_y = 0$$
$$C''_{y\,max} = 2\,ma \cdot \omega^2 \cdot r \cdot \lambda \cdot b \neq 0$$



F       = forces
C       = moments
'       = refers to first order forces and moments
"       = refers to second order forces and moments

ma      = reciprocating mass connected to each crank
$\omega$       = angular speed
$\lambda = \dfrac{r}{l}$   = ratio of crank radius to connecting rod length
b       = distance between two adiacent cranks

Fig 6    Free forces and moments resulting with a crankshaft having two pairs of cranks phased at 180°

# C42/81

# Pre-assembly of a large compressor set for a land based petrochemical plant

P E SIMMONS, BSc (Eng)
ICI Petrochemicals Division, Wilton

SYNOPSIS     A large (21MW) motor driven air compressor complete with two stages of power recovery has recently been built in Germany for ICI's Petrochemicals Division at Wilton, England.  The machine was completely erected on its steel foundation in the maker's works, fully tested and then shipped to site in one piece with the minimum of disassembly; shipping weight 270 tonnes.  The advantages of this arrangement are discussed and the various factors which came together to make it possible.  The method of transporting the unit is described together with the techniques developed to reproduce accurately on-site the alignment of the complete unit which existed on the test bed.

## THE DUTY

In 1977 a requirement arose for a large motor driven air compressor for a new plant being built for the Petrochemicals Division of ICI at Wilton, England.  The machine train was also to include a two stage power recovery turbine:  the two stages being required because the gas was not available at a sufficiently high temperature to allow expansion down to atmospheric pressure without reheat.

In round numbers the duty was :
to compressor 120 000 $m^3$/hr (40kg/s) of air from ambient pressure to 32 bar with 30kg/s of gas (mostly nitrogen) at 24 bar, $200^{o}$C available for re-expansion.

## THE MACHINE SELECTED

After consideration of various combinations of axial and centrifugal compressors with different arrangements and numbers of intercoolers the arrangement shown in Fig 1 was selected.

In this arrangement the 21 MW squirrel cage induction motor is located in the middle of the train.  From one end it drives, through a single helical gearbox, the axial low pressure compressor, which has one final radial stage.  From the other end the motor drives, through a second single helical gear, the centrifugal high pressure compressor, which is in two sections.  There are only two intercoolers; one after the axial compressor and the second between the two sections of the centrifugal machine.

The two stages of the power recovery turbine are separated and each comprises a single wheel overhung on the outboard end of the appropriate compressor.

## THE CONCEPT OF PRE-ASSEMBLY

From past experience, the author has a preference for steel rather than concrete foundations for a machine which is elevated above grade.  Two smaller machines had already been erected at Wilton on steel foundations with no unforeseen problems.  The enquiry for this machine had therefore requested that a steel foundation be offered as an option.

During the various discussions with the selected vendor, GHH, Sterkrade, the proposal was made by the author that the machine should be completely erected on its steel foundation in the maker's works, tested and then, if possible, shipped in one piece as a complete unit.  The overall dimensions of the unit are shown in Fig 2,  the total weight being approx. 270 tons.

Initially this proposal was received with considerable incredulity but in a relatively short time the vendor came to appreciate the advantages and from then on, co-operated in developing the concept, with almost as much enthusiasm as the author.

On the whole project, it was the declared intention of the project management  to minimise the on-site work by doing the maximum of off-site pre-assembly and pre-fabrication.  There was therefore little or no resistance within ICI to this novel proposal and  ready  acceptance of the inevitable increase in the order value of the machine set.

## ENABLING FACTORS

In this case a number of conditions or factors fortuitously coincided, which enabled this proposal to be developed without any major difficulties as regards the basic concept.

These factors were :

(i)    The acceptance within ICI and by the project management in particular, of the concept of pre-assembled units.

(ii)   A preference on the author's part for a steel foundation and acceptance by the other parties concerned of the author's viewpoint.

(iii)    A compressor vendor (GHH) with a great
deal of experience with steel foundations, who
was also able to offer a compressor that was
acceptable in all other aspects.

(iv)    The fact that this vendor had recently
developed a large new test facility which was
not so heavily loaded so that the space and time
could be made available for the erection and
testing of the machine.

(v)    Close proximity of the vendor's works to a
canal which connected into the river Rhine and
the existence of a large crane alongside the
canal with which the complete set could be
loaded into a barge.

(vi)    The fact that this vendor was accustomed
to handling large loads, as the same factory
also produced large pressure vessels for the
nuclear power industry.

(vii)    Reasonable road access to ICI's works at
Wilton, England, for oversize loads from the
local dock.

It is possible that had one of these
factors not been present, the concept would have
been dropped.

DEVELOPMENT OF THE PRE-ASSEMBLED UNIT

Having made the decision to proceed with
the concept of pre-assembly the next stage was
to ensure that maximum advantage was gained.  To
realise this aim a number of objectives were
identified :

(i)    To include in the pre-assembled unit all
the compressor ancilliaries and other associated
equipment in so far as it was reasonable to do
so.

(ii)    To support all these ancilliary items
from the main steel foundation to minimise the
number of separate supports to be provided on
site.

(iii)    To have all work within the unit
completely finished before shipment including all
instrument and electrical cabling, painting,
insulation (thermal and acoustic) etc.

(iv)    To develop a method of accurately
measuring the levels throughout the steel foundation and a system of adjustment so that the
levels existing on the test bed could be
reproduced on site to minimise the work of
realigning the machine train.

Achieving several of these objectives was
considerably assisted by having a large scale
model (scale 1 : 10) of the complete unit built
(See Figs  3 & 4).  Construction of this model
was commenced as early as possible and was used
as a design tool in the development of the unit.
The model was jointly reviewed by the compressor
vendor, the contractor and the customer at
various stages as the detailed arrangement
evolved.

The oil system serving the train was
completely built on the steel foundation.  The
oil reservoir and the pumps are located under the
main platform but supported from it.  The oil
coolers, filters and pressure regulators etc,
are located on top.

The intercoolers, which are of plain tube
construction with the cooling water on the shell
side, are suspended from the steel foundation as
are the large interconnecting pipes.

The only significant item which could not
be included in the pre-assembly, was a heat
exchanger which reheats the off-gas before it
expands through the second power recovery turbine.  It was not possible to find an arrangement for this exchanger and the associated
pipework which was sufficiently flexible to
keep the machine flange loadings down to
acceptable levels, and which was acceptable to
ICI.

To reproduce on-site the precise alignment
of the set which existed during the works tests,
the following procedure was adopted.

(i)    The feet of the steel foundation were
designed to be supported on sole-plates with
shim plates between these sole-plates and the
undersides of the feet (See Fig  5).

(ii)    On the top surface of the foundation, a
reference flat was located directly over each
leg (See Fig  6).

(iii)    On the test bed, the relative levels of
these ten references flats were accurately
measured using a precision water gauge   (see
Fig  7).

(iv)    On-site the levels were checked using the
same instrument, and final adjustment made by
varying the thickness of the shim plates.

In the event the test bed levels were reproduced
within 0.2mm.

TESTING

A complete performance test was carried
out, in the maker's works, in accordance with
VDI 2045.

In addition a mechanical test at full
speed was carried out on both the main and
spare rotors.

For these tests the machine was driven by
passing steam through the expansion turbines and
exhausting into the permanent test bed condenser.

The instrument panel associated with the
machine, was shipped from the U.K. where it was
built, to Germany, so that the supervisory
instruments, anti-surge control gear and safety
systems could also be proved on test, as far as
possible.

SHIPMENT

Prior to removal from the test bed, the
steel foundation was stiffened by the addition
of massive, temporary steel beams between the
feet and a number of diagonal braces.  Also, a
few items that projected beyond the agreed
transport limits were removed.

The whole unit was lifted by four hydraulic

jacks to a sufficient height so that a large
multi-wheeled trailer could be driven under it.
(See Fig 8).

The unit was moved on this trailer to the
nearby canal where it was lifted into a barge in
which it was transported to Rotterdam. (See
Fig 9).

At Rotterdam the machine was loaded on to the
deck of a special ship, using the ship's own
derrick, for the journey across the North Sea
to Tees-side. (See Fig 10).

At Tees dock the ship's derrick lifted the
unit on to a similar multi-wheeled trailer upon
which it was towed into the Wilton Site and
into the part-built compressor house. (See
Fig. 11).

Approximately in its final position the
machine was again raised by jacks so that the
trailer could be removed and the machine lowered
on its foundation block. Final adjustments to
its lateral position were made using traversing
jacks. (See Fig 12 ).

ADVANTAGES

In the author's opinion four main advantages accrued from the pre-assembly of this
machine.

Firstly, the machine was tested in a
situation which was much closer to the eventual
operating situation on-site. Following that
test, virtually no dis-assembly of the machine
was necessary.

Secondly, even though the manufacturer
under-estimated the time and effort required in
his works to carry out the work he had been
persuaded to accept, and the final shipment was
several months late, nevertheless there has
undoubtedly been a considerable overall saving
in time and effort and therefore in cost.

Thirdly, pre-assembly of this machine has
contributed to a reduction in the peak labour
requirements on-site and hence reduced the
problems of managing a very large transient
labour force.

Fourthly, there has been an improvement in
the quality of workmanship in some areas. For
example, the machine has forty vibration
transducers, more than fifty thermocouples,
numerous pressure switches etc. Routing all
the cables to these instruments has been done
under the control of the compressor maker and
hence in such a way as to interface much less
with the maintenance access to the machine than
would typically have been the case, had all
these cables been run on-site by an independent
instrument contractor.

LESSONS FOR THE FUTURE

The pre-assembly of this machine is
considered to have been an unqualified success.
However, this was the first machine of this
size to which these principles have been
applied. Hence lessons have been learned and
in some small ways the next job would be done
differently.

It is the author's firm opinion that with
a complex piece of equipment such as the subject
compressor, every item should be individually
tested and proved before it is built into the
whole. Whilst in this case, the motor and gearboxes were individually test run in their
respective maker's works, the compressor casings
were not. In the event, a critical speed
problem came to light on the test bed. Fortunately, the problem was solved by a simple
modification to one bearing and only delayed
the combined test by a few days but it might
have been much more serious.

In order to provide sufficient flexibility
in the main process pipework within the unit
to avoid exceeding the acceptable flange loads
on the machine, bellows units had to be built
into several of the lines. Due to a long
history of failure of such bellows units, the
author's company prefers to avoid them and would
not have accepted them had the compressors been
handling a flammable or toxic gas. The author
believes that had the layout within the unit been
designed from the outset with piping flexibility
as the prime consideration, the use of these
bellows could have been avoided.

The shipment of the unit proceeded without
any problems and the accuracy within which the
test bed levels were reproduced on-site was well
within the target limits. However, had the
method of shipment been decided at the outset,
including such details as the dimensions of the
transport trailer, the foundation could have
been designed with this in mind. As a result
the whole process of stiffening the unit,
lifting it, transporting it and finally locating
and levelling it on-site could have been
simplified.

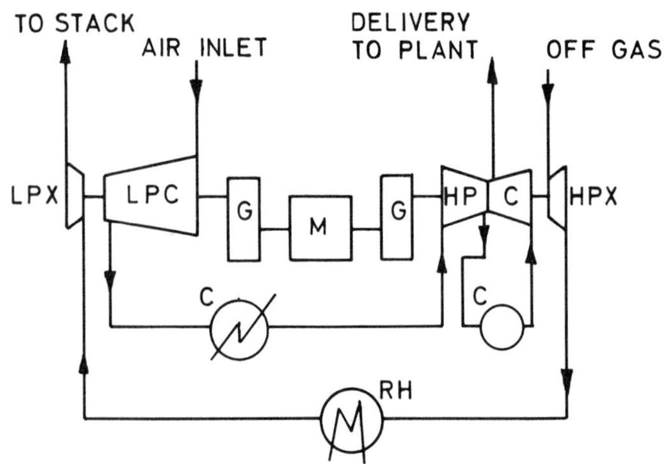



Fig 1    Diagrammatic arrangement of compressor set

M      Motor
G      Gear
LPC    Low pressure compressor
HPC    High pressure compressor
HPX    High pressure expander
LPX    Low pressure expander
C      Intercooler
RH     Reheater

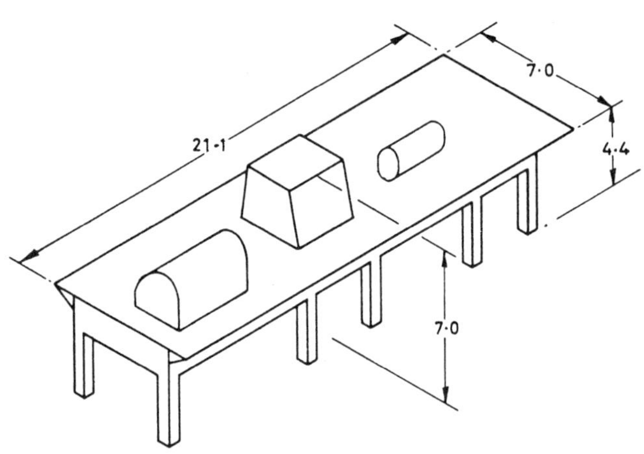



Fig 2    Overall dimensions of compressor set as shipped (dimensions in metres)

Fig 3    Model of compressor set: East side

Fig 4    Model of compressor set: West side

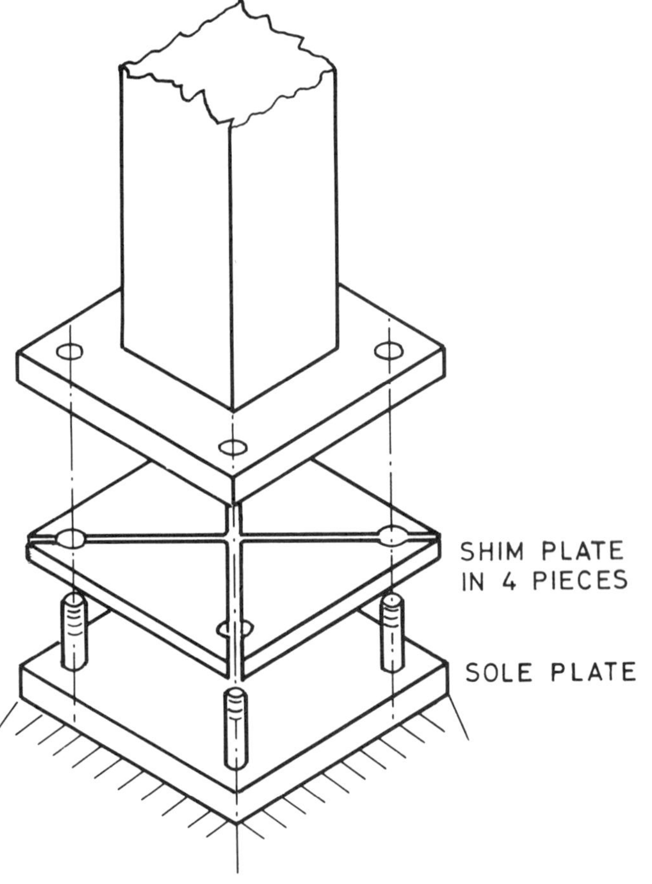



Fig 5    Arrangement at feet of steel foundation

Fig 6    Reference plate

Fig 7    Precision water gauge in use

Fig 8    Inserting trailer under unit

Fig 9    Loading unit into barge

Fig 10    Lifting unit on to special ship

Fig 11    Compressor unit in part-built compressor house

Fig 12    Traversing jack

# C49/81

# The application of turboexpanders for energy conservation

J HOLM, Member ASME
Rotoflow Corporation, Los Angeles, USA

SYNOPSIS   Substantial energy can be recovered using low grade waste heat, process gas, or waste gas pressure letdown.

Centrifugal (radial inflow) turboexpanders are well adapted to such energy conservation schemes and with recent developments which have increased their reliability, are suitable for unattended service on a 24-hour, 7-day week operational basis.

Some of these recent developments presented herein include shaft seals, thrust bearing monitoring, and control devices, etc.

The turboexpander's qualifications to meet the requirements of energy conservation are enumerated below.

INTRODUCTION

Forty years of development in turboexpander technology has resulted in highly-efficient machines which can be applied in the profitable recovery of energy from waste heat sources, and gas pressure letdown.

With increased demands and with energy sources being depleted at an ever increasing rate, the need for conservation and for the recovery of energy from sources once thought unprofitable, is emphasized.

In the past, the use of the turboexpander as an energy recovery device was limited because of the following:

A.   The return on capital investment did not justify a power recovery system unless more than several thousand horsepower was recovered.

B.   Finding a market for recovered power, when there is not immediate use for it within the plant was difficult.

C.   Continuity and reliability of this energy source was required if it were used as 'base load', which required standby equipment, spares, and constant operator attendance.

D.   Lack of confidence in new power recovery schemes which were not yet proven made both government and private industry reluctant to invest in these systems.

At present, these conditions, limiting the use of turboexpanders for power recovery, have improved, as follows:

A.   With increasing cost of power, the return on capital investment has vastly improved.

B.   A more favorable attitude by the utility companies toward returning electricity to their grid has made a number of power producing schemes practical.

C.   High efficiency of expanders, and their relatively short pay-back period make smaller units economically attractive.   A curve showing power recovery versus cost is shown in Figure 1.   This is based on a unit of medium pressure (up to 60 Bars), and temperatures up to 350 Deg. C.

D.   The present high degree of reliability of expanders, proven by hundreds of units which have been in continuous uninterrupted service for many years, has removed the concern for back-up equipment and has demonstrated that unattended operation is feasible.   Several demonstration projects are now operating successfully.

This paper briefly summarizes turboexpander applications, the present state-of-the-art, and the features incorporated in the turboexpander design which enable it to meet most power recovery requirements.

Expander Applications

The turboexpander has been used for many years in cryogenic processing plants to provide low temperature refrigeration (Ref. 1).   Power recovery has been of secondary importance. Since the expander's efficiency determines the amount of refrigeration produced and, in gas process plants, the amount of product usually depends on the available refrigeration, there is a large premium on efficiency and, of course, on reliability.

The main market for the turboexpander has been
in low-pressure air separation plants, expanding
down from 5 Bars, and in hydrocarbon processing
plants, expanding natural gas from as high as
200 Bars.  The air separation expanders are
roughly divided into two types:  the first type
ranges from a few horsepower up to 100 hp, where
the expander power is too small to be economically recovered, and is absorbed by an oil brake
or other similar device; the second type ranges
from 100 hp to over 2000 hp, where the power is
used to drive electric generators or process
booster compressors.

The hydrocarbon expanders range in the order of
100 hp, up to 8000 hp (Figure 2)(Ref. 2).
These expanders are usually designed with a process compressor directly driven by the expander
for power recovery.  The gas is usually expanded
from a 100 Bar to 50 Bar range, down to a 50 to
15 Bar range, resulting in an expansion ratio of
2:1 to 4:1, a very suitable expansion for a
single-stage expander with high efficiency in
the 84% to 86% range.

There are numerous, large turboexpanders operating in the pressure range of 200 to 130 Bar,
most in well-head natural gas service (Ref. 3).

Expanders are also used for the purification of
gases, such as $H_2$ or He, by condensing contaminants.  These are usually small units, 5 to 50
hp, operating at speeds from 45 000 to 70 000
rpm, and not usually considered economical for
power recovery.


Power Recovery Turboexpanders

The number of applications where the recovery of
power is important is growing, and they are being explored and developed to an increasing
degree.

Several small demonstration plants are operating,
or are about to begin operation, to minimize
potential problem areas for new, large power
plants in the planning stage, utilizing such
sources as ocean-thermal energy conversion
(solar heat), geothermal, waste heat, natural
gas, or waste gas pressure letdown, etc.

The cycles in these power recovery applications
are relatively simple (Figure 3).  They involve
the removal of solids or liquids ahead of the
expander, and often the incoming stream is heated
so its temperature will not reach its frost
point at the discharge (Ref. 4).  This addition
of heat also increases the amount of available
power.  Some examples of this application are
expansion of waste gas, waste products of combustion in oxidation processes, waste carbon
dioxide, and expansion of high-pressure synthetic
gas streams.

In applications with impulse or axial reaction
turbines, care must be taken to discharge just
above the dewpoint of the expanded gas.  If gas
enters the turbine at, or near its dewpoint, the
turbine will operate in the condensing range,
resulting in two-phase flow in the turbine outlet.  This condensate has caused severe erosion
problems in ordinary turbines; however, the
design of the radial inflow turbine solved these
problems, as will be discussed later.

A 1200 kW power recovery expander-gear-
generator designed to be installed in parallel
with a natural gas pressure letdown station is
illustrated (Figure 4).  The expander shown
receives the process gas at 11 Bar and 42 Deg.
C , and expands it to 5 Bar.  In this case, the
temperature at the discharge is calculated to be
1 Deg. C , and since the gas contained water
vapour, it will condense in the expander.  This
will bring the gas to a suitable dewpoint, and
droplets are removed in a separator downstream
of the expander.

Another application for turboexpanders is in
power recovery from various heat sources utilizing the Rankine cycle.

The heat sources presently being considered for
large scale power plants include geothermal and
ocean-thermal energy, while small systems are
directed at solar heat, waste heat from reactor
processes, gas turbine exhaust and many other
industrial waste heat sources.

Some of these systems are discussed below in
greater detail:

There are two general geothermal resources, dry
(steam) fields and wet (brine) fields.  More
than 800 MWe is being produced from such dry
geothermal steam fields in Northern California
and several other flash stream power plants are
under construction.

The wet fields usually cannot  be utilized in
this manner and Rankine cycle-type systems
(binary plants) are being considered at such
locations.

At such wet fields as we find in the Imperial
Valley in Southern California, the geothermal
fluid is a 250 Deg. C  brine which does not lend
itself for use in conventional steam turbines.
A typical binary cycle is illustrated (Figure 5)
(Ref. 5).  The power recovery is accomplished by
pumping the hot water or brine from underground
wells through heat exchanger equipment to boil
a working fluid maintained in a closed cycle.
The resulting vapour is expanded to drive the
turbine-generator and then recondensed and
pumped back into the heat exchanger to repeat
the cycle.  This expansion of the vapour produces
saleable power, so efficiency is at a premium.
Several working fluids are suitable for binary
cycles, and include iso-butane, iso-pentane,
propane, and hydrocarbon mixtures.

Without further development, suitable turboexpanders with high efficiency, reliability, and
seal systems to meet the geothermal requirements
are presently available.

A study of this type of application, to develop
a conceptual design for a radial reaction turbine, was completed by Rotoflow for EPRI (Electric Power Research Institute in Palo Alto,
California)(Ref. 6).  The result of this study
program was a 65 MWe gross output turboexpander
operating at 3600 rpm, for direct coupling to a
synchronous generator.

The turbine design has a double (back-to-back)
rotor, 122 cm in diameter, placed between the
bearings with a single inlet port and double
discharge ports.

A hydrocarbon mixture was selected as the working
fluid and the vapour at the inlet to the turbine
was 33.3 Bar at 143 Deg. C , and expanding to 5
Bar, a condenser temperature of 63 Deg. C.

Since this plant was to be located in the Southern California Desert, the condensing was to be
done with air, hence the selection of high
expander discharge temperature. A comprehensive
study was made to determine the effect of the
machine design with the large change of ambient
temperatures found at this location, which can
vary from a high of 50 Deg. C, to well below
freezing from summer to winter, or from day to
night. Such wide swings may cause extensive
condensation in the turbine, but this can be
efficiently and safely handled in a modern turboexpander, as discussed later.

One of the problems that complicates plant design
in the wet geothermal field is the corrosive
characteristics of the brine.

The system described in the foregoing involved
pumping the brine to the plant, and then from the
plant into the ground, thus keeping the brine
from flashing and causing severe scaling in casings, pipes and heat exchangers.

To circumvent this problem a pilot plant was
constructed by Daedalean Associates in Maryland
under the sponsorship of the U.S. Department of
Energy (DOE), utilizing direct-contact heat
exchangers (Ref. 7). In this design the working
fluid, in this case iso-pentane, is sprayed in
direct contact with the geothermal brine and
vapourized. The fluid and water vapour at 66 Deg.
C is expanded from 3 Bar to 1 Bar in a 100 kw
expander/integral-gear/generator unit (Figure 7).
The first series of tests have been successfully
completed and it is interesting to note that only
1 ppm of the iso-pentane was absorbed in the
'boiler' brine.

Much attention is also being given to solar
energy. It does not appear that direct solar
heat is economically feasible for large power
plant energy source; however, this resource has
great potential for several process and heating
applications.

One form of solar heat does offer interesting
possibilities and is referred to as OTEC (Ocean-
Thermal Energy Conversion). The OTEC power plant
principle uses the solar heat of ocean surface
water to vapourize ammonia as a working fluid in a
Rankine cycle. After the fluid is expanded in
the turbine, it is condensed by the 22 Deg. C
colder water pumped from the ocean depths.

The Mini-OTEC platform was designed and constructed with funding from several private companies,
and has a 50 kW ammonia turbine/gear/generator
unit which expands the ammonia from 7 Bar and 21
Deg. C , to 6.5 Bar at 10 Deg. C (Figure 7).
Both the boiler and the condenser were designed
for a 5.5 Deg. C temperature approach, using the
27 Deg. C surface water for heating and 4 Deg.
C water pumped from 663 meters below the surface.

The platform has operated since August 1979, in a
location 2.5 km off the west coast of the Big
Island of Hawaii (Figure 8).

We recently received a letter informing us of
the unqualified success of the initial 600 hours
of operation. Just before the barge was pulled
into the harbour for examination, the plant ran
smoothly and continuously despite the fact that
the barge rolled 10 deg. to 20 deg. due to large
ocean swells. This would not create any problems on a full-scale, floating OTEC Power Plant,
since most of it would be below the surface, as
shown in the 'Artist's Conception' of a 160
Megawatts Power Plant (Figure 9).


Power Absorption Methods

Turboexpanders, presently in refrigeration processes, develop power, but recovery of this
power has been secondary. The usual methods of
absorbing power are described below. All of
these are directly applicable to energy recovery
expanders.

1. Compressor, Direct-Connected (Figure 10)

    The most popular method of absorbing turbo-
    expander power is by means of a single-stage
    or two-stage centrifugal compressor, mounted
    directly on the expander shaft. In a cryo-
    genic process, there is nearly always a place
    where this compression energy can be used.
    The addition of a compressor load to the sys-
    tem is inexpensive because the turboexpander
    bearings also support the compressor impeller,
    and therefore only an impeller, case and seal
    need to be added.

2. Gear and Generator

    If a compressor is not applicable and power is
    of value, a gear speed reducer and electric
    generator is a widely used and reliable method
    for the recovery of energy.

    This generator arrangement usually consists of
    a high-speed gear, couplings and a generator.
    Other rotating machinery, such as a pump, may
    also be used to absorb the energy.

    A simplification of this arrangement is an
    expander with integral speed reducing gear,
    in which the pinion gear is on the turbo-
    expander shaft. The pinion gear directly
    engages the low-speed master gear and reduces
    the speed of the available power to 3600 to
    3000 rpm, as required (Figure 11). This
    arrangement has several advantages. It permits
    easy application of a mechanical seal on the
    low-speed shaft that hermetically seals the
    expander gearbox. These integral gear boxes
    are designed to be pressurized up to 10 Bar.
    Effective sealing is especially important in
    the expansion of poisonous or explosive gases
    like helium or freon.

    The integral gear also eliminates the power
    losses in the high-speed pinion gear bearings
    of an external reduction gear and the windage
    of the high-speed coupling, as well as elimi-
    nating the alignment and noise problems of the
    latter.

## Turboexpander Qualities

From the preceeding applications and from many hydrocarbon applications, it is apparent that a turboexpander is a special turbine which should be designed with quality features to meet the following requirements:

* Maintain high efficiency with varying flow;

* Toleration of dust or condensation of gas stream;

* Bearing strength to avoid damage if the rotor should be unbalanced by ice deposits, or damaged by erosion;

* High efficiency (usually requiring high speed);

* Proven reliability;

* Positive shaft seals or other special seals;

* Wide range of sizes.


## Variable Flow Control

A high quality turboexpander has variable flow control nozzles capable of withstanding the total pressure and acting as the flow control for the main gas stream through the plant. The variable nozzle should be matched with a rotor to give high efficiency over a wide range of flows (usually from 50% to 120% of design or wider)(Ref. 9)(Ref. 10)(Figure 12). They should be designed for negligible blow-by and for durable performance, even if constantly moved as by a pressure-controller or other controlling signal.


## Expansion of Condensing Streams

As to the utilization of turboexpanders for condensing streams, this is solved by a very interesting development in which the blades in the expander rotor are so shaped that their walls are at every point parallel to the vector resultant of the forces acting upon suspended fog droplets (or dust particles)(Ref. 11)(Ref. 12). Then the suspended fog particles do not drift toward the walls, collect and interfere with performance and erode the blades. Over 100 turboexpancers are in successful operation involving condensing liquids.

Dust-laden streams can also cause operational problems. A turboexpander that can efficiently process condensing streams (gas with fog droplets suspended) can usually handle a stream with suspended solid particles, as long as the particle size does not exceed 2 to 3 microns. A new design has been developed for reducing erosion of an expander back rotor seal by disposing of the dust that accumulates at the seal and discharging it through the balance holes in the expander rotor (Figure 13)(Ref. 13). We believe large expanders can be designed to handle dust or particles up to 10 microns.


## Thrust Bearing Force Meters (Figure 14)

Machines with an expander inlet pressure of the order of 10 Bars carry thrust loads usually within the capabilities of the thrust bearings. At higher pressures it is essential to carefully balance the thrust loads against each other. Thrust loads, even though originally correctly balanced, may change greatly, and exceed the thrust bearing load-carrying capacity. This imbalance of thrust loads may be caused by either erosion of a seal, icing or off-design operating conditions. This problem has been solved by a force measuring meter on each thrust bearing, and in some cases, a thrust control valve which controls the thrust by control of pressure behind the thrust-balancing drum (Ref. 12).

Because of features such as these, the reliability of turboexpanders is exceptionally good. Operation for several years without repair is most common.


## Shaft Seals

Virtually all turboexpander seals are of the close-fitting, labyrinth-type, because the high velocities prohibit the use of rubbing-face seals. With close-clearance seals it is important that the shaft be closely maintained in its rotating position, thus flexible shaft design (operation above the first lateral critical) is usually not acceptable. Bearings which maintain the closest alignment of the shaft are obviously the best for such applications, and for this purpose, a close-clearance journal-type bearing is used.


## Available Sizes of Turboexpanders

Turboexpanders, currently in operation, range in size from about 1 hp to above 10 000 hp.

We recently delivered five cryogenic expander-compressors, rated at 15 000 hp, to the US Air Force (Figure 16).

In this enormous scale-up, it is interesting to observe the size ranges where various problems are dominant. In the small sizes, the problems are miniaturization, Reynolds Number effects, heat transfer, seals and mechanical problems, such as bearings and critical speed.

In intermediate sizes, these problems become less significant, but bearing rubbing speeds and vibration become increasingly important.

Vibration becomes critical in the intermediate ranges because structural members are relatively less massive and the speeds are high enough to match the resulting natural frequencies in some cases. Thorough testing of rotors is essential, and extensive work has been done in this area (Ref. 15).

In intermediate and larger sizes the thrust bearing problem requires more attention, but it has been effectively solved recently by the introduction of the thrust force meter and thrust force adjustment valve, described earlier.

Presently, designs for radial inflow turboexpanders in sizes up to 70 MWe are available for
use in geothermal power plants.


Conclusion

Turboexpander processes have evolved as a result
of available turboexpanders with the required
high performance capabilities and reliability;
and conversely, the development of these processes has pointed the way toward further improvements in turboexpanders.

In summary, the state-of-the-art of modern turboexpanders has reached a very high level of reliability and efficiency.

Mechanical designs of low-temperature, high-speed
machinery are routine.

Stiff shaft designs have eliminated shaft and
bearing criticals in the entire operating range.

Rotor resonance problems are well known to the
designers and are, in most cases, totally
eliminated.

Thrust bearing problems (the weakest area in
high-speed machinery) can be accurately monitored
and controlled.

Condensing streams and some dust in gas can be
handled without erosion.

These are the most important features that make
the use of turboexpanders ideal for recovery of
power from the vast resources of pressurized gas
streams.


REFERENCES

1.  J.S. Swearingen 1971 AIChE Annual Institute
Lecture.

2.  8,000 hp NG Expander-Compressor, installed
at Petro-Canada Exploration, Inc., Empress Plant
at Empress, Alberta, Canada.

3.  Chemical Engineering 75, No. 11, Pg. 146
(May 24, 1968).

4.  Stodola:  Steam and Gas Turbines, Peter Smith
Publishers, Pg. 312 (Steam Quality).

5.  Diagram from paper by The Ben Holt Co.
"Advanced Binary Cycles for Geothermal Power
Generation".

6.  E.P.R.I. Contract No. RP928-3.

7.  Daedaleon Associates, Springland Research
Center, Woodbine, Maryland.

8.  Lockheed Missiles and Space Co., Sunnyvale,
California, Sub-contract YF80C1230A.

9.  Swearingen U.S. Patent 3,232,581 (February
1966).

10. Swearingen U.S. Patent 3,495,921 (February
1970).

11. Swearingen U.S. Patent 3,610,775 (October
1971).

12. Swearingen Japanese Patent 617,904
(September 1971).

13. Swearingen U.S. Patent applied for.

14. Swearingen U.S. Patent 3,828,610 (February
1974).

15. Swearingen, J.S. and Mafi, S.: Experimental
Investigation of Vibrations in High Speed
Rotating Machinery, ASME Paper No. 69 VIBR-58
(1969).

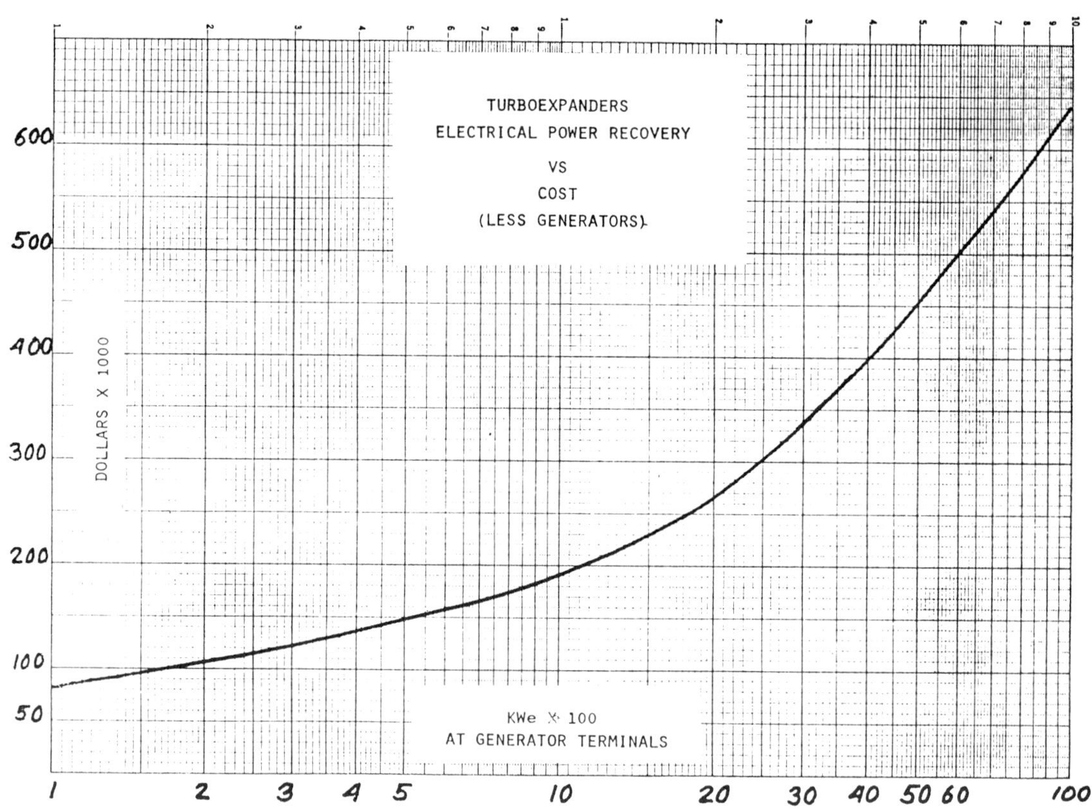


Fig 1    1980 cost of turboexpanders *vs* horsepower

Fig 2    8000 h.p. natural gas expander-compressor

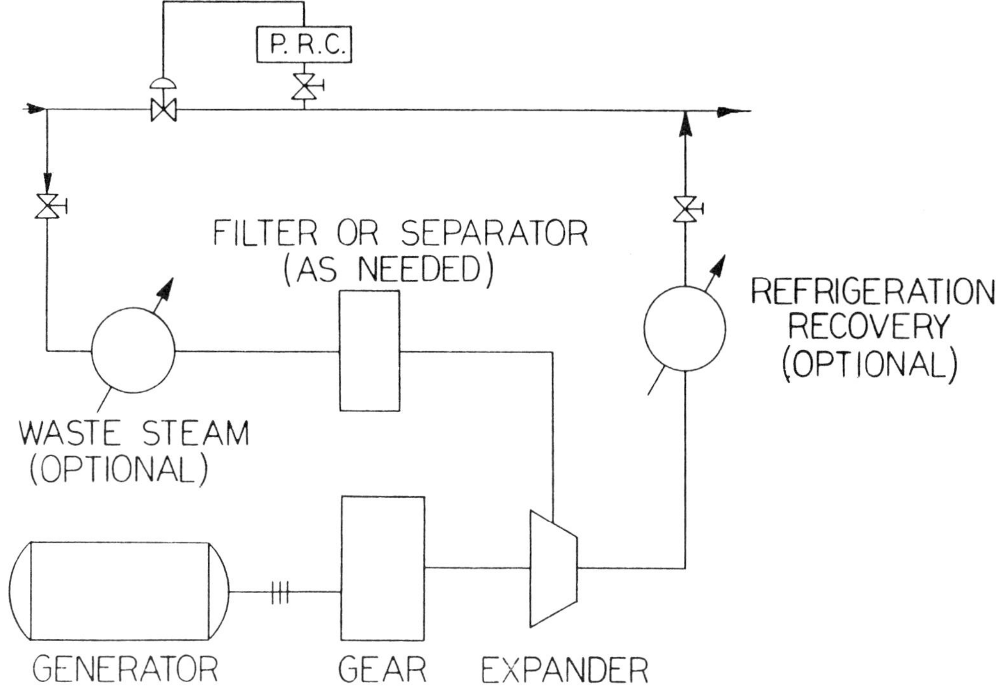


Fig 3    Gas pressure letdown power recovery schematic

Fig 4    1800 kW natural gas pressure letdown expander-gear-generator

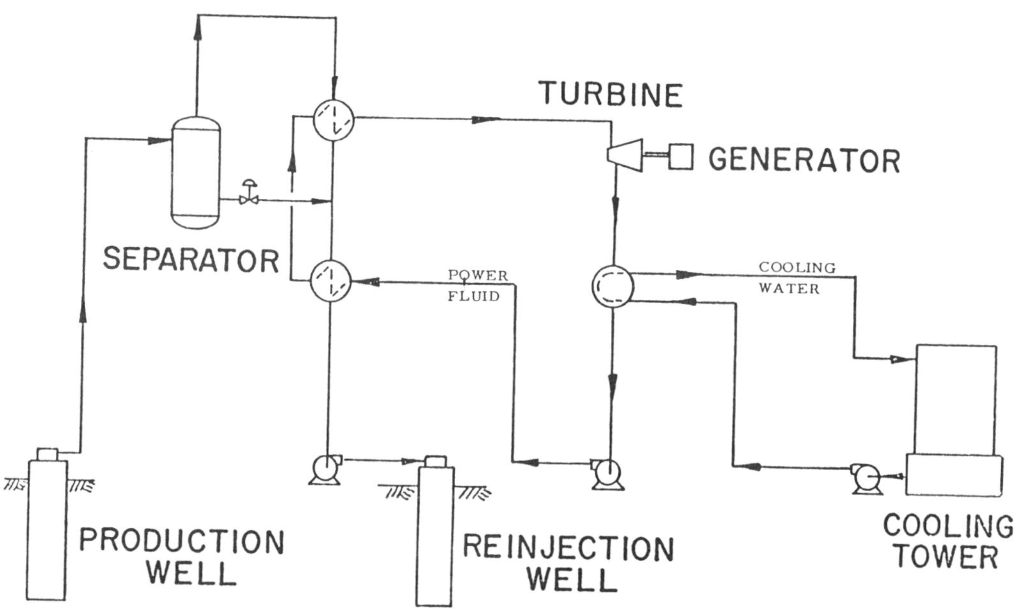


Fig 5    Typical binary geothermal cycle

Fig 6    100 kW iso-pentane expander-gear-generator

Fig 7    MIni-OTEC ammonia expander/gear with 50 kW synchronous generator

Fig 8    Mini-OTEC platform anchored off Hawaii

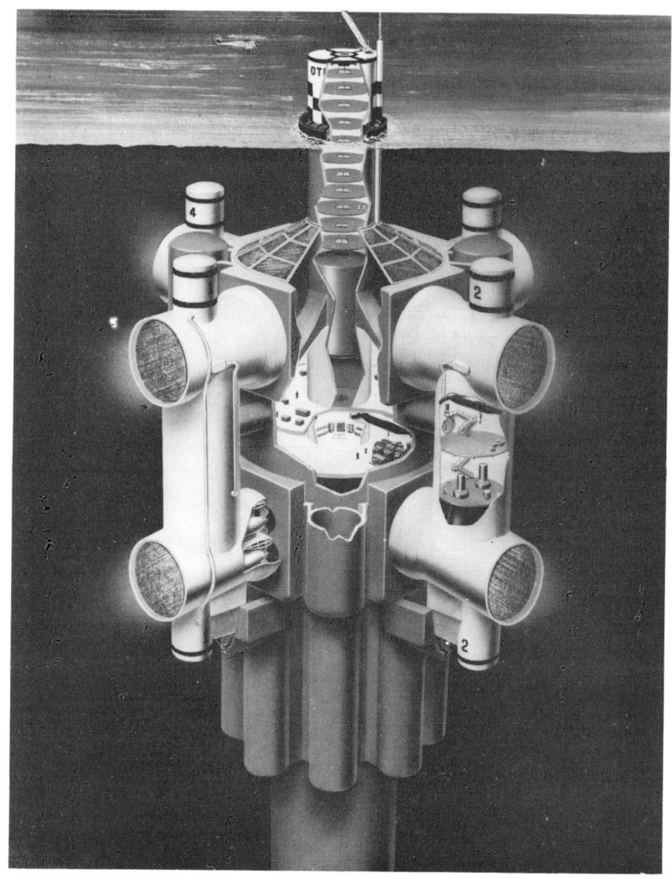

Fig 9    Conceptual design of a 160 MWe floating OTEC power plant

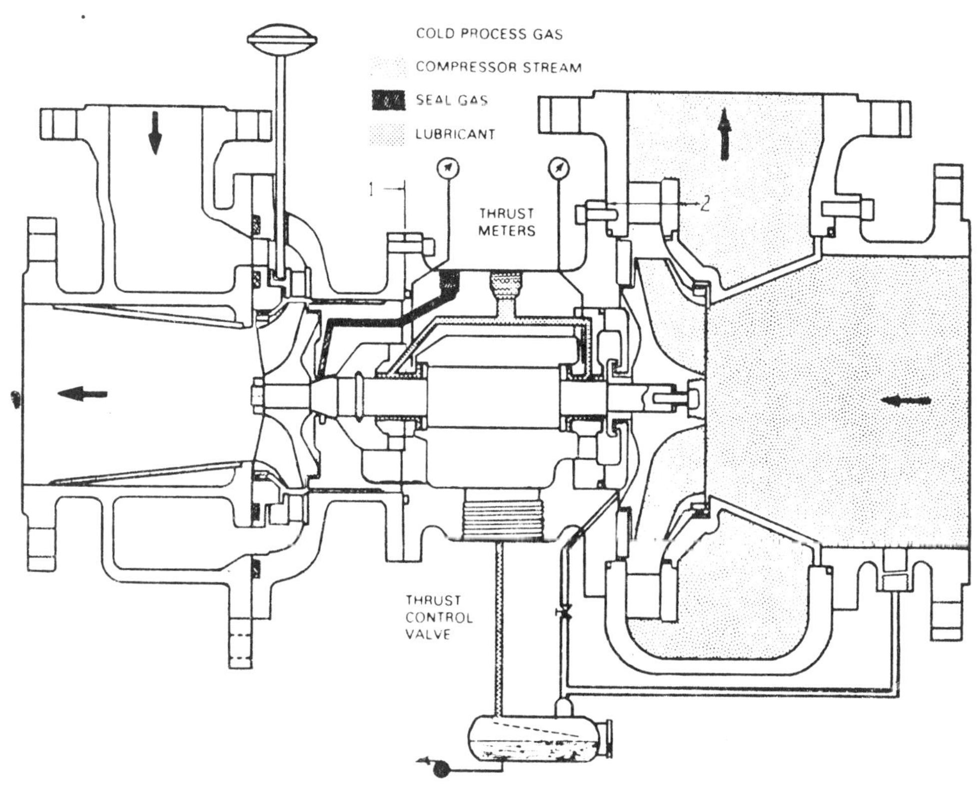


Fig 10    Cross-section of an expander-compressor

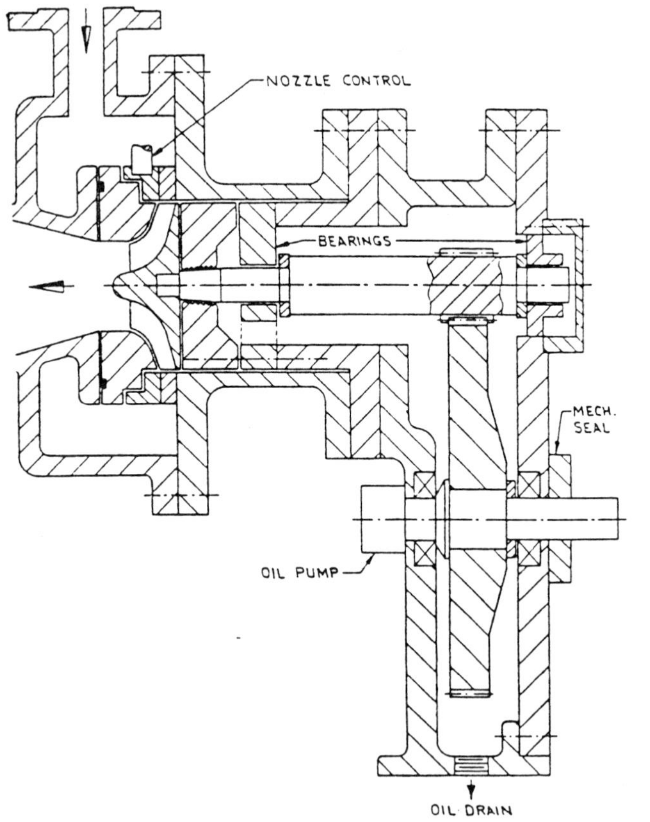



Fig 11  Cross-section of an expander with integral gear for power recovery

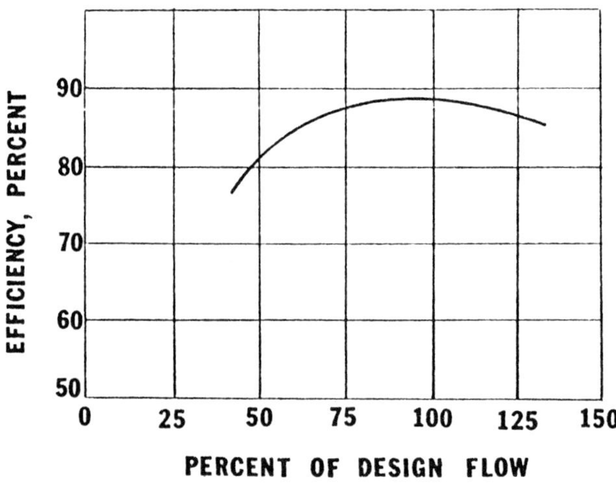



Fig 12  Turboexpander efficiency *vs* flow

Fig 13  Variable inlet guide nozzles for turbo expanders

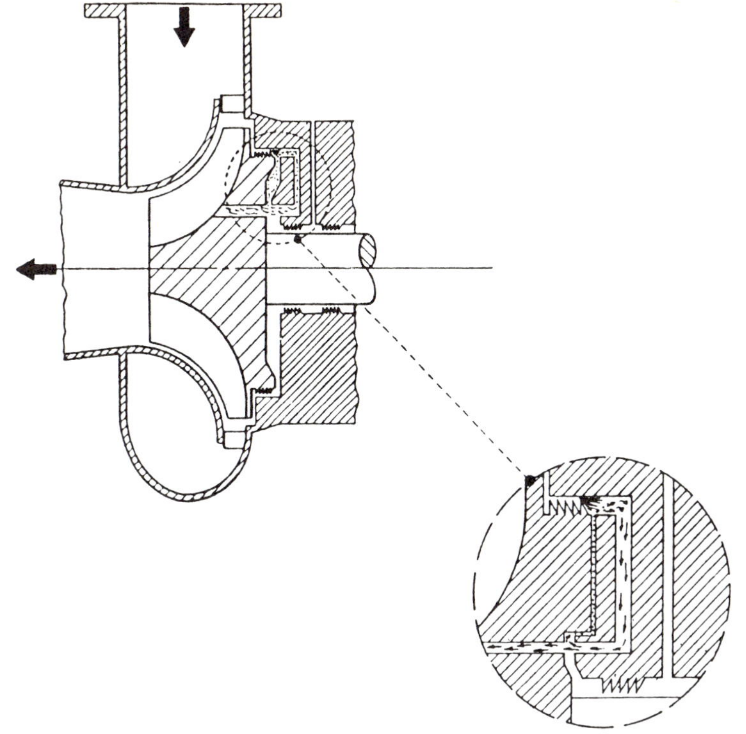

Fig 14    Rotoflow's dust-free expander seal design

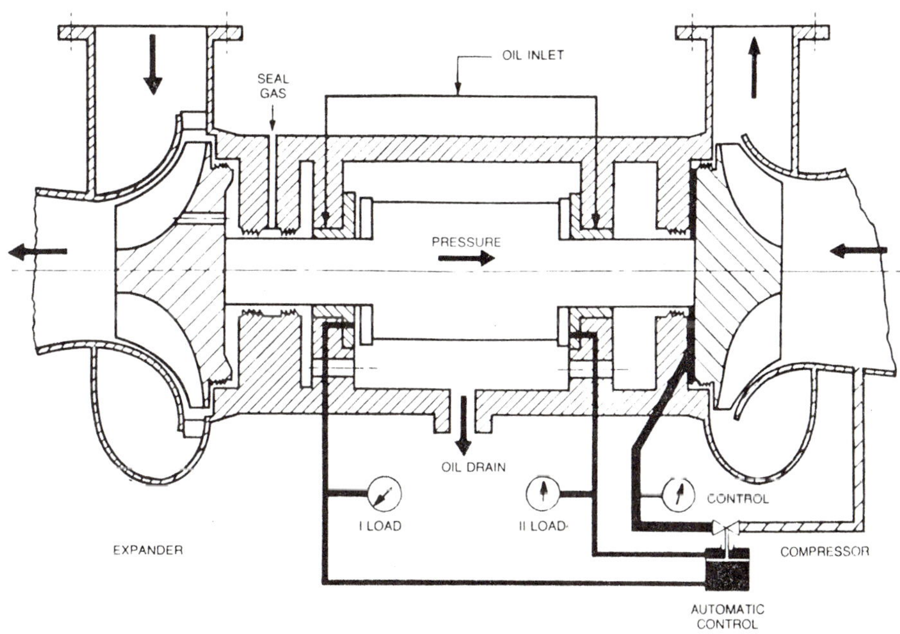



Fig 15    Rotoflow's thrust meters and thrust control

Fig 16    11 000 h.p. expander-compressor for US Air Force test facility

# C44/81

# Thermodynamic design of improved efficiency small steam turbines for industrial applications


J WAGGOTT
Turbodyne Corporation, Wellsville, New York



SYNOPSIS    This paper presents the statistical and thermodynamic studies involved in designing improved efficiency small steam turbines.  The power range considered is 373 to 4474 kilowatts (500 to 6000 horsepower) and the prime objective was to improve efficiency by 20 points over the near 50 percent efficiencies of currently available small steam turbines.  The resulting designs use established technology, have economic viability and are suitable for use in traditional industrial applications.  Capabilities of specific designs are compared with the requirements of complex ranges of application using a concise method.  This method should be of increasing value in this time of rapidly changing energy conversion policies.


NOTATION

$W$    mean blade speed, m/s
$V_o$    isentropic spouting velocity, m/s

INTRODUCTION

Small industrial steam turbines in the power range from a few hundred to several thousand kilowatts have traditionally been relatively simple durable machines. In the past, the prevailing energy conversion economics have induced designs that frequently sacrifice efficiency in order to achieve the necessary reliability at relatively low capital cost. Increased energy costs have changed the economic evaluations making improved efficiency desirable, providing that it is cost effective and can be achieved without sacrificing reliability and durability.

This paper describes the statistical and thermodynamic evaluations leading to designs with desired efficiency improvements. The work was carried out as an integral part of a contract with the U. S. Department of Energy (DOE) calling for the complete design and development of these improved efficiency small steam turbines.

OUTLINE

The thermodynamic design of these turbines was achieved through four major steps listed below:

a.  A review of historical data to establish the range of thermodynamic requirements.

b.  Preliminary staging specification to satisfy performance requirements and mechanical and thermodynamic criteria.

c.  Comparison of competitive configurations on the basis of efficiency, application coverage and design difficulty, resulting in a selection of the most desirable configuration.

d.  Developed shaft seal specifications that gave an acceptable efficiency penalty without compromising the need for rotor stability.

The material which follows contains more detailed descriptions of the procedures and results of the above steps.

Establishing thermodynamic requirements

The specific objective was to design and develop improved efficiency steam turbines in the 373 to 4474 kilowatts (500 to 6000 horsepower) range. Beyond this basic specification, it was necessary to establish probable ranges of steam conditions and the distribution of power requirements within the specified range. This was done by studying the information contained in Turbodyne's commercial data base, which is a detailed computerized record of bookings back to 1966. The data sample was first compared with the records of the National Electrical Manufacturers Association (NEMA) and determined to be representative of total U. S. domestic populations.

The computerized nature of the data allowed complex data reduction procedures. Figure 1 was derived from the data and shows histograms of turbine population versus power (kW) and available energy (J/g). It can be seen that approximately two-thirds of the population has between 200 and 600 J/g available energy and below 1350 kW. Many of the turbines with above 700 J/g or over 2500 kW are relatively efficient multistage machines and there would be little value in designing the new system to cover such applications. The average available energy is 448 J/g or 397 J/g, depending on whether complete or abridged populations are considered.

Preliminary staging specification

Stage selection is based on the use of axial flow stages because of the wide range of flow capacities that can be achieved within a single frame configuration.

Radial inflow stages did not appear to offer as wide a range and could not conveniently handle the wet steam that is frequently present during normal or abnormal operation. It was recognized from the onset of this design exercise that there is an inherent conflict between desired turbine shaft speeds and traditional driven unit speed. Therefore, the design and development of suitable gearing was accepted as an integral part of the project and relative freedom in selecting turbine shaft speeds was assumed.

To control stresses, a maximum allowable blade speed of 320 meters per second was established. This blade speed limit is higher than normally used on standard designs but is well within the range used for specifically engineered designs. The limit selected is felt to be the highest speed consistent with achieving a reliable and universal rotor system (Disc/Attachments/Blades).

It was clear that the new turbine designs would typically have short blades and/or partial admission and thus require low reaction (impulse) stages in order to avoid excessive leakage. This, in turn, requires 0.45 to 0.50 velocity ratios $(W/V_o)$ for maximum efficiency. The blade speed (W) limit and desired velocity ratio lead to a maximum of approximately 230 J/g/stage for maximum efficiency and 280 to 350 J/g/stage with only a modest decline from optimum. Lower blade speeds could be used for lower available energy per stage.

The maximum stage capabilities discussed above, together with the requirements shown in the histograms of Figure 1, indicate that a two-stage design is desirable. This will give a very high efficiency for applications with up to 460 J/g while still offering acceptable efficiency for the relatively small number of applications with 600 to 700 J/g. Clearly, larger numbers of stages could be considered, but this would not be consistent with the implicit need for a relatively simple and versatile system.

This has established that two stages are desirable but has not, at this point, established mean diameter or shaft speed.

Selection of final stage configuration

To establish the most desirable configuration, three alternative combinations of mean blade diameter and shaft speed were compared. These were:

200 mm Dia./30 000 rpm
300 mm Dia./20 000 rpm
450 mm Dia./13 333 rpm

Blading was assumed to be conventional impulse type, thus allowing well established performance prediction methods to be used. The flow capacity of each stage was varied by changing blade height or arc of admission. To limit stresses and allow the use of constant section blades, the radial height of the flow path was limited to give a minimum hub/tip ratio of 0.8. This established a maximum flow path size for each configuration and also achieves the geometric similarity implied when using blade speed as a stress criterion.

The application coverage of each configuration was determined using the diagram shown in Figure 2. The abscissa in this diagram is the product of kilowatts and isentropic exhaust specific volume. This parameter strongly influences flow path size and was therefore called 'size parameter'. 'Available energy' essentially completes the thermodynamic specification for a particular application and is used as the ordinate in this diagram. The reference population of turbine applications was plotted on this plane and fell within the shaded area of Figure 2. Conventional turbine performance calculations were then executed at various available energy levels for each of the competitive configurations. The second stages were fixed at the maximum allowable flow path size while first stages were sized for optimum efficiency at each energy level. The output powers from these calculations allowed 'size parameter' capabilities to be computed at several energy levels. These maximum capability lines are plotted on Figure 2. A specific configuration cannot satisfy applications which lie to the right of its maximum capability line. The percentages of the reference turbine population which can be satisfied by each size increment are shown on the diagram.

Performance calculations were done for a wide range of applications for each configuration. These calculations indicate that between 2 and 6 percentage points of efficiency penalty are incurred with each size increment. Averaging these penalties for the reference turbine population shows an average penalty of 3.4 percentage points for increasing from 200 to 300 mm mean diameter and 4.4 percentage points for the 300 to 450 mm diameter increase.

With all of this information pertaining to configuration, application coverage and efficiency available, a final configuration selection was made. The 300 mm/ 20 000 rpm configuration was judged to give the best compromise of application coverage and efficiency, while presenting an acceptable gearing and rotor dynamic challenge. By comparison, the 200 mm diameter configuration would cover fewer applications and present a severe gearing and rotor dynamics challenge, and the 450 mm diameter configuration would incur unnecessary efficiency penalties.

Shaft seals

Complete turbine design concepts were developed around the flow path specifications defined above and an overhung configuration was evolved. To minimize axial thrust, and therefore thrust bearing or balance piston losses, the flow direction was away from the bearings. This configuration caused the end gland leakage to bypass the second stage, as shown schematically in Figure 3, and thus subtract from the power output. Rotor stability requirements tended to restrict the space available for the seal, and it was necessary to establish the minimum requirements from an efficiency standpoint.

It was arbitrarily decided that a median efficiency penalty of one percentage point would be acceptable. Using a similar method to that used in Figure 2, the median second stage nozzle area was determined to be 14.2 cm$^2$. This is illustrated in Figure 4. With the size and therefore flow capacity of the second stage nozzle specified, the flow capacity of an end gland seal that would depress efficiency by one percentage point was derived. With shaft diameter, seal clearance, tooth geometry and flow capacity defined, the number of labyrinth points was determined. The resulting seal configuration is 110 mm diameter and has 15 points in a 45 mm length.

This established the number of labyrinth points before the first leakoff which is assumed to return flow to the exhaust stream. Any additional space available will accommodate as many labyrinth points as possible to minimize the flow lost to an end gland condenser in high exhaust pressure applications.

## CONCLUSIONS AND DISCUSSION

With the complete flow path configuration defined, a wide range of performance estimates were calculated and are illustrated in Figure 5. This figure has the same abscissa as Figure 2 and by entering with size parameter for a particular application and moving up to the appropriate available energy line, efficiency can be read on the ordinate. The performance estimates include windage losses, partial admission effects, interstage leakage, end gland leakage, inlet losses, exhaust losses and assume that the maximum allowable first stage admission is 80%, to allow for additional flow capacity under a 'hand valve'. These estimates did not include mechanical losses or the effects of moisture in turbines operating with wet steam.

There are some obvious questions about the merits of using historical data rather than future predictions to assess the field of applications for these turbine designs. This topic should be viewed in the context that many of the application specifications are inherent to the working fluid and do not change significantly with time. In addition, it is anticipated that many future applications will, in fact, be the replacement of existing units. The author's conclusion is that an incomparably more difficult and less accurate future prediction approach would have yielded little additional insight into the potential field of applications.

The system used in Figure 2 is intended to be a concise, easy to use and versatile method of comparing the capabilities of particular turbine configurations with complex fields of potential applications. A particular merit of this system is that the plotted field of application points is defined quite independently of the expanders that are to be considered. The capability lines for particular turbines or expanders are derived from actual performance calculations and therefore include efficiency effects. Virtually any type of expander can be considered, including positive displacement. The system will, in fact, work moderately well with different working fluid applications plotted on the same diagram.

The simplifying approaches used in this procedure may introduce some inaccuracies in the assessments of application coverage and efficiency. However, when considering such inaccuracies it should be recognized that size evaluations covered two decade ranges, making small percentage errors in assessing size requirements relatively insignificant. Any marginal inadequacies in flow capacity for a given configuration could be accommodated by changing the blade geometry from the fixed standards or

limits assumed in the original calculations. Most of the inaccuracies resulting from the simplifications are quite small, but failing to consider moisture effects may introduce significant errors.

A relatively simple improvement could be implemented by first establishing an effective quality as an average of the isentropic quality at the middle and at the end of the expansion. The original size parameter would then be divided by this effective quality to give a modified size parameter for plotting on the coverage diagram (Figure 2). This same modified size parameter could also be used for efficiency estimates from Figure 5. The efficiency read from Figure 5 should then be multiplied by the effective quality to obtain an efficiency estimate corrected for moisture.

Using isentropic quality values in the above method may appear to be an exaggerated correction. But this approach is intended to provide an additional correction to compensate for the larger expansion ratios that accompany a given energy drop in wet steam. These larger expansion ratios typically require slightly less efficient staging than is implied by the available energy parameter. This inefficiency, and therefore increased flow requirement, is additional to that induced directly by the moisture, thus justifying the slightly exaggerated correction method. This method will be qualitatively correct for most expanders but is more specifically intended for the two-stage axial flow turbines being considered here.

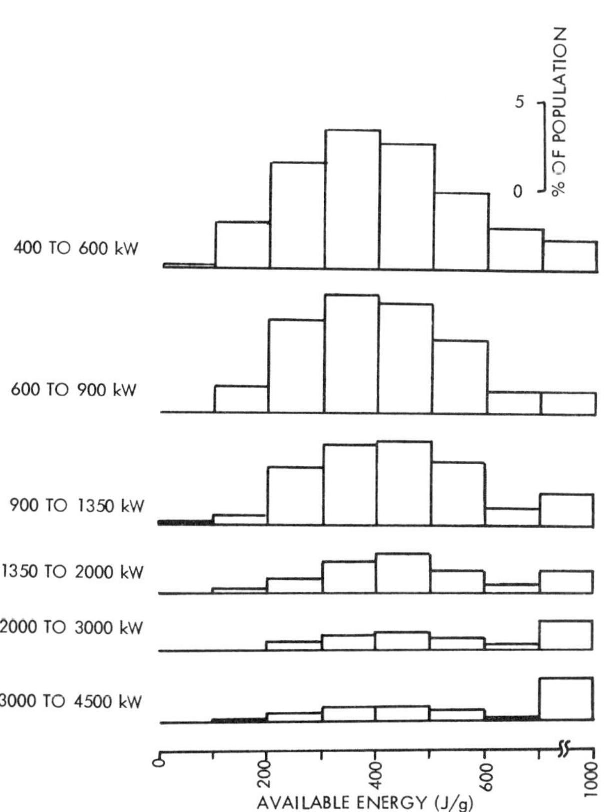


Fig 1  Turbine populations versus power and available energy

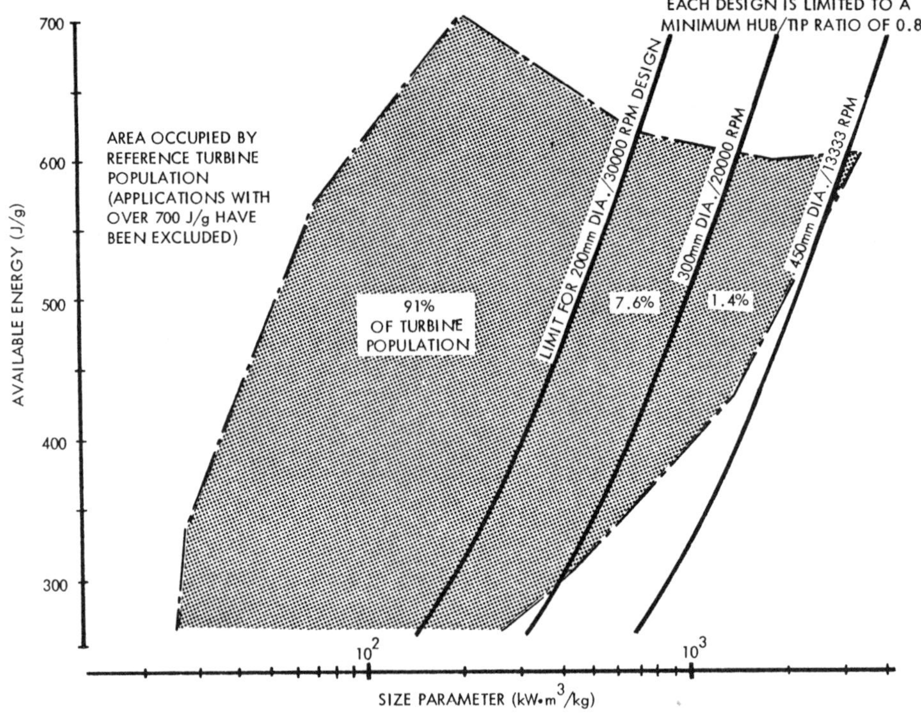


Fig 2   Coverage achieved by competitive two-stage designs

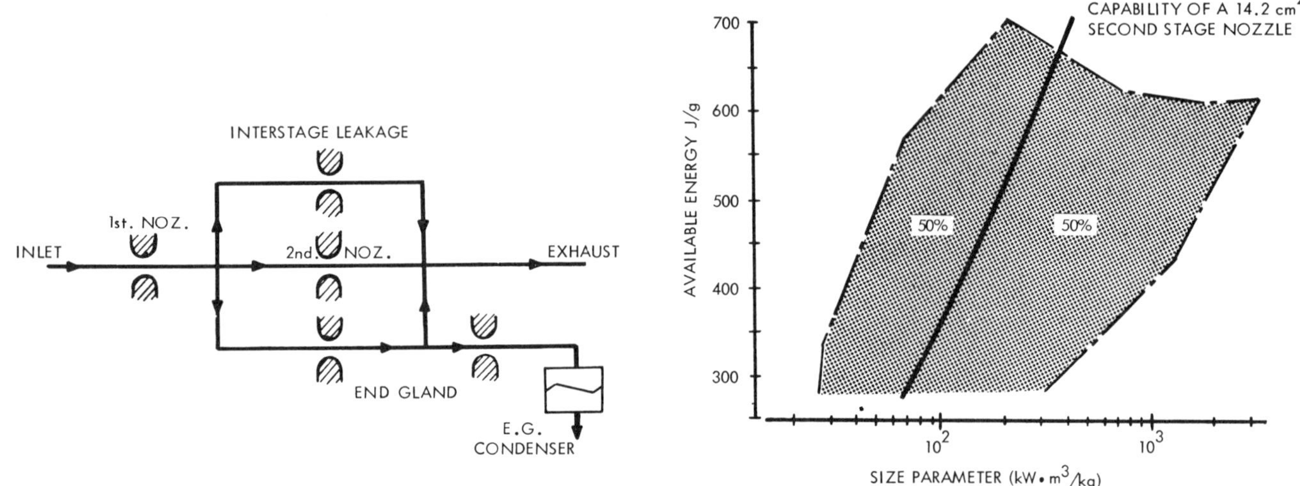


Fig 3   Flow schematic                              Fig 4   Establishing median nozzle area

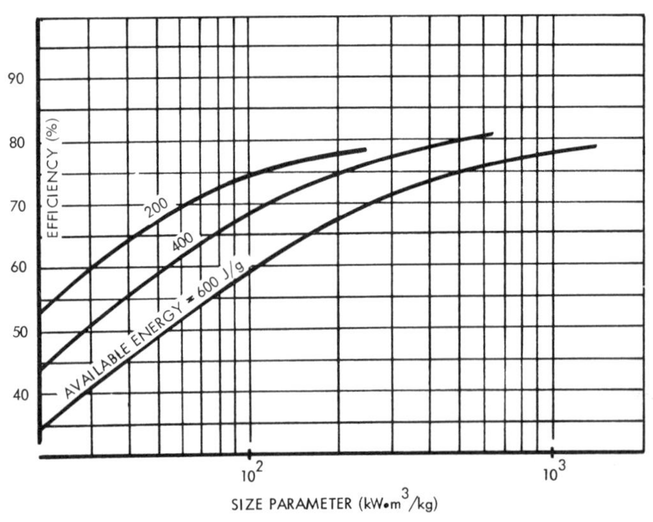


Fig 5   Estimated efficiency of final design

                                   © IMechE 1981   C44/81

# A case history of carbonate solution pumping problems

R H WATKINS, CEng, MIMechE, MIMarE
Bechtel Great Britain Limited

SYNOPSIS    A standard, heavy duty, refinery type pump was selected for a potassium carbonate service in a gas liquefaction plant.  The operational and maintenance history of this pump is described and used as a basis for indicating refinements of the construction of both the pump and the associated system piping which would have eliminated or significantly reduced the problems encountered.

## INTRODUCTION

Nine 2200 kW motor driven pumps installed in the $CO_2$ removal units of a LNG liquefaction plant had the following problems during the first eighteen months of jobsite operation.

(a)  Premature failure of bearings

(b)  Accelerated deterioration of wear rings

(c)  Malfunction of shaft mechanical seals

(d)  Rotor imbalance

(e)  Shaft malalignment

(f)  High level of vibration

To resolve these problems an assessment of the mode of operation was made at the jobsite;  the compatibility of the design of the pump and the system was reviewed;  a vibration survey was implemented.

The service conditions were as follows:

| Liquid | lean carbonate solution |
|---|---|
| Temperature, $^{o}C$ | 118 |
| Specific gravity | 1.25 |
| Flow rate, $m^3$/hr | 950 |
| Discharge pressure, bars | 62 |
| Inlet pressure, bars | 5.5 |
| Rotation speed, rpm | 2960 |
| Material of pump | 18-8 stainless steel |

The pumps were located in three identical process units in which two pumps were normally operated in parallel with one pump on standby. The pumps as depicted in figure 1 were a two stage design, with the impellers arranged opposed in a radially split casing, centreline mounted, between bearings.  The shaft was supported radially in plain sleeve bearings and

located axially by a tilting pad thrust bearing. The bearings were lubricated by a pressurized oil supply.  The main oil pump was driven by the carbonate pump shaft.  An auxiliary motor driven oil pump was used for pre-lubrication prior to the start up of the carbonate pump and was arranged to start automatically on low oil pressure.  If the oil pressure was reduced even lower the carbonate pump motor was tripped out. Each journal bearing had an oil ring to assist lubrication during rundown.  A supply of high pressure, clean water was provided for flushing each mechanical seal housing.  The shaft vibration was continuously monitored by two proximity probes, spaced 90° apart, adjacent to the pump journal bearings.

Observations made at jobsite confirmed that in general the pumps were operated between 90% and 110% of rated duty as indicated by the differential pressure across the pump of 53 to 60 bars and the motor current of 270 to 290 amps. The fresh water flush was supplied at near design flow rate of 0.45 $m^3$/hr to each seal of the two pumps in service but was reduced to below 0.25 $m^3$/hr to each seal of the pump on standby.  The standby pump was not maintained in the hot condition although a bypass around the discharge check valve had been installed for this purpose.

## PREMATURE FAILURE OF BEARINGS

There were several occasions on which the journal and thrust bearings of the carbonate pumps had been damaged or subjected to accelerated wear due to inadequate lubrication. The main cause of lubrication failure could be attributed to reverse rotation of the non-operating carbonate pump, either on standby or during rundown, resulting from the malfunction of the check valve in the discharge piping of the carbonate pump.  When reverse rotation occurred the shaft driven oil pump discharged in the opposite direction to normal but reverse flow was prevented by the check valve in the oil pump suction and discharge piping.  The evidence indicated that the carbonate pumps had been allowed to run in the reverse direction for an excessively long period of time which

caused the oil retained in the bearing housings
to overheat and thereby reduce its lubrication
properties.  The lubrication oil system did not
have provision in the piping arrangement to
allow the circulation of oil to the bearing
housings when the oil pump was running in the
reverse direction.  Also, the electricity supply
to the auxiliary motor pump was isolated when
the motor of the carbonate pump was de-energised.

As an interim solution the auxiliary motor
driven oil pump was started manually before the
carbonate pump was shut down and left running
continuously while the carbonate pump was on
standby.  Thus some additional time was made
available in which the lubrication oil piping
arrangement could be modified and the malfunction of the check valve in the carbonate
pump discharge piping rectified without causing
an unscheduled shut down of the plant.


ACCELERATED DETERIORATION OF WEAR RINGS

Some corrosion/erosion damage at the high
velocity areas of the casing, impeller and wear
rings occurred due to entrained pipe scale and
carbonate crystals.  In particular the damage
to the wear rings resulted in an accelerated
deterioration of the close clearance gap between
the rotating and static components.  The
consequential increase in stuffing box pressure
reduced the preset throughput of flushing liquid
flow and allowed the build up of carbonate crud
which impaired the performance of the mechanical
seals.  In addition there were several occasions
on which the wear rings had worked loose.  The
explanation of this was not readily apparent
because the material of the wear rings (type 316
stainless steel) had nominally the same
expansion/contraction characteristics as that of
the material of the impeller and casing (type
CF8 stainless steel).

The impellers and casings were repaired by
welding and ground smooth.  The damaged wear
rings were replaced in type 304 stainless steel
without a stellite deposit.  The only cause of
concern with this choice of material was the
probability of galling should running contact
occur, but with the estimated shaft deflection
and clearance gap in use the risk involved was
deemed to be acceptable.  To prevent the wear
rings becoming loose in service those on the
impeller were preheated and shrunk on with an
interference fit of 0.13 to 0.25 mm and those
on the casing were spot welded at three points.

A study of the operators log indicated that
the normal procedure was to isolate the standby
pump and allow it to cool.  In consequence,
there was no outlet path for the flushing liquid
and it was reasonable to presume that carbonate
crud could accumulate in the wear ring
clearance.  The advantage of maintaining the
standby pump at operating temperature was the
elimination of thermal shock when it was brought
into service and a reduction in carbonate crud
is generally associated with hot recirculation.
Therefore, instructions for standby conditions
were issued recommending that the bypass around
the discharge check valve be left open, the
design flushing flow rate of 0.45 m3/hr to
each seal be maintained at all times, and the
casing drain valve be opened to provide an
outlet for these inflows.


MALFUNCTION OF MECHANICAL SEALS

There were several occasions on which the
mechanical seals had malfunctioned and allowed
heavy leakage.  The seal components examined
indicated that the carbon and tungsten carbide
contact faces were in good condition with an
even wear pattern.  Although some of these two
components were fractured in a radial plane
there were no apparent signs of excessive heat
generation.  However, there was severe
contamination by carbonate crud with a
particular build up around the static 'O' ring
seals and spring locations.  Hence it was
concluded that the malfunction of the seals was
mainly due to insufficient flushing by the
external supply of clean water.  The radial
fractures were probably the result of physical
loads applied during the disassembly procedure.
The flushing liquid was injected radially at
the contact faces and at a single position.
Complete loss of the flushing supply could not
be ruled out and a significant reduction in the
supply due to an increase in the stuffing box
pressure caused by accelerated deterioration of
the wear rings was suspected.

To prevent the build up of carbonate crud
on the seal components the number of flushing
liquid inlets were increased to three ports,
120 degrees apart at each seal.


ROTOR IMBALANCE

Difficulty was experienced in maintaining
an acceptable level of rotor balance as
indicated by an increase in the shaft vibration
amplitude at running frequency.  Some
corrosion/erosion of the impellers was present
and there was reasonable cause to suspect that
the output data from the workshop balancing
machine was either incorrect or being mis-
interpreted.

Attempts were made to carry out in service
balancing of the rotors but this solution was
not entirely successful.  Although the weight
and dimensions of the shaft coupling were
relatively small compared to those of the motor
and pump rotor, the spacer part of the coupling
was connected by flexible membrane to the shaft
hubs.  This coupling spacer assembly appeared
to be the greatest source of the problem.  The
addition of balance weights at the coupling to
correct an imbalance at the impeller did not
always produce satisfactory results.  It is
generally accepted that rotor imbalance results
in a vibration component at a frequency
corresponding to the speed of rotation and
malalignment results in a vibration component
at a frequency corresponding to twice the speed
of rotation, but this is an over simplification
of a very complex dynamic problem.  A study of
the shaft vibration on an oscilloscope
confirmed that very unstable shaft dynamics
were present.  Variation in shaft alignment
with time was suspected.

Therefore the assistance of a specialist from the balance machine manufacturer was obtained to check the output data of the machine, review balancing procedures and supervise the rebalance of the rotors under controlled workshop conditions.

SHAFT MALALIGNMENT

The inability to maintain the correct shaft alignment was a continual problem and the situation was confused by the presence of rotor imbalance.  There was no doubt that the relative positions of pump and motor shafts moved in service by a greater amount than that caused by the predicted thermal growth between cold and hot service conditions.  The geometry of the pump and motor assemblies indicated that the cold alignment should be set with the centreline of the motor rotor 0.1 mm low in the vertical direction at both inboard and outboard ends. Any deviation in the horizontal direction should have been within the permissible limits of the spacer coupling.  Routine alignment checks verified that this situation was not valid. The combined motor and pump assemblies appeared to be subjected to external loads other than the predictable thermal growth causing shaft mal-alignment of up to 0.4 mm.  The most likely source of these external loads was the forces and moments applied to the pump nozzles by the inlet and discharge piping.  Despite repeated attempts to reposition and adjust the supports of the piping only marginal success was achieved.

It was concluded that the difference in physical size between the pump and the piping was too great and the permanent solution to the problem was to either select a physically larger and heavier pump or reduce the size and change the method of support of the inlet and discharge piping.

HIGH LEVEL OF VIBRATION

During the initial pump start up on jobsite the displacement measurements from the installed proximity probes indicating the relative movement of the pump shaft to the bearings was similar to those recorded during the manufacturer's works tests, (approximately 0.05 mm).  However, the vibration velocity measured on the pump bearing housings was a cause for concern (approximately 9 mm/s at 50 Hz, 7 mm/s at 250 Hz and 4mm/s at 350 Hz).  The 50 Hz component corresponded to the shaft rotation speed, the 250 Hz component corresponded to the blade rate passing frequency of the first stage impeller and the 350 Hz component corresponded to the blade rate passing frequency of the second stage impeller.  Because the flow conditions, fluid concentrations and operating parameters were not firmly established nor fully controlled at this time, no immediate remedial action was taken but a programme of periodic monitoring of the shaft and bearing housing vibration was introduced.

After some three months of operation, steady state operating conditions had been attained, minor adjustments had been made to the inlet and discharge piping support, and

allowance had been made for variations between cold to hot shaft alignment.  During this period the shaft to bearing displacement remained nominally constant but the bearing housing vibration had increased by 50%.

Hence, a comprehensive vibration survey was implemented which consisted of recording external vibration spectra at various points along the pump and piping, shaft vibration as transmitted through the installed proximity probes, and pulsation spectra at selected points on the inlet piping, discharge piping and internal to the pump casing.  Some typical results of the tests made on jobsite are illustrated graphically in figures 2, 3, 4 and 5.

Figure 2, depicts the relationship between the pressure pulsations in the piping and the vibration signatures taken from the pump which illustrates the sensitivity of the pump bearing supports at the impeller blade rate passing frequency excitation.

Figure 3, depicts the response of the piping, pump bearing housings, pump rotor and motor rotor to shock loading applied by a piece of 4 x 2 timber and thereby identifies the resonant frequencies of these components.

Figure 4, depicts the response of the piping and pump cross over passages to transient changes in pressure during shutdown and thus gives an indication of their pressure amplification characteristics".

Figure 5, depicts the mode shapes which verifies the longitudinal flexibility of the inlet piping and the pumps at the blade rate passing frequency of the first stage impeller.  Similar, but less pronounced mode shapes were obtained at the blade rate passing frequency of the second stage impeller.

The tests included:  normal pump operation; startup and shutdown;  impact of pipe, pump and bearing housing;  and flow variation.  Also, the result of modifying the piping con-figuration, baseplate stiffness, different impeller designs and the application of a dynamic absorber were compared with the results obtained from the original installation.

The following deductions were made:

(a)	The predominant vibration responsible for the high seismic readings measured at the bearing housing was due primarily to inlet piping pulsations and vibrations that were generated by the impeller vanes passing the cutwater, resulting in five and seven times running speed excitation.  In general, the pulsation levels were highest in the inlet and cross-over passages reaching levels in excess of 10% of the static pressure.  The pulsations were amplified by the inlet piping and also excited a shell resonant mode of the thin wall Schedule 20 pipe.  This flexibility of the inlet shell appeared in the form

of an egg or multi-lobal mode shape.  The
axial inlet piping arrangement provided
a mechanism for coupling the pulsation
into vibration and excitation of the pump
bearing housing in the vertical plane.
Vibratory motion along the top of the
pump at 250 Hz and 350 Hz indicated that
the bearing brackets flexibility, combined
with the rocking action of the pump casing,
was responsible for the high level of
vibration at the bearing housing.

(b)  Vibration levels at running speed
frequency could be contained within
acceptable limits provided the rotor was
properly balanced.  The shaft vibration
and bearing housing vibration moved
together and there was sufficient bearing
clearance to withstand the 250 Hz and
350 Hz vane passing frequency excitation.
Shaft fatigue, bearing support fatigue,
bearing metal fatigue, bearing wear, or
mechanical seal failure that could be
attributed to the high vibration levels
were not present.

(c)  The inlet and discharge piping as
originally installed had one diameter
taper pieces and tight radius elbows at
the pump nozzles.  Also the inlet piping
arrangement immediately upstream of the
bend introduced an axial approach component
into the first stage, double entry impeller.
The piping configuration was modified to
provide additional distance between the
elbows and the pump nozzles with connecting
taper pieces four nozzle  diameters long.
A possible reduction in vibration level of
approximately 20% was attributed to this
modification but it was difficult to
substantiate because of the varying
conditions.

(d)  The increase in stiffness of the baseplate
was achieved by the addition of extra grout
and pedestal bracing but no significant
improvement was observed.

(e)  The vanes of the first stage impeller were
staggered to reduce the shock losses and
hence pressure pulsations at the blade
rate passing frequency.  This modification
achieved a 30% reduction in the vibration
level at 250 Hz.  The outlet tips of the
impeller vanes were notched in the shape
of a 'V' but did not show any improvement.
A first stage impeller with seven vanes
was also tested but as one would expect
the vibration level at 250 Hz was replaced
by a much higher level at 350 Hz.  The
diameter of the second stage impeller was
reduced slightly to increase the radial
gap between the vanes and the cutwater
which resulted in an apparent reduction
in the 350 Hz vibration level but the
amount was difficult to quantify.

(f)  The application of strain gauges on the
bearing brackets and inlet piping did not
reveal any strain levels of any concern.

(g)  There was some evidence of gas/vapours
coming out of solution in the inlet piping.
The pressure pulsations varied in magnitude

within a period of 15 minutes.  The
maximum pulsation reached 6.9 bars peak
to peak on a 6.9 bars static pressure and
indicated a cavitation type wave form.
This dynamic degassing of the fluid in the
inlet piping was probably a contributing
factor to the corrosion/erosion of the
casing, impellers, and wear rings.

(h)  The application of a dynamic absorber to
the bearing housing reduced the 250 Hz
vibration level but increased the 350 Hz
vibration level.

(i)  Impact tests on the pump casing in the
ready to run condition gave resonant
frequencies between 240 Hz and 250 Hz.
Similar tests on the pump shaft indicated
that the critical speed was between 3900
rpm and 4200 rpm.

(j)  Some of the original pump shafts, in type
316 stainless steel material, were found
to be bent (from 0.13 mm to 0.2 mm TIR)
and were replaced by shafts in 17-4 ph
hardened steel.  Obviously the vibration
level at 50 Hz was reduced by this change
but the pumps had been dismantled several
times and the root cause of the shafts
bending could not be identified.


It was concluded:

(a)  The installed proximity probes measuring
the relative motion between the pump shaft
and the bearings was a better indication
of potential damage than the seismic
readings on the bearing housing.
Therefore, provided the shaft displacement
did not exceed 0.075 mm the pumps could
continue to operate.

(b)  Periodic monitoring of the bearing housing
vibration would continue and be analysed
for trends.

(c)  During the first plant overhaul the inlet
piping would be replaced by thicker
schedule pipe to avoid the thin wall
resonant frequency.


POST MORTEM

While the six problem areas have been
discussed separately in the previous sections
they occurred simultaneously.  Therefore the
level of confidence in the pumps was severely
diminished and a major plant shutdown was only
avoided by an abnormally high maintenance
effort (2 000 manhours per pump).

Although discharge check valves were
installed to prevent reverse flow the
probability and consequences of malfunction
should have received more consideration.  The
provision of a clean water flush to the moving
parts would have reduced, if not eliminated,
the number of occasions that the check valve
failed to close fully .  Similarly, the addition
of a simple cross-connection between the
suction and discharge piping of the shaft
driven oil pump would have prevented the
lubrication oil supply failures.

The accelerated deterioration of the pump wear rings was not a major problem but the axial pressure balance across the pump was altered with a consequential increase in the thrust bearing load which would eventually reduce the effective operational life of the bearing. Also, the contribution it made towards the mechanical seal failures fully justified greater care in the selection of the material and the provision of an external flushing supply of clean water to the wear ring annulus.

Although the leakage of carbonate solution was not considered a safety hazard it occurred in sufficient quantities to create very unpleasant working conditions. In hindsight a double seal assembly of two seals back to back with the space between flushed by a supply of clean water at approximately 1 bar higher than the stuffing box pressure (i.e. 10.5 bars) would have been a worthwhile investment. Thus all the working components of the seal assembly are contained within the flushing liquid and isolated from the carbonate solution by the inner seal. Should the inner seal fail the outer seal prevents leakage to the atmosphere until the pump can be withdrawn from service. The quantity of flushing liquid mixing with the carbonate solution in the pump is reduced from 0.9 m$^3$/hr to 0.06 m$^3$/hr provided the inner seals are in good condition.

An adequate attainment of rotor balance and shaft alignment are prerequisites to any complex vibration analysis and the difficulty experienced in maintaining these fundamental requirements was a major factor in the time, effort and cost of identifying the cause of the high vibration at the pump bearing housing. The selection of a pump driven by a four pole motor would have resulted in a heavier and physically larger machine. This selection might have been a more stable base for the attachment of the large bore carbon steel piping but would have cost approximately 50% more to purchase and still not, necessarily, avoided the impeller blade rate passing frequency excitation. Piping design philosophy generally dictates that a completely flexible system should be used to reduce the forces and moments to acceptable values but this results in an infinite number of degrees of freedom. However, the selection of smaller bore, stainless steel, inlet and outlet piping would have facilitated changes to the supports and configuration thereby increasing the options available for changing the resonant frequencies of the piping. The ideal marriage of pump and piping in terms of size, weight, hydraulic flow and vibration characteristics may not always be possible but one can only hope that the recording of this experience will help to avoid them being so completely divorced in future installations.

The sensitivity of the pump bearing support arrangement to the impeller blade rate passing frequencies was a major design fault. Why this phenomenon remained undetected on a pump reputed to have a good track record was not explained and was a reflection of the adequacy of the pump development programme. The selection of a diffuser casing design as depicted in figure 6 would have reduced the magnitude of the pulsations and increased the excitation frequency by a factor equal to the number of diffuser flow passages. Similarly, a double casing design would have been a heavier and more rigid construction capable of transferring the piping loads to the pump baseplate without affecting shaft alignment. Also, any high velocity erosion damage to the stator components would be restricted to areas within the refurbishable cartridge and not to the pump casing.

SUMMARY OF RECOMMENDATIONS

Refinements to the piping design features.

(a)  Corrosion/erosion resistant material such as type 316L stainless steel should be selected to permit the use of higher flow velocities and hence the use of smaller bore pipes.

(b)  The supports should be adequate and sufficiently rigid to attenuate unavoidable excitation frequencies.

(c)  Consideration should be given to the hydraulic characteristics generated by the piping configuration equal to that currently given to the mechanical and thermal loads.

(d)  The centreline of the inlet and discharge piping should be at right angles to the pump shaft centreline from the pump nozzles to the second bend with at least a length equivalent to four pipe diameters between the changes in direction.

(e)  Moving parts in the discharge check valve assembly should be flushed with a clean water supply.

Refinements to the pump design features.

(a)  A diffuser casing design should be selected for all high pressure services with a differential pressure above 40 bars.

(b)  A double casing design with a cartridge assembly of the internal components should be selected.

(c)  The wear rings should be flushed continuously with a clean water supply.

(d)  Double mechanical seals should be used and flushed continuously with a clean water supply.

(e)  The bearing lubrication oil system should be suitable for prolonged running with both forward and reverse rotation of the pump.

(f)  The inlet pressure should be increased to permit the use of a single entry impeller in the first stage of the pump.

(g)  The pump development programme should include a vibration survey and spectra analysis to identify the critical resonant frequencies.

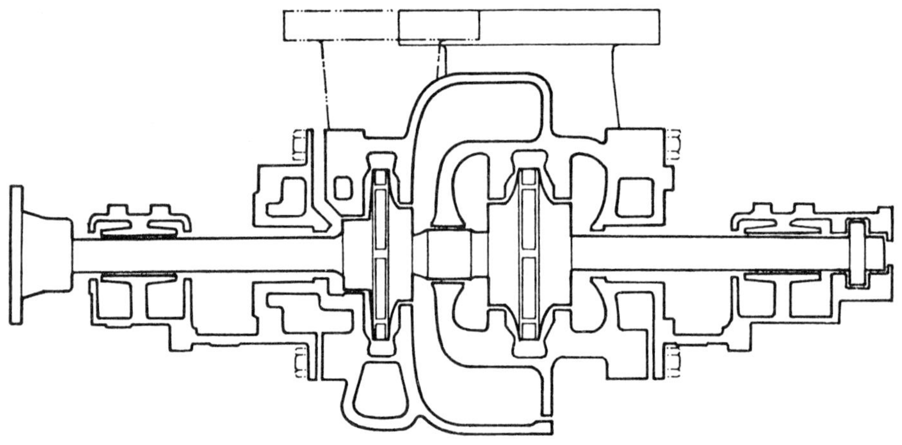

Fig 1    Volute casing pump

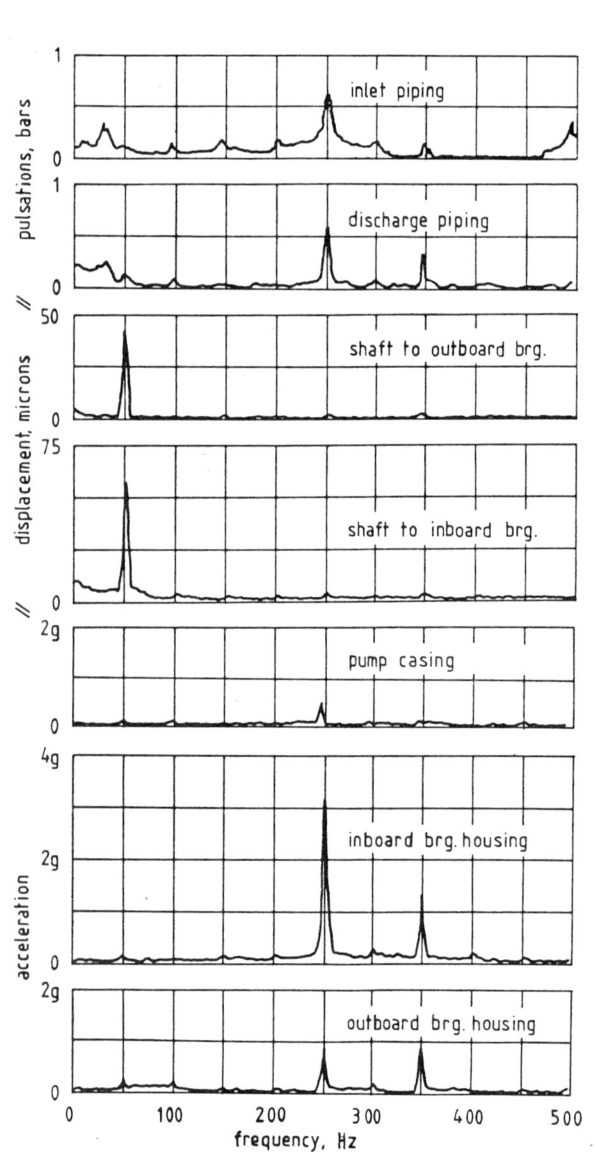


Fig 2    Normal operation test results

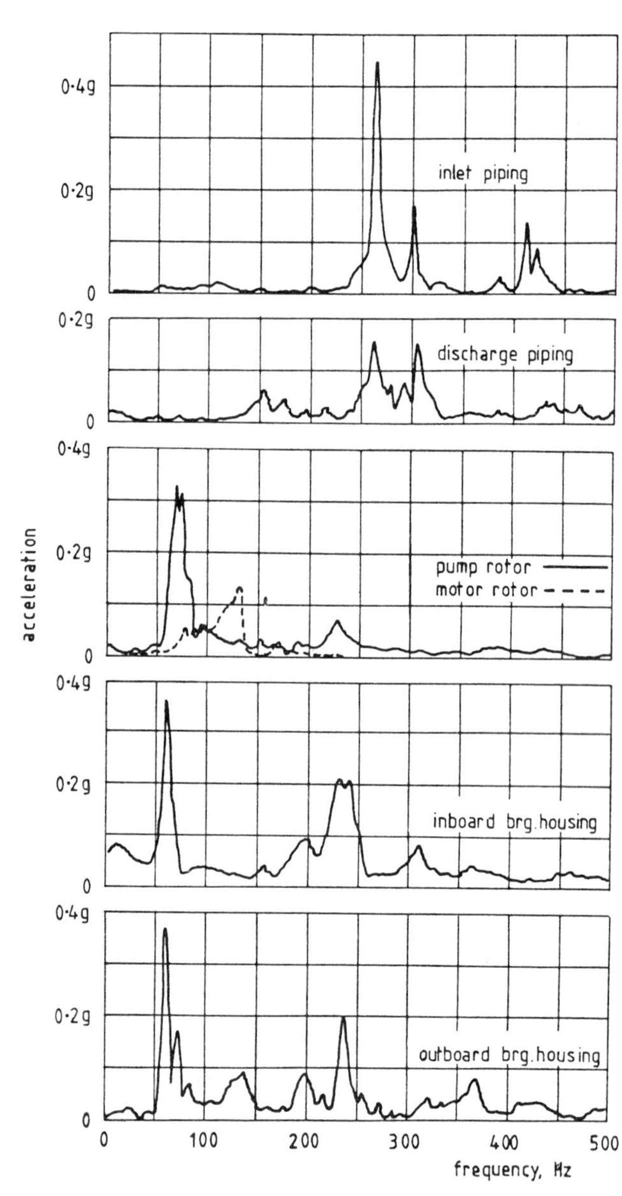


Fig 3    Impact test results

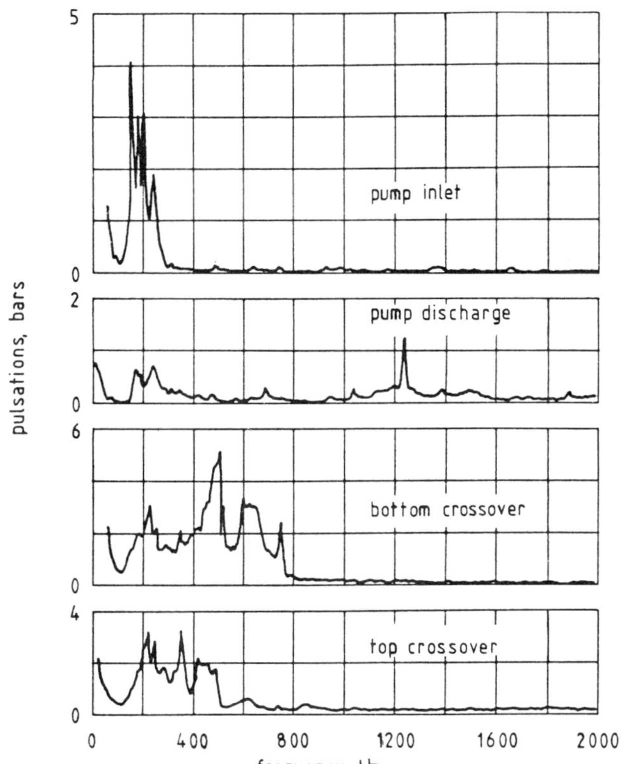



Fig 4    Shutdown tests

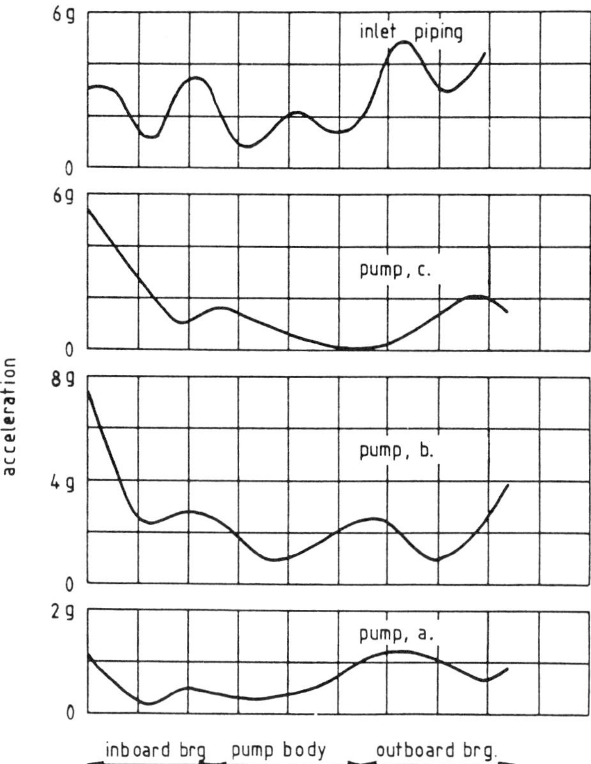



Fig 5    Mode shapes at 250 Hz

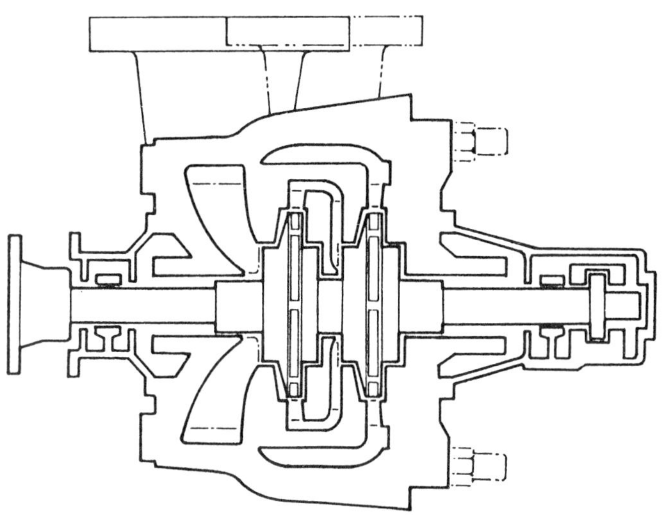

Fig 6    Diffuser casing pump

# C52/81

# The misapplication of centrifugal pumps for refinery and petrochemical duties

A P SMITH, Imperial Chemical Industries Limited, Middlesbrough
I C MASSEY, CEng, MIMechE, Mather and Platt Limited, Manchester

SYNOPSIS  Due to the inadequate selection procedures for pumping equipment to serve refinery and petrochemical duties, the application of centrifugal pumps, particularly multi-stage machines, is viewed with apprehension by many maintenance and commissioning engineers.  The problems experienced arise from the inability of the plant designer to specify his requirements and commercial pressures upon the pump supplier to stretch equipment to meet diverse requirements.  A further contributory factor to the problems of the maintenance engineer is associated with the long commissioning periods encountered in complex plants.

It is the purpose of this paper to present typical user experience incidents and relate these to the design factors adversely affecting machine performance in order to promote an understanding of the parameters to be considered in the selection of pumping equipment.

## INTRODUCTION

Most experienced maintenance engineers on refinery and petrochemical plants will have at least one horror story associated with multi-stage centrifugal pumps.  They are also likely to have experienced either noise or excessive vibration problems on single stage units.

The adverse experience with multi-stage pumps has led to an increasing tendency to use high speed single-stage pumps in the area of application associated with the smaller multi-stage units and a significant section of the market has now been lost.

A major source of these problems lies in the running conditions.  The centrifugal pump is of fixed geometry design and therefore can only develop optimum performance at a discrete flow rate for a given speed.  Most pump manufacturers, particularly those supplying large multi-stage units, developed their initial design for power and water undertakings where normal practice is to run the pump close to its design point.  In the refinery and petrochemical markets, start-ups, plant changes and operational problems often lead to pumps being run for long periods at significantly different flow rates to those for which the design is optimised.  Long running periods are experienced at leak-off flow rates which had been traditionally considered to be an infrequent condition.

To counteract this problem it has been necessary for pump manufacturers to call for larger bypass circuits.  In order to reduce the tendency of the system to surge when running machines designed with high head coefficients at low flow rates, these bypass circuits are often modulated with the attendant high costs and high maintenance, generation of noise and large bore pipework required to accommodate the leak-off flow.

When manufacturers are asked for evidence why such large bypasses are required, replies are often vague.  Rarely is quantitative evidence produced for sizing the bypass; normally one is told that it is from background experience.

A further major contributory factor to the problems experienced is the demand to achieve lower NPSH. Design philosophies employed to achieve exceptional NPSHR at high flow rates can often lead to severe problems at reduced flow-rates if fundamentals of design are overlooked. This is often the case where standard machines are utilized to cover a multitude of duties and applications.

Despite the diverse range of operation which may be encountered by machines, most specifications concentrate on guaranteeing performance at a perceived duty point.  The fact that specifications do not embrace the machine operating environment fully, leaving the obligation of the equipment unclear, may result in long time delays in resolving problems which are experienced.  In such cases the user may be left with a pump which is a continuous maintenance or noise problem; the manufacturer is left with a customer who is un-satisfied and will avoid buying further products from his Company.

## CASE ANALYSIS

### 1. Symptoms

The performance of two large (2MW) single-stage double suction circulating pumps running at 1500 rpm exhibited severe rotor shuttling problems, generated excessive airborne noise and vibrated in a random manner.

Performance analysis on the manufacturers test bed indicated that the noise and vibration pattern was random and that the head-flow characteristic contained a kink at a flow rate of approximately 85% of design flow.

Operating at reduced speed the response to the excitation was tolerable but the shape of the characteristic curve was identical to its full speed counterpart.

After installation on site with large piping and control valves fitted far away from the machine, the situation was barely tolerable at full load and intolerable at partial load. Severe hydraulic vibration with no set pattern was experienced in the pump suction developing high accelerations and pressure excursions in excess of 2 bars peak-to-peak. Cavitation damage was severe.

Investigation into the pedigree of the machines indicated that performance had been extrapolated from 1200 rpm to 1500 rpm with no experience of operation at the higher speed. The NPSH available at duty flow rate on site was low and therefore the eye of the impeller was of special design to achieve the required suction specific speed.

The problem was clearly associated with the suction side of the pump and therefore steps were taken to modify the impeller inlet design. The system flow requirements had been over-specified and it was possible to reduce the duty point to 80% of the original requirement result-ing in a less onerous suction condition. The re-design of impeller utilised a smaller eye which was adequate to achieve the new NPSH, although the breakdown performance was inferior to the original design.

The introduction of the new impellers transformed the behaviour of the machines, especially at low flow rates with a dramatic reduction in hydraulic noise in the suction of two orders of magnitude to 30 millibars peak-to-peak. Airborne noise levels reduced by 10 dB and cavitation erosion was eliminated.

Analysis

The application of performance criteria from one duty to another often has undesirable results if the conditions and philosophies relevant to each application are not clearly understood. Inspection and scaling of performance characteristics is insufficient to guarantee acceptable performance, particularly when applying such techniques to suction performance.

It is notable that many design techniques employed to achieve superior cavitation break-down performance were developed for such applica-tions as free level extraction pumps. Under normal conditions of operation cavitation was used to control flow rate. The development of cavities under conditions of significant head decay is generally neither damaging to the impeller nor is excessive noise experienced. It should be noted that cavity collapse occurs well into the impeller under such conditions of operation away from vane and shroud surfaces.

Design philosophies adopted employed large impeller eyes, low vane numbers and over corrected angles to improve the vane-to-vane area which is the most significant factor when considering breakdown performance. The applica-tion of classical theory to a large eye leads to low vane angles which restrict the area between vanes thus limiting the ability of the impeller to continue to generate head. To enhance performance many designs employ an incidence addition increasing the vane-to-vane area and inhibiting performance breakdown. Additionally, foundry capability may severely restrict the minimum angle which may be practically considered.

The consequence of this philosophy is to develop an inlet design which is incompatible with the required duty of the pump leading to the early onset of incipient cavitation with the associated noise and reduced impeller life. Inspection of Figure 1 shows the consequence of the above design approach. The impeller suction is sized for capacity in excess of the required design flow and the incidence addition results in severe cavitation over most of the operat-ing range at what may be considered to be an adequate level of NPSH.

The mechanism of reverse flow into the impeller suction (1) may exacerbate the noise problem as cavitation is often developed due to the high rate of shear between outgoing and incoming flow to the impeller. It should be noted that cavitation of this type rarely leads to physical damage in the impeller but may exhibit itself in the form of cavitation damage in the inlet guide vanes. Such damage may be observed in Fig.2.

The observed kink in the head characteristic supports the prognosis. Past work (1) has related this event to the onset of secondary flow in the eye of the impeller.

Operation of the machine in the area of head breakdown may lead to severe surging in the suction or discharge of the system at low frequency. This mechanism was particularly common when condensate extraction systems moved from free to fixed or controlled level.

The NPSH problems outlined above are generally a result of commercial pressures where pumping equipment is generally assessed on capital cost rather than technical merit. To be competitive it is necessary to increase head per stage which tends to be synonymous with increas-ing unit speed. A lack of understanding of the dangers to be encountered by the purchaser may often lead to designs which may generate considerable problems in the field.

Axial shuttling is a common phenomenon with large centrifugal pumps incorporating bearings with a significant axial clearance. The phenomenon usually develops at low flow rates and is an indication of the unstable conditions which may be generated at an impeller inlet or outlet. The forces generated at the bearing can lead to reduced life and the use of a flexible spacer coupling and unlocated drive rotor may lead to large resonant axial excur-sions. For large machines of the double entry type it is necessary to design a thrust bias into the configuration of sufficient magnitude to overcome the transient forces described, but in the worst cases this may be difficult to achieve.

The use of a high flow recirculation around the pump is clearly a solution to the noise problem caused by reverse flow from the impeller. The lower the flow rate the larger the reverse flow length into the suction generating large shear zones. Anti-rotation vanes in the suction branch reduce the pre-rotation associated with the secondary flow, and make a contribution to a reduction in levels of generated noise. The reduction in the pre-rotation has a stabilizing influence on the pump performance in the surge regime; however, such vanes must be presented well into the impeller eye and maybe subjected to the attack described above.

2. Symptoms

A process plant has two pairs of typical small (45 kW) ten-stage boiler feed water pumps running at 3000 r pm with adequate NPSHA and operated close to the duty point. The pumps were designed as 100% units, therefore no interference was possible between the pumps.

A rash of failures were experienced with these machines, taking the form of a build-up of synchronous whirl leading to severe wear and eventual failure. A consequence of the high vibration levels experienced was a bent shaft which was detected on a number of occasions following machine failure.

Due to poor delivery from the pump supplier it had been found necessary to purchase shafts locally. Investigation after an unusually high number of consecutive failures showed that these shafts were often outside tolerance. The genuine spare impellers were also poor quality and often had to be re-machined to bring them to within the tolerance required by the manufacturer. This machining upset the static balance provided at the time of original manufacture.

A satisfactory unit was not achieved until all the tolerances were tightened to design and a multi-plane dynamic balance was achieved by adding two impellers at a time and making the appropriate correction to the added impellers.

The units have operated satisfactorily since the precautions described above were introduced.

Analysis

Clearly the fit between the impeller and shaft is important when operating in the higher speed regimes.

Loose impellers due to excessive clearances in the bore, which significantly grow due to centrifugal force, will lead to excessive vibration. It is often the case with small multi-stage pumps that the fit of impellers is not closely controlled.

The smaller multi-stage machines invariably have their first resonant frequency in air below the running speed and therefore rely on the stiff-ness and damping developed in the fine clearances at the eye and back hub of the impeller to provide the qualities of stiff shaft operation . Once wear is generated in the machine due to excessive excitation from out of balance rubbing, casing hogging or out of balance hydraulic radial forces, there is a danger that the pump may run close to, or at a resonant frequency with a low level of damping in the system, thus tending to whirl.

A series of typical response curves for a multi-stage machine are shown in Fig.3 which relates response frequency to internal clearance.

The technique applied to the dynamic balance of rotors is important particularly when correcting for secondary forces. It is often found convenient to apply corrections at the coupling or thrust block to reduce the levels of material to be removed. This may lead to shaft mode shapes during operation which lead to excessive wear due to shaft bending, (Fig.4).

When testing a pump for hydrocarbons duty, water testing may be misleading. The stiffness and damping effects described above are Reynolds number dependent and therefore if care is not taken to compute the critical speed in the fluid to be pumped, a machine which performs well on test on the manufacturer's testbed may prove to be troublesome on site, (Fig 5).

The choice of stationary and rotating vane numbers also has an effect on the stability of the rotor. Certain combinations lead to high amplitude response at impeller blade passing frequency and high radial loads due to rotor eccentricity (2).

A lack of attention is also common at the selection stage, and with the maintenance, of drive couplings. These are a common source of problem under both test and site conditions. Coupling hubs and transmission units often represent a significant mass eccentricity if poorly designed or maintained, and may have a detrimental effect on rotor performance. Spigots and fitted bolts must be treated with great care. To reduce the incidence of such problems light-weight units should be employed but again commercial pressures often lead to an unsatisfactory compromise.

3. Symptoms

A set of 450 kW boiler feed pumps supplying feedwater to 900 psi boilers were initially fitted with packed glands. However, even under these conditions after a period of two to three years, an increase in vibration level was experienced and not only was the packing severely damaged but heavy rubs were experienced between impeller necks and neck rings. As the previous example the damage had the characteristics of wear due to synchronous whirl.

For safety reasons the packed gland was changed to a mechanical seal; with this configuration the period from commissioning to failure was as little as 3 to 4 months. The pattern of failure was generally the same; throughout a majority of the period of operation vibration levels were low and were only seen to increase in a dramatic manner shortly before failure.

An exacerbating factor was the fact that the boilers being supplied were often on a standby duty to take sudden load changes on the system so that the machines spent considerable time operating at approximately 40% load. Low flow bypasses were fitted; however, these were of the on-off type operating at slightly less than 20% of duty flow rate.

Initially these pumps operated singularly and had a very flat head curve over most of the operating range. Re-design of the system necessitated running the pumps in parallel requiring a re-design of the hydraulics to improve stability. The impellers were close to full diameter and therefore room for manoeuvre was small. New impellers were installed to achieve a stable characteristic although the final clearance between impeller and diffuser was small.

The combination of impeller and diffuser vanes was chosen to give the minimum disturbance at flows down to leak-off.

Analysis

The problem of surging due to interaction between pump operation in parallel with unstable head curves is well known (3). Many designs have curves which, if not unstable, are flat leading to paralleling difficulties due to variation in dimension between one machine and its sister. The pressure is on the pump designer to increase the head coefficient of a design to achieve maximum head per given speed and diameter, this tends to produce such instability.

In the current climate of commercial appraisal the more head per given stage diameter and speed the more competitive the machine, although long term this may not be the case after involvement in site problems. Delivery delays may also be incurred due to problems in the manufacturers test facility.

The case for conversion from packed gland to mechanical seal clearly requires careful consideration as the packed gland forms a stiff bearing with inherent damping which is beneficial to rotor response. The problem described above, however, can be exacerbated by casing hogging which may occur in boiler feed pumps under both operating and standby conditions depending upon the design features employed. In the hogged situation the packed gland helps the shaft accommodate the casing or barrel distortion without inflicting excessive wear on the pump internals. Conversion to mechanical seals may lead to accelerated wear from hogging in a manner similar to that described in the case. Excessive pipe loads may also lead to significant distortion of casing or barrel.

With the development of internal wear in the machine where a relatively flexible rotor is employed, rotor excursions due to out of balance forces and unbalanced hydraulic forces generated from the positioning of the impeller in the diffuser will increase. These forces are offset by the stiffening effect of the internal clear-ances. However, with excessive wear the diffuser eccentricity forces will predominate leading to metal-to-metal contact under normal conditions of operation, further excessive wear and the development of sub-synchronous whirl due to friction effects. (2) (4). This phenomenon tends to exhibit itself in the form of high levels of vibration shortly before failure.

It should be noted that the diffuser forces described above are flow dependent and operation at part load exacerbates the problem.

CONCLUSIONS

Benefits of pumps stable at low flows

What benefits can be expected if we design pumps specifically to be stable at low flow rates. Are these large enough to affect sales ?

For normal, relatively small single stage centrifugal pumps, where the situation is generally tolerable, the benefit will usually be inadequate unless noise specifications are unusually stringent. However, on large centrifugal pumps and, even more, on multi-stage pumps, it is felt that the advantages are so significant that they might even to some extent overcome the present objection by many users to multi-stage pumps. In detail, the benefits are :

a) For the User – a pump with greatly reduced maintenance and better reliability. This would also be an encouragement for the user to fit mechanical seals instead of packed glands. It can also be expected that a pump of this type can be assembled and be expected to work first time.

There should also be reduced maintenance on bypass circuits.

b) For the Contractor Building a Plant :

advantages on the maintenance will weigh less with the contractors as they normally do not quantify this in their bid comparisons. However, stability at low flows allows the bypass circuit to be reduced in size or, in some cases, eliminated; in others made a simple on/off circuit. This reduces the overall cost, complexity and size of the pipework associated with the pump.

At present, some of these bypass circuits can cost about 50% of the cost of the pump itself and also be a potent source of noise, which will need more cost and complexity to counteract. The overall advantages in capital saving are obvious.

Another factor is that such pumps should be better performers on the Works' test bed. We know of one case with a multi-stage centrifugal pump where 29 separate test runs were required before the pump was acceptable. (Such a pump should be rejected outright on the grounds that we could never re-build it to a satisfactory condition afterwards). Such prolonged test work must inevitably have repercussions on delivery.

c) <u>For the Manufacturer</u>, advantages appear
more dubious. However, once a manufacturer has committed himself to the larger
casing necessary for a pump with a stable
characteristic, he should not need to
increase size further to achieve stability
at low flow rates. What in some cases may
be required is a new family of impellers
with geometry carefully chosen for this
type of stability. Once he has these,
he will obviously be required to show on
the test bed with fairly complex instrumentation, showing hydraulic and mechanical vibrations, that the requirements for
stability have been met. However, most
pump manufacturers already have such
equipment or can hire it relatively
easily. The actual changes to the
geometry of the wheels are not particularly complex and should be achievable by
a good hydraulics engineer without much
difficulty or development time.

In exchange for this expenditure, the
manufacturer should have a product which
he can show is manifestly superior
to any competitor who has not taken
similar action. He should therefore, be
able to command a larger slice of the
market and indeed, with greater confidence in multi-stage pumps; the size of
the market itself may well increase.

However, in view of the costs, the
manufacturer may well feel hesitant in
going ahead until needs have been clearly
specified.

SPECIFICATION

a) <u>On Mechanical Balancing</u> : the requirements for the new specification are
relatively clear cut and should in
part be covered by the new ISO/DIS 5406
for balancing of flexible rotors now in
course of preparation, though, in its
present draft, it will clearly require
considerable amplification to be
adequate.

b) <u>On restricting extreme designs for low
NPSHR</u> : the requirements are less clear
cut but should be met as an outcome of
the process of meeting the essential
requirements for stability at low flow
rates.

c) <u>On specification on Stability of Low
Flow</u> : we are in yet further difficulty in that there is not even a clear
cut vocabulary on which we can start to
work. However, the ultimate aim of
this specification is clear; what we
want to achieve, as stated above, is a
pump which can achieve flow rates - say
20% of best efficiency point - without the rotor coming in contact with
the static parts due to either mechanical or hydraulic out-of-balance forces.
We can establish this experimentally on
the test bed by measuring the actual
displacement of the shaft in the bearings and seeing that this does not

exceed the set clearances. However,
equipment to measure this is somewhat
complex and difficult to install inside
a production pump. A suggested
solution is to build a prototype pump
on which the necessary measurements
are carefully made - preferably with
independent witnessing by consultants.
From this set of tests may then be
derived measurements which can be more
easily made on the production pump to
show that the pump as built will behave
in an acceptable manner as the prototype did.

Noise measurements and adequate
frequency analysis, preferably with
real time analysers, may be useful
tools in the derived readings. However
before they can be used with confidence,
the measuring procedures on many test-
beds would have to be much improved.
In particular, noise measurement tests
carried out from the test beds at
present are farcical.

In closing, a pump which the manufacturer
himself can send to his own test beds with
confidence would also be a significant advantage to himself.

REFERENCES

1. MASSEY I.C.  The Suction Instability
problem in Rotodynamic Pumps.  International
Conference on Pumps and Turbines, N.E.L.
Scotland, September 1976.

2. HERGT P & KRIEGER P.  Radial Forces in
Centrifugal Pumps with Guide Vanes.  Institution  of Mechanical Engineers, April, 1970.

3. STEPANOFF A.J.  Centrifugal and Axial Flow
Pumps Theory, Design and Application.  John
Wiley & Sons Inc.

4. BLACK H.F.  Dynamics of Rotors.  International Union of Theoretical and Applied
Mechanics Symposium, Lyngby 1974.

ACKNOWLEDGEMENTS

The Authors would like to thank the Directors
of Mather & Platt Limited and Imperial Chemical
Industries Limited, for permission to publish
this paper.

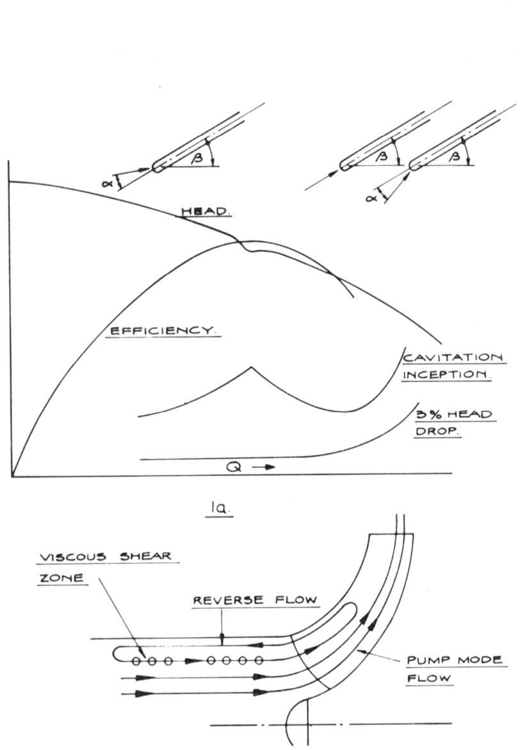


Fig 1    The influence of reverse flow on cavitation inception characteristic of centrifugal impeller

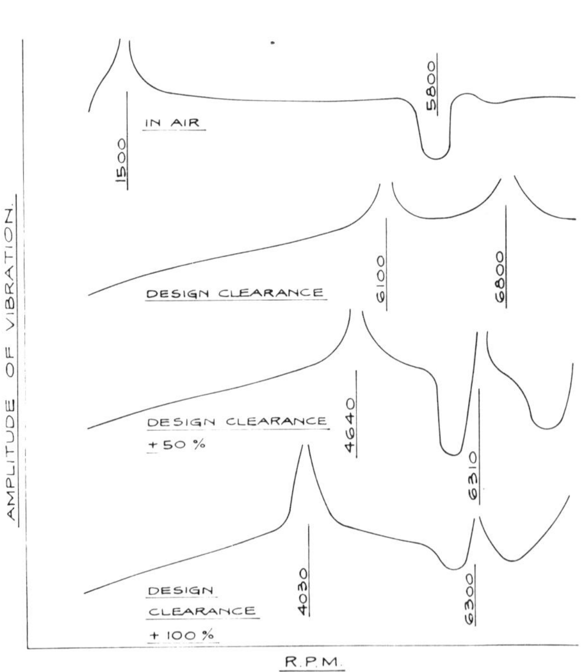

Fig 2



Fig 3    The influence of clearance on rotor resonant response frequency 8/10in 8 stage axially split pump

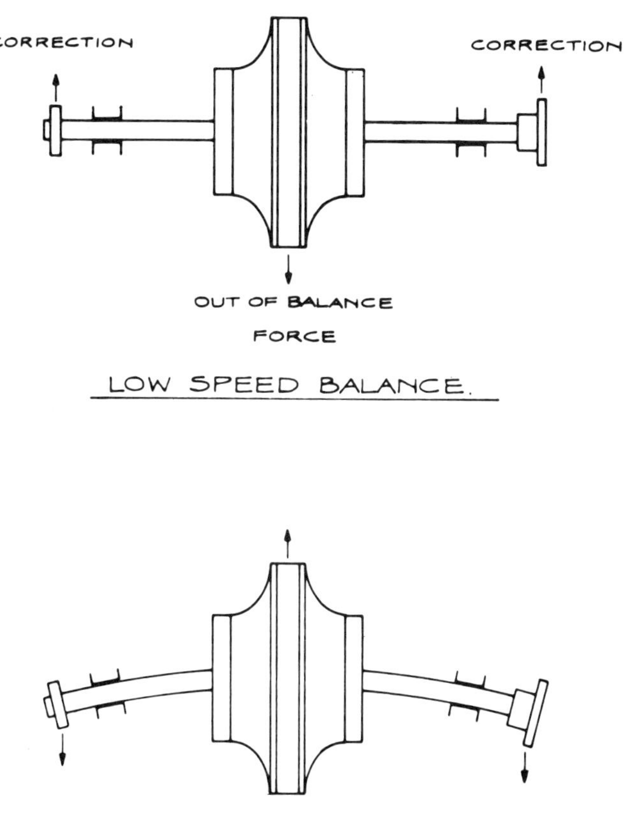


Fig 4   Effect of incorrect balancing procedure

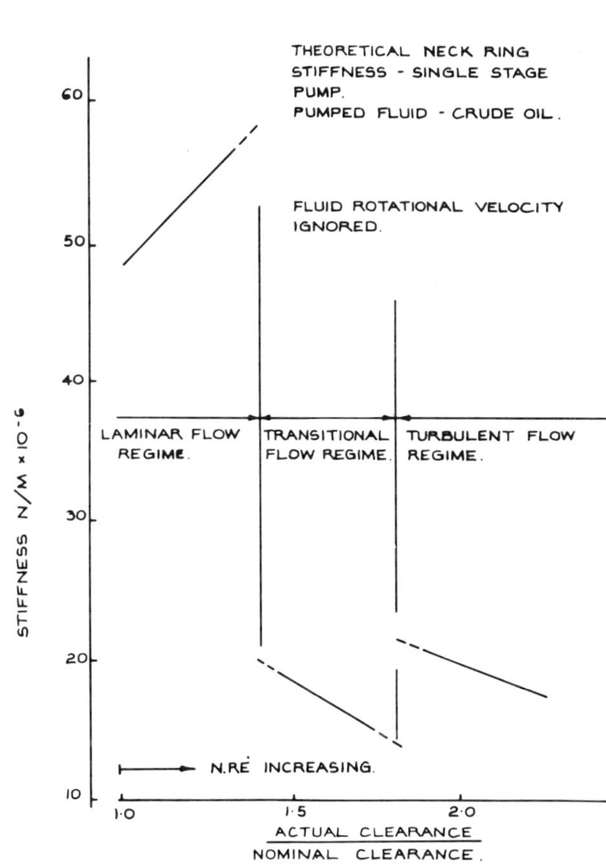


Fig 5   Reynolds number effect on internal stiffness
values

# C47/81

# Diffuser or volute pump? A comparison of performances


J H BUNJES
Stork Pompen BV, Hengelo, The Netherlands



SYNOPSIS    Within the constraints of International and Pump User's Standards, a designer has some
freedom to choose the basic hydraulic concept of a pump. Various solutions are possible to meet
specified requirements depending upon the starting point chosen. However, in each case the hydraulic
design more than any other factor determines the end result of his work. This result is then judged
by the user on the basis of performance, investment, operation costs, reliability and maintainability.

The volute and radial diffuser concepts embrace most of the possible design solutions for the single
stage pumps; a number of hydraulic aspects of these concepts are compared with regard to operation at
best efficiency and part load conditions.


## INTRODUCTION

The author's company has manufactured pumps of
widely diverse types for more than 80 years. Most
of these pumps have been shop tested to check
operation and guarantee performances, thereby
providing an enormous amount of vital information
for today's designers.

A data bank has been created to assemble this
test information together with the related pump
design data in such a way that rapid analysis is
possible. Many systematic studies have been conducted using this information to obtain empirical
data for those phenomena which influence relationships between theory and practice.

One of these studies concerned a comparison of
two conventional pump types:  the radial diffuser
and volute single stage pumps. About 250 pumps of
each type were included having various dimensions
and specific speeds. All were of the single stage
overhung impeller type. The comparison considered
their basic design concepts, the flexibility of
the concepts to accommodate design and performance adjustments, the influence of major construction features on hydraulic behaviour, hydraulic
forces and hydraulic performance characteristics.
The results of this comparison are summarised in
this paper and amplified with data from literature.

Mechanical construction aspects of the two pump
types offer interesting comparisons as well, particularly those relating to cost, quality and
reliability. However, these aspects can only be
considered objectively for specific cases and
therefore it would be incorrect to try to genera-
lise. Mechanical construction aspects are therefore not covered.

## NOTATION

| | |
|---|---|
| $a_d$ | Diffuser throat area |
| $D$ | Impeller diameter |
| $D_v$ | Impeller vane diameter |
| $E$ | Efficiency ratio    $\eta/\eta_{opt.}$ |
| $g$ | Acceleration of gravity |
| $H$ | Pump head |
| $K_s$ | Suction specific speed    $\dfrac{\omega \sqrt{Q}}{(g.NPSH)^{3/4}}$ |
| $n$ | Number of rotations per minute |
| $n_\omega$ | Specific speed    $\dfrac{\omega \sqrt{Q}}{(gH)^{3/4}}$ |
| $Q$ | Capacity |
| $Q_c$ | Capacity at point of contact between enveloping curve and efficiency curve (see fig. 9) |
| $\delta$ | Specific diameter    $\dfrac{\psi^{1/4}}{\phi^{1/2}}$ |
| $\eta$ | Efficiency |
| $\phi$ | Capacity coefficient    $\dfrac{Q}{\omega D^3}$ |
| $\psi$ | Head coefficient    $\dfrac{gH}{(\omega D)^2}$ |
| $\omega$ | Angular velocity    $\dfrac{\pi.n}{30}$ |

Subscripts

| | |
|---|---|
| d | Diffuser |
| opt | Magnitudes at best efficiency point |
| s | Volute |
| t | Cutwater |
| v | Vane |

BASIC DESIGN CONCEPTS

## 1. Hydraulic design

The basic constructional difference between the
diffuser and volute pump is clearly defined in
their names. In the diffuser pump, shown in
fig. 1, the basic hydraulic and pressure containment functions are substantially separated: the
impeller and diffuser fulfilling the major hydraulic functions with a circular collecting chamber
performing only a minor hydraulic role.

The volute, illustrated by the sectional drawing
of a heavy process pump (fig. 2), provides a combination of hydraulic and pressure containment
functions:  flow diffusion after the impeller
being achieved in the divergent channel of the
volute.

Clearly, separating the hydraulic and pressure
containment functions permits pumphouse standardisation and also optimisation of both components
for their principal roles. The influence of this
optimisation on pump behaviour is discussed later
in the paper.

## 2. Design flexibility

Demands to produce higher efficiency are justi-
fiably always present, particularly when standard
pumps are selected for particular duties, or
tuned for a duty on a production testbed. The
pump hydraulic concept determines the facility
the designer has to make adjustment to a pump to
optimise the match between pump performance and
system requirement.

With a volute design, the flexibility for adjustment is really limited to the impeller modifications, whilst with a diffuser design a very power-
ful additional adjustment facility is provided.
If the diffuser has an open side, then substantial changes can easily be introduced to the area
of the diffuser 'throat'. The influence of the
dimension of this area on the hydraulic performance characteristic is illustrated in figure 3.
Here, the characteristics of an impeller and diffuser combination are shown from tests with the
'throat' area reduced to 47% and then 25% of the
original area. This illustration makes it particularly clear, that at part load improved efficiency can easily be obtained with an altered
diffuser design. Looking at the power consumption
this means a reduction at low flows; the reduction decreases to zero at $Q \sim 0,5\ Q_{opt}$. At larger flows the power consumption of the pump with
modified diffuser is equal to that of the pump
with the original diffuser.

The effect of impeller diameter variation  is
shown in figure 4a. Nominal maximum diameter, reduced diameter and vane diameter only reduced
illustrate the 'turn down' effect with a diffuser
pump of rather low specific speed. A trade-off of
efficiency for stability is clearly demonstrated.
The example of the diffuser pump has been chosen
(the impeller diameter reduction of a volute pump
generates a similar effect) to compare the characteristics with those of an extreme variant where
the diffuser has been removed (fig. 4b). Here the
best efficiency has shifted to about 20% larger capacity; the capacity - head curves show a
quite stable character, even for the trimmed impeller, albeit at a lower head level. The capacity
- power curve coincides with the curve of the
pump with the original diffuser. Although unusual,

this alternative appeared to be very useful in
special cases.

With the previous examples, it is clearly demonstrated  that the diffuser pump has an increased
adjustment flexibility from the performance optimisation point of view.


FEATURES WHICH INFLUENCE PUMP MECHANICAL
BEHAVIOUR

The mechanical behaviour of a pump is largely a
consequence of the hydraulic design. Noise,
vibrations, seal and bearing life, stresses and
erosion are some areas in which the behaviour is
strongly influenced by the pump hydraulic concept.

## 1. Tip clearance effects

The clearance between the impeller tips and the
diffuser or volute cutwater, plays a major role
in the energy released to initiate mechanical
instability. Small gaps generate noise and vibrations, particularly at off-design flow conditions when erosion of the cutwater or diffuser
tips can also occur. In this respect the volute
and diffuser designs are equally sensitive. When
the phenomenon is taken into account in the pump
design, problems can be easily avoided by choos-
ing a reasonable radial gap. Indications of
dynamic stresses related to the ratio between the
impeller diameter and the cutwater and the diffuser inlet diameter respectively are given in (3),
(6) and (7).

Increasing the radial gap reduces noise and vibrations, but also reduces the efficiency. The
latter must be taken into account particularly
when applying a Process Industry Standard. Here
a reserve capability of 10% head is required,
which implies the use of an impeller with reduced
diameter.

## 2. Noise and vibrations

In the last 15 years considerable work has been
done in the field of pump noise analysis. Various
publications present these experimental investi-
gations e.g. (9), (10).

In the author's company and in the hydraulic
laboratory of Vmf-Stork / FDO Engineering Consultants, flow induced vibrations and noise have
been studied and a lot of tests have been performed. Test results give the frequency analysis in
whole or one third octave bands. As a part of
this work the hydraulic noise was measured with
various impellers operating in volute casings as
well as with radial diffusers in a collector. For
these measurements use was made of either a micro-
phone, a hydrophone or an accelerometer. Unfortunately for the present comparison, where an impeller was tested in either a volute or with a diffuser, different noise measuring methods were used
or the measurements were performed in only one
pump type. It is not possible therefore, to show an
objective comparison with just the volute or diffuser constructional elements as parameters from
own noise data. An overall comparison is, however,
possible.

Analysing the noise readings in the lower frequency range (< 1 kHz) for the pump operating at
design conditions, we can conclude that the test
results agree very well with the data presented
in (9). For both pump types the shaft rotational
frequency and the impeller blade frequency plus
their first harmonics are the dominant frequen-

cies. The noise levels at these frequencies mainly determine the general noise level of the pump.
From the test data of diffuser pumps it appears
that a diffuser blade component is not apparent;
even with the one third octave band frequency
analysis, a peak was not detected at a frequency
which could be related to the number of diffuser
blades.

From the data in (9) and (10) it appears that the
noise level (dBA) at best efficiency point depends
on the hydraulic concept: the level is about
3 to 5 dB lower for double volute and diffuser
pumps, compared to a volute pump for the same
duty.

Noise originates from pressure fluctuations, and
pressure fluctuations are due to unsteady flow.
Reduction of the intensity of the unsteady flow
results, therefore, in a lowering of the noise
level. This has to be taken into account when
comparing the noise levels of the two pump types
at off-design conditions.

The noise level characteristic as a function of
the pump capacity has a comparable shape to that
found for radial loading. Near shut valve conditions the volute pump can generate a noise level
up to 10 dB higher than at best efficiency point;
for the diffuser pump this increase is only about
3 dB. Both figures can be reduced substantially
by increasing the gap between the impeller and
the cutwater or the diffuser.

It can be concluded that a significant noise
level difference can  be expected between the
volute and diffuser pump types; this difference
becomes more unfavourable for the volute pump
when operating at off-design conditions.


## 3.  Hydraulic forces

The static and dynamic pressure distribution surrounding an impeller results in both axial and
radial forces and couple loads on the shaft end.
Frequently, consideration of these forces is restricted to the static component only, whilst it
is well known that the unsteady forces can be of
a magnitude almost equal to the static force. The
consequence of these forces on bearings and shaft
seals is often too apparent.


### (a)  Axial thrust

At first sight the generation of axial thrust
might be considered identical for both diffuser
and volute pump alternative designs. Axial thrust
being the net result of the pressure fields in
the axial spaces on both sides of the impeller.
However, the pressure field surrounding the impeller periphery is not uniform, particularly for a
volute pump. These pressure differences result
in considerable secondary flows, asymmetry and
instability of the pressure field which creates
additional axial thrust.

Comparing the two pumptypes of great significance
is the reduction of the secondary flows, asymmetry
and instability of the pressure field surrounding
the impeller of the diffuser pump. Because of the
multivaned construction of the diffuser the circumferential pressure distribution is clearly improved. Furthermore, the small radial gap which
normally exists between the impeller and diffuser
shrouds also substantially reduces the communica-
tion between the radial and axial pressure fields.
Secondary flow, asymmetry and instability of the
pressure in the axial gap is therefore reduced.
Reference (8) also describes this influence.

The unstable flow pattern in the axial clearance
space between the impeller shrouds and the casing
is significantly increased at part loads. Clearly,
such a condition can also become critical with a
diffuser pump because of the relatively smaller
axial space adjacent to the impeller shrouds.
Dramatic effects with unstable flow conditions
can result, as it is well known that small changes to the impeller shroud axial gaps can have
remarkable effects on axial thrust, particularly
in a multistage pump.


### (b)  Radial thrust

The radial thrust constitutes a major effect on
the difference between the volute and diffuser
pump concepts. Adopting a double volute goes some
way to improving the symmetry of radial loading,
but the multivaned diffuser provides the ideal
answer.
In practice the ideal answer is rarely achieved.
Some residual radial load remains with the diffuser concept, partly due to the flow in the circular collecting chamber and also to eccentricity
of the impeller and diffuser hydraulic centres
created by manufacturing inaccuracies (see ref.6).
A comparison of radial load for the alternative
designs is illustrated in figure 5.

Reference 7 also refers to manufacturing effects
on radial forces. The influence of the wearing
rings is considered here to be relevant if they
are able to act as bearings and offer some radial
support. Increasing wearing ring clearance is
sometimes considered to increase pump reliability
but this can cause a counter effect which increases bearing load and shaft deflection also resulting in seal problems.


HYDRAULIC PERFORMANCE

The hydraulic performance of pumps can be compared by the use of non-dimensional quantities:
specific speed $n_\omega$, specific diameter $\delta$, the head
and capacity coefficients $\psi$ and $\phi$, the efficiency $\eta$ and the suction specific speed $K_s$.
The specific speed indicates the pump type and
unit speed, the specific diameter reflects the
pump dimensions.

Pump cavitation behaviour will not be discussed
here. The impeller intake configuration combined
with the pump inlet geometry dictate the cavitation performance and these are independent of the
volute or diffuser configurations being compared.

For an analysis of the hydraulic performance of
volute and diffuser pumps, use was made of the
test results from about 240 diffuser pumps in the
specific speed range $n_\omega \leq 0.8$. From about 250
available volute pumps, 130 pumps appeared to be
in the same specific speed range, so their test
results could be used for the comparison. From
all these pumps the test curves (about 5 per pump)
have been interpreted to obtain the data for the
nominal impeller diameter at design and off-design
conditions. The basic geometrical dimensions of
the impellers, volutes and diffusers were also
available in the data bank, so that analysis of
relations on performance and geometry could easily
be performed.

The first aspect of the comparison is shown in
figure 6: the relation between specific speed and
specific diameter. For both pump types the same
relation was found, also with about the same
scatter.

In the detail plottings a decrease of pump numbers
with a stable head-capacity curve in the specific speed range $n_\omega < 0.35$ was found. The specific
diameter of these pumps appeared to be about 10%
larger than the comparable value of the pumps
with unstable curves. This confirms the opinion
that at low specific speeds a stable curve can be
obtained at the cost of the head factor and so of
a relatively larger impeller.

The second aspect of the comparison concerns the
efficiencies: here significant differences for
the two pumptypes are revealed.

For each pumptype the peak efficiencies have been
plotted versus the specific speed in limited ranges of the impeller diameter (roughly ± 5%) and
speed. In these various plots enveloping curves
were determined and from these curves a smoothed relation between efficiency and diameter with the
specific speed as a parameter was obtained. This
relation is considered to be the best attainable
efficiency for the pumptype in question.
Having these figures for both the volute and diffuser pumps the comparison is a simple task. Fig.
7 gives the result and shows the proportional
differences as a function of the impeller diameter and specific speed. From the peak efficiency
point of view the diffuser scores over the volute
pump in the specific speed range $n_\omega < 0.6$. It is
surprising to learn, that here also the impeller
diameter plays a role. At increasing diameters
the advantage of the diffuser decreases. This
phenomenon can only be explained by a detailed
analysis of the various hydraulic losses.

The results of the peak efficiency analysis moti-
vated a study of the comparable relations at off-
design conditions. For this investigation we
decided to use only a part of the experimental
data available. Only those characteristics having
a peak efficiency close to the initial enveloping
curve were used: i.e. $n_{opt} \geq n_{env} - 3$. This
should exclude wrong interpretations. More than
half the original pump data was retained in this
way for further analysis.
From these test curves the efficiency ratio E was
calculated for various ratio's $Q/Q_{opt}$ and plotted
versus the specific speed at best efficiency
point. Comparison of these figures for the two
pumptypes also showed substantial differences,
now mostly in favour for the volute pump. The
result is indicated in figure 8, the proportional
differences at three off-design capacities related to the specific speed at the best efficiency
points. Analysis of the effect of the impeller
diameter on these figures did not give a reliable
relationship.

To complete a total overall impression about the
most favourable area of application for each
pumptype, the results at design and off-design
conditions have to be combined. Bearing in mind
the diameter influence on the peak efficiency
differences (fig. 7) it is clear that such a comparison cannot be performed with non-dimensional
figures only. Therefore, efficiency curves were
calculated for a fixed impeller diameter and
speed for both types of pumps and compared subsequently. Now the full area where each pumptype is
best becomes clear. For the diffuser pump:
$n_\omega < 0.6$; for the volute pump: $n_\omega > 0.6$, see fig.
9. In this figure an enveloping curve is drawn
for each pumptype indicating the best attainable
efficiency for a given capacity with a pump impeller diameter and speed as indicated.
Further information, which can be obtained from

fig. 9, include the points of contact of the various efficiency curves with the enveloping curve
for the volute and diffuser pump respectively.
The capacity at these points related to the capacity at best efficiency point ($Q_c/Q_{opt}$) has been
plotted versus the specific speed and is shown in
figure 10. This empirical relation illustrates
the optimal relative capacity of a pump (for a
given impeller diameter and speed) for a particular specific speed when the efficiency is the
optimisation criterion.

## CONCLUSIONS

For each of the important aspects considered the
advantages and disadvantages of the design differences for the volute and diffuser pump became
clear. It was a conscious choice to confine the
analysis to the hydraulic performance and mechanical behaviour to obtain an objective comparison.
Reviewing the aspects considered, the conclusions
can be summarized as follows:

1. The diffuser pump affords significantly increased flexibility for performance adjustment
   and component standardisation.

2. It appears that the hydraulic noise of the diffuser pump at design conditions is comparable
   to the noise of the double volute pump and
   positively lower compared to the single volute
   pump noise.
   At off-design conditions the volute pump shows
   a substantial increase of the noise; this is
   not the case for the diffuser pump.

3. The mechanical loading due to static hydraulic
   forces is lower in the case of the diffuser
   concept especially due to the low radial thrust.
   This is very substantial at part load conditions.
   The volute pump can be less sensitive to the
   dynamic part of axial hydraulic forces. The
   importance of this aspect increases with the
   pump speed.

4. The efficiency of the diffuser pump is better
   than that for the volute pump in the specific
   speed range $n_\omega < 0.6$. This applies for full
   load as well as for part load conditions.

## ACKNOWLEDGEMENTS

The author thanks the management of Stork Pompen
B.V. for permission to publish this paper.
Special thanks are due to Mr. D.J. Burgoyne and
to the colleagues of the Hydraulic Engineering
Dept. for their advice and help in preparing the
paper.

REFERENCES

1. STEPANOFF, A.J., Centrifugal and axial flow pumps. 1957 (J. Wiley & Son, New York).

2. PFLEIDERER, C. and PETERMANN, H. Strömungsmaschinen. 4. Auflage, 1972 (Springer Verlag).

3. WELDON, R., Boiler feed pump design for maximum availability. Convention on advanced class boiler feed pumps, pp 18-35, London: Instn. Mech. Engrs., 1970.

4. AGOSTINELLI, A., NOBLESS, D. and NOCKRIDGE, C.R., An Experimental investigation of radial thrust in centrifugal pumps. J. Engng Power, Trans. ASME, 1960, 82A(2), pp 120-126.

5. BIHELLER, H.J., Radial force on the impeller of centrifugal pumps with volute, semivolute and fully concentric casings. J. Engng. Power, Trans. ASME, 1965, pp 319-323.

6. HERGT, P. and KRIEGER, P., Radial forces in centrifugal pumps with guide vanes. Convention on advanced class boiler feed pumps, pp 101-107. London: Instn. Mech. Engrs.,1970.

7. SCHWARZ, D. and WESCHE, W., Radialschub an doppelflutigen, einstufigen Kreiselpumpen. Technische Rundschau Sulzer, 2/1978, pp 63-68.

8. MAKAY, E. Better understanding of sources of feed pump damage booster performance, reliability. Power, 1979, pp 72-74.

9. SIMPSON, H.C., MACASKILL, R. and CLARK, T.A., Generation of hydraulic noise in centrifugal pumps. Convention on vibration in hydraulic pumps and turbines, pp 84-108. London: Instn. Mech. Engrs., 1966.

10. SAXENA, S.V., WONSAK, G. and NAGEL, W., Geräuschemission von Kreiselpumpen. Bundesanstalt für Arbeitsschutz und Unfallforschung Dortmund, Forschungsbericht MR 184, 1978.

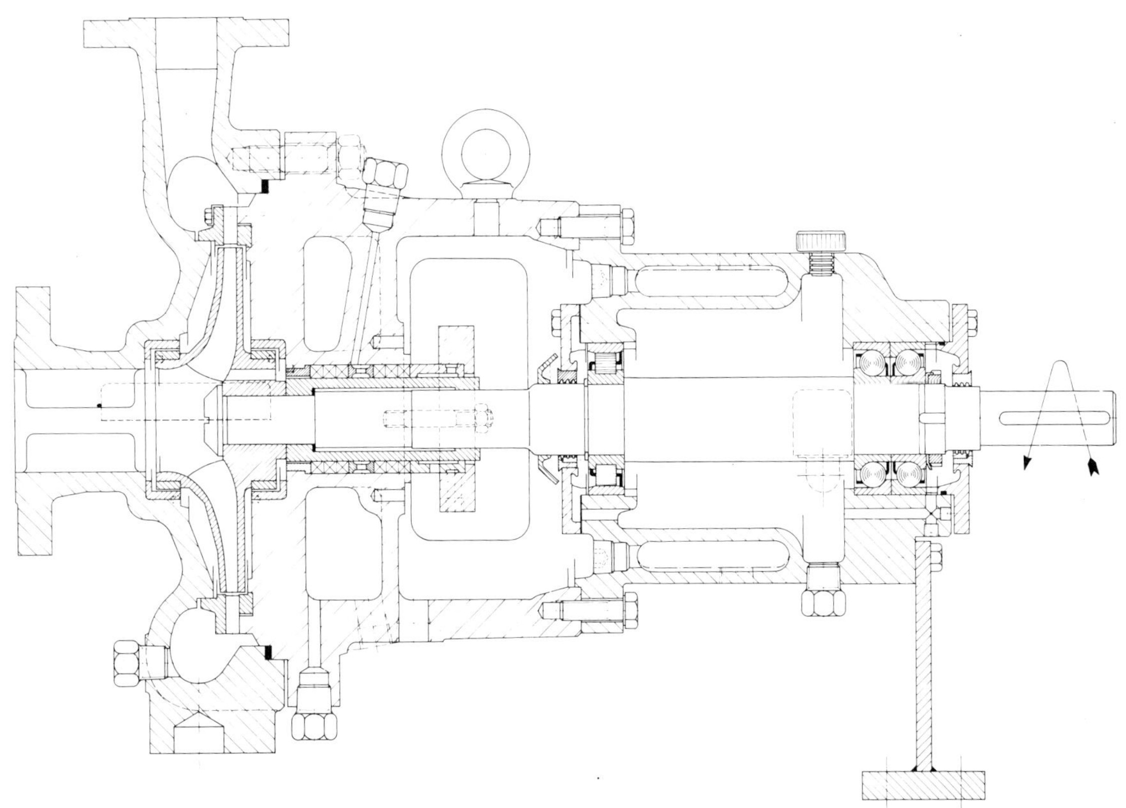

Fig 1   Heavy duty diffuser type process pump, Stork  type PHL

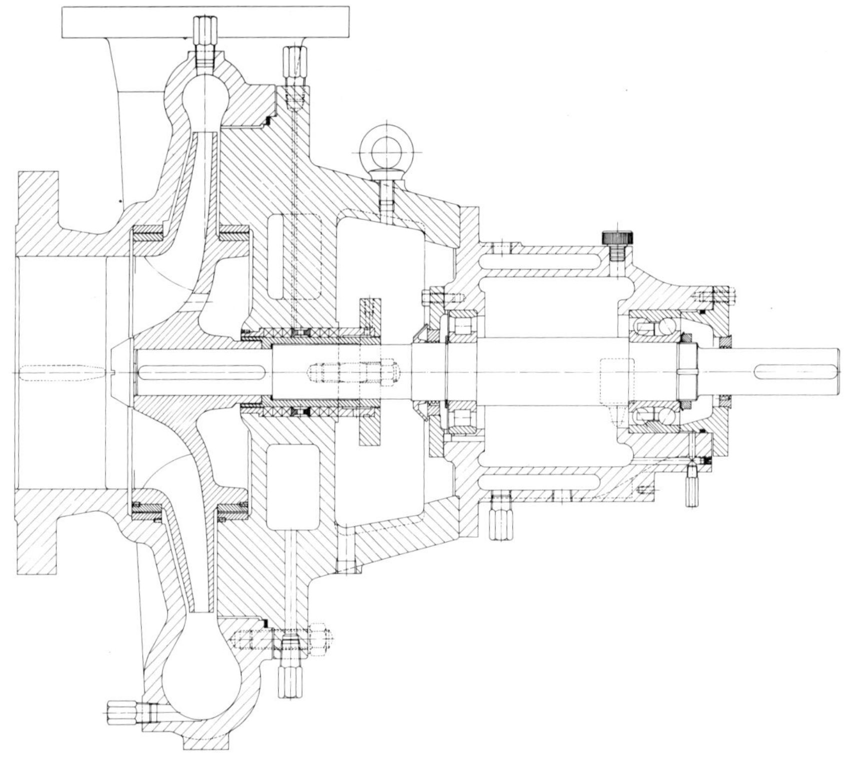

Fig 2   Heavy duty process pump with volute casing,  Stork type PH

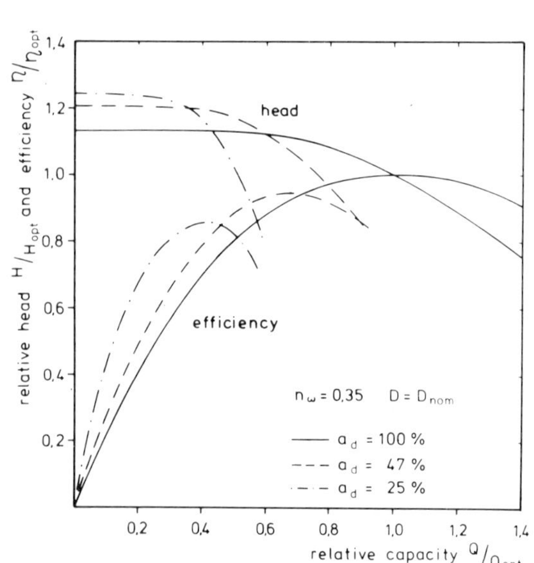

Fig 3   Influence of the diffuser throat area on the hydraulic pump performance

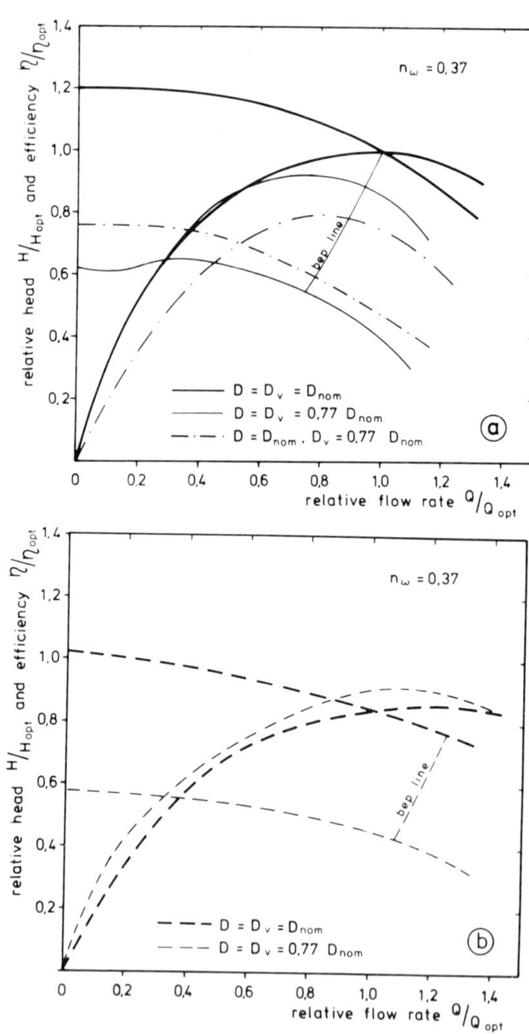

Fig 4   Hydraulic performance of a diffuser pump with $n_\omega = 0.37$
  (a) Pump with full and reduced impeller diameter
  (b) Pump with full and reduced impeller diameter, diffuser removed

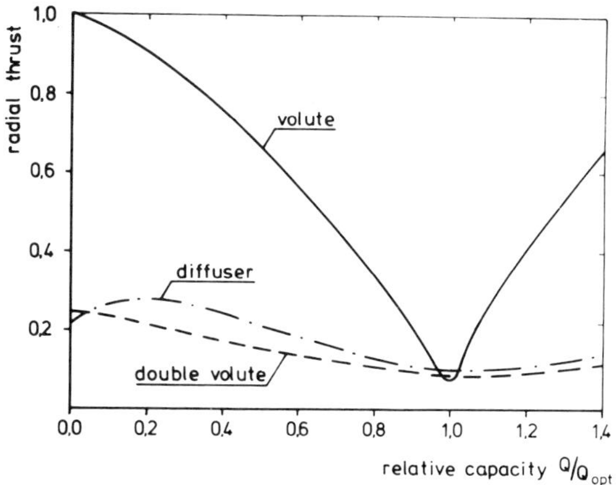

Fig 5   Comparison of radial forces as percentage of force at zero flow related to the relative flow rate

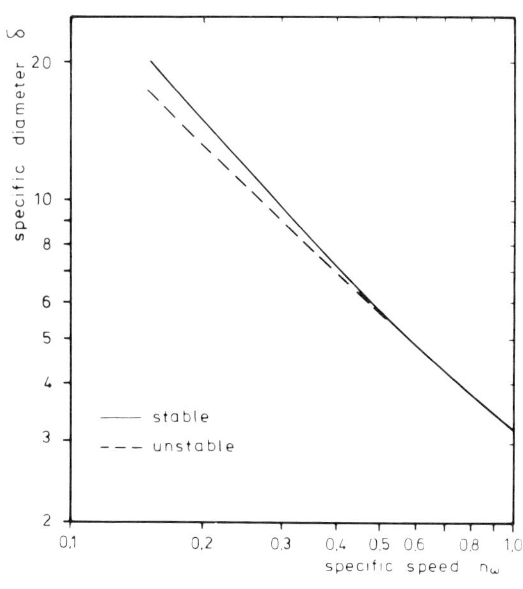

Fig 6   Specific diameter versus specific speed for volute and diffuser pumps

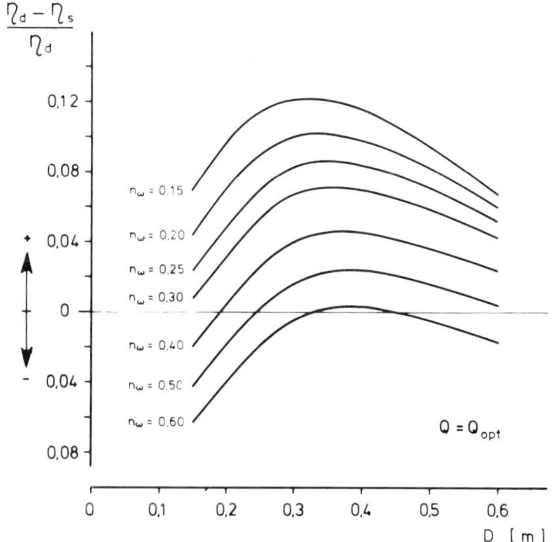

Fig 7   Comparison of hydraulic efficiencies at best efficiency point of diffuser and volute pumps related to the impeller diameter and specific speed

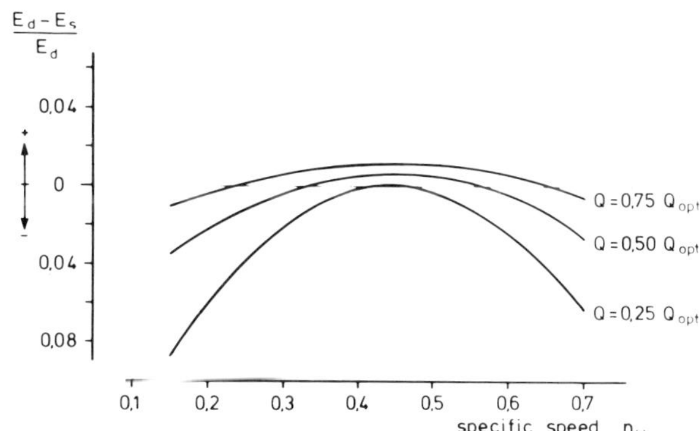

Fig 8   Comparison of hydraulic efficiencies at off-design conditions of volute and diffuser pumps related to the specific speed

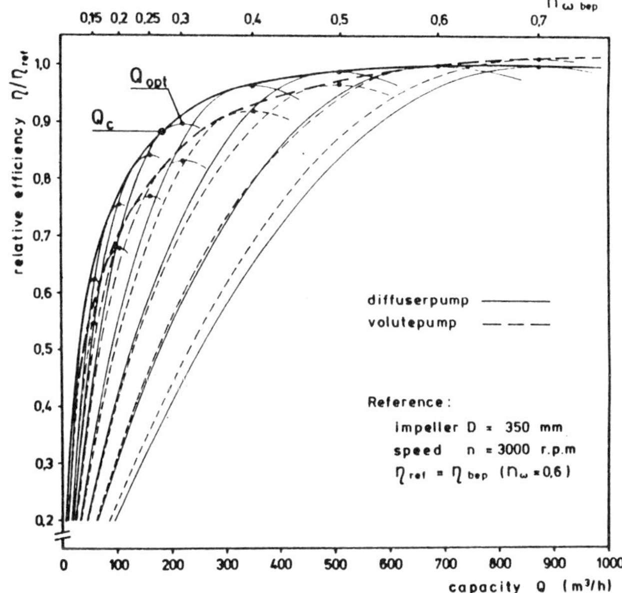

Fig 9   Calculated efficiency curves for volute and diffuser pumps with impeller diameter $D = 350$ mm and speed $n = 3000$ rev/min

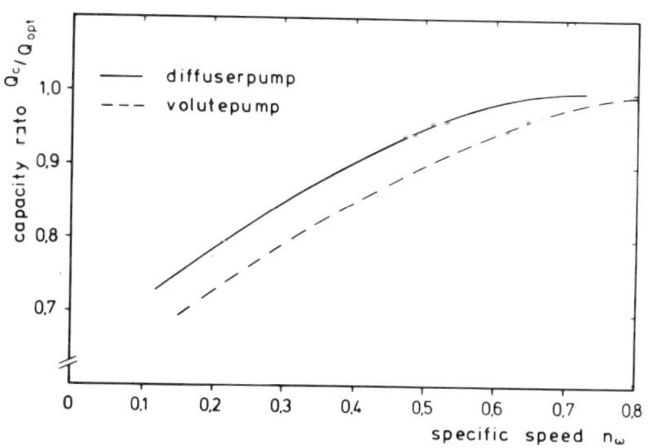

Fig 10   Relative capacity for optimum hydraulic efficiency as a function of specific speed

# C45/81

# Design aspects of baseplates for oil and petrochemical industry pumps

E J BUSSEMAKER
Stork Pompen BV, Hengelo, The Netherlands

SYNOPSIS    A study has been made of requirements which are specified for oil and petrochemical industry pump baseplates. Thermal and pipe loading effects on coupling alignment and other aspects of structural  design, such as thermal expansion of pedestals, and pedestal stress, are considered.

Various baseplate styles have been evaluated of grouted and non-grouted types. Finite elements and experimental analyses are presented and discussed in relation to specified requirements.

## INTRODUCTION

In this paper attention is given to different aspects of baseplates for oil and petrochemical pumps. Baseplates for horizontal pumps with overhung impeller are specifically considered.
The baseplate construction normally used is a welded structure which supports the pump as well as the driving electric motor. Normally the support of the pump is at centreline height using pedestals which are welded or bolted to the base structure.
Usually the baseplate structure is grouted, the grout being poured into the structure via grout openings after the baseplate has been bolted down and a first alignment check has taken place. The function of the grout is to provide extra stiffness and also to fill spaces in which gas or liquid might collect.

The main functions of a baseplate can be summarized as :
- to keep the pump and motorshaft acceptably in line, whilst forces and moments, caused by mechanical, weight, thermal expansion and hydraulic piping thrust effects, are working on the pump and motor structure.
- to drain leaked pump liquid via a raised lip or pan construction.
- to minimize and damp vibrations of pump and motor.
- to assure that natural frequencies of the total structure do not interact with exciting frequencies from the pump or piping.
- to offer cooling possibilities for pedestals or other features which provide correct alignment of the pump and motor shaft when the pump is pumping hot liquids, say with temperatures above 350°C.

In this paper attention is given to a number of design aspects, particularly to the following subjects:
- demands from the most important pump specifications.
- pump shaft deflection under pipe loading conditions.
- influence of pump casing pressure rating.
- pedestal stress.
- pedestal temperature.
- vibration behaviour.

No specific attention is paid to stresses or temperatures in the pump structure itself.

## SPECIFICATIONS

### API 610 baseplate requirements

The current 5th edition of the API standard 610 is dated March 1971 but will shortly be renewed. This important standard for oil refinery pumps is used by most firms which purchase pumps for oil refineries and petrochemical plants. The allowable forces and moments which are specified in the 5th edition differ substantially from those in the 6th edition final draft. Attention is therefore paid to both versions.

In the 5th edition (chapter 26) attention is given to baseplates for horizontal pumps. Figures are given for baseplates which support pumps having 4 inch or smaller discharge size. For fully grouted baseplates the shaft deflection measured at the coupling is limited in the specification to 0.010 inch (= 0.25 mm) in any direction when the pump is subjected to summed moments $M_x = 3.0$ W, $M_y = 2.0$ W and $M_z = 1.5$ W, in which W is the pump weight, with minimum values of 3000, 2000 and 1500 ft lb  (= 4067, 2712 and 2034 Nm). These moments

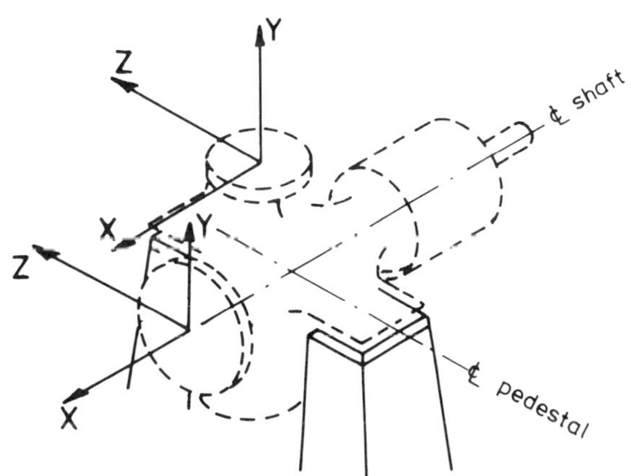



Fig 1    The coordinate system

are assumed to act at the intersection of the X-, Y- and Z-axis (see figure 1). Actual forces and moments considered to act separately on the suction and discharge branches must be transferred to the intersection of the X-, Y- and Z-axes and the resulting summed moments compared with the maximum moments mentioned earlier. The influence of the forces as pure forces is not taken into account.

This last remark is in contradiction with the baseplate requirements as formulated in the 6th edition. Here the forces are also taken into account as real forces and not only as causing a moment by the influence of the arm distance of the branches to the pump XYZ-centre.

The 6th edition data also cover a wider range of pump sizes with suction branches of 12 inches and smaller. This normally relates to discharge branches of 10 inch and smaller, which is a much broader coverage than the 4 inch maximum in the 5th edition.

For the pump baseplate and pedestal support assembly the shaft deflection measured at the coupling is limited to 0.005 inch (= 0.13 mm) in any direction when the suction and discharge branches are simultaneously subjected to the forces and moments mentioned in table I.

In both API 610 editions the maximum shaft deflections must not be seen as allowable figures during pump operation, but as a measure of the stiffness of the assembly for design purposes only. A correct alignment for operating temperatures is always recommended.

When a comparison is made between the specified forces and moments in the 5th edition and those in the 6th edition we come to the figures as shown in table II.

The sum of the moments in the new edition has increased substantially by a factor between 1.72 and 2.17. This means an average increase factor of 1.95. Furthermore the forces are also taken into account, however their influence is relatively small.

Summarizing, the 6th edition API 610 draft specifies a baseplate and pedestal support which is twice as rigid as that in the 5th edition.

The 6th edition also mentions the possibility of specifying a support structure which allows for a further doubling of the forces and moments. This means a stiffness increase of 300 % compared with the current 5th edition.

## Table 1

(Source: API 610 final draft - 2.1.1980 - table 1)

| Force (pounds)/Moment (foot-pounds) | | Nozzle nominal flange size in inches | | | | | | |
|---|---|---|---|---|---|---|---|---|
| | | 2 | 3 | 4 | 6 | 8 | 10 | 12 |
| Top nozzle | $F_x$ | 100 | 240 | 320 | 560 | 850 | 1200 | 1500 |
| | $F_y$ (compression) | 200 | 300 | 400 | 700 | 1100 | 1500 | 1800 |
| | $F_y$ (tension) | 100 | 150 | 200 | 350 | 530 | 750 | 920 |
| | $F_z$ | 130 | 200 | 260 | 460 | 700 | 1000 | 1200 |
| End nozzle | $F_x$ | 200 | 300 | 400 | 700 | 1100 | 1500 | 1800 |
| | $F_y$ | 130 | 200 | 260 | 460 | 700 | 1000 | 1200 |
| | $F_z$ | 160 | 240 | 320 | 560 | 850 | 1200 | 1500 |
| Each nozzle | $M_x$ | 340 | 700 | 980 | 1700 | 2600 | 3700 | 4500 |
| | $M_y$ | 260 | 530 | 740 | 1300 | 1900 | 2800 | 3400 |
| | $M_z$ | 170 | 350 | 500 | 870 | 1300 | 1800 | 2200 |

## Table 2

| Loads on pump for maximum 0.010 inch deflection | | API 610 5th edition | | API 610 6th edition final draft | | Ratio 6th/5th edition | |
|---|---|---|---|---|---|---|---|
| | | 2"x3"x13" | 4"x6"x10" | 2"x3"x13" | 4"x6"x10" | 2"x3"x13" | 4"x6"x10" |
| $M_x$ | (Nm) | ± 4 067 | ± 4 067 | ± 6 368 | ± 7 916 | 1.57 | 1.95 |
| $M_y$ | (Nm) | ± 2 712 | ± 2 712 | ± 4 874 | ± 6 154 | 1.80 | 2.27 |
| $M_z$ | (Nm) | ± 2 034 | ± 2 034 | ± 3 878 | ± 5 024 | 1.91 | 2.47 |
| $M_x+M_y+M_z$ | (Nm) | ± 8 813 | ± 8 813 | ± 15 120 | ± 19 094 | 1.72 | 2.17 |
| $F_x$ | (N) | – | – | ± 7 116 | ± 9 074 | | |
| $F_y$ | (N) | – | – | + 5 340 / – 7 120 | + 5 872 / – 7 650 | | |
| $F_z$ | (N) | – | – | ± 6 784 | ± 7 296 | | |
| $F_{hor}=F_x+F_z$ | (N) | – | – | ± 13 900 | ± 16 370 | | |
| $F_{vert}=F_y$ | (N) | – | – | + 5 340 / – 7 120 | + 5 872 / – 7 650 | | |

## Europump and ISO baseplate requirements

A Europump working group has also worked on the subject of allowable pipe loads on end suction pumps built to the standard ISO 2858; the so-called 16 bar pumps. A number of pump manufacturers have tested the various pump baseplates assemblies and presented the shaft deflections caused by forces and moments in a specification; for example Second Draft Proposal ISO/DP 5199 'Technical specifications for centrifugal pumps'.

The tests were carried out on several pumps and as a result the shaft deflections were judged to have a limit of 0.15 mm, 0.20 mm and 0.25 mm for pumps with shaft diameters of 24 mm, 32 mm and 42 mm respectively. In comparison with refinery and petrochemical pumps it is interesting to look at the demands for steel case pumps on grouted baseplates. For a number of pumps the maximum allowable horizontal and vertical forces and moments are reproduced in table 3.

When we compare the allowable moments for 4 example pumps following API 610 5th edition with the Europump figures, we see a difference of a factor 0.45 up to 0.90 for pumps with discharge branches of 4 inches and smaller. In this comparison it must be realised that the Europump allowable shaft deflections are also smaller: 0.15 mm, 0.20 mm and 0.25 mm compared with 0.25 mm.
Furthermore, the Europump draft proposal for a specification includes the total horizontal and vertical forces on the pump structure, whilst the current edition API 610 specification does not account for such forces in relation to baseplate stiffness.

## PIPE LOADING ANALYSIS

### Shaft deflection test results

In the authors firm a number of tests have been carried out on steel end-suction refinery pumps installed on welded construction steel baseplates and pedestal assemblies.
The test results will be given here as shaft displacements in the X- and Z-direction caused by moments $M_x$, $M_y$ and $M_z$, see also figure 1.

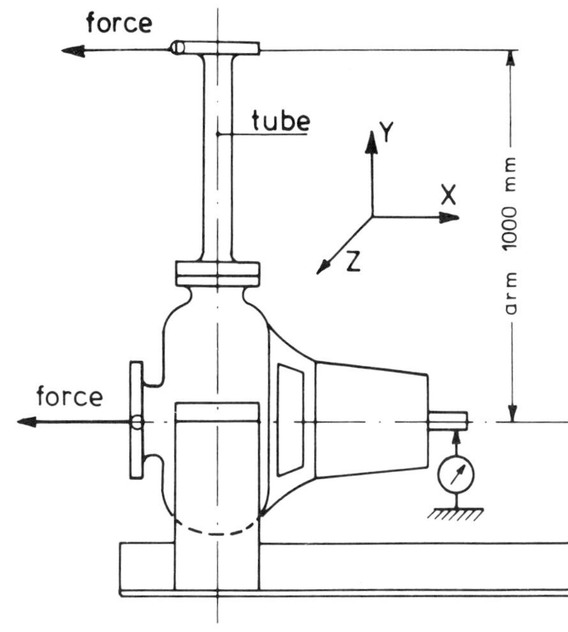


Fig 2    Pump loading situation

Table 3

| Pump type | ISO 2858 identification | Dimensions (mm) | | | | Maximum Europump Loads [1][2] (total loading on both nozzles) | | |
|---|---|---|---|---|---|---|---|---|
| | | A | B | C | d | $(F_x+F_z)_{max.}$ = $F_{hor.}$ (N) | $F_{y,max.}$ [3] = $F_{vert.}$ (N) | $(M_x+M_y+M_z)_{max.}$ (Nm) |
| 2 x3 x13 in | 80- 50-315 | 125 | 280 | 500 | 32 | 4 000 | 5 500 | 1 600 |
| 3 x4 x13 in | 100- 65-315 | 125 | 280 | 530 | 42 | 5 000 | 6 500 | 2 400 |
| 4 x6 x10 in | 125-100-250 | 140 | 280 | 530 | 42 | 8 000 | 9 800 | 4 600 |
| 8 x6 x10 in | 200-150-250 | 160 | 375 | 530 | 42 | 12 000 | 13 000 | 7 100 |

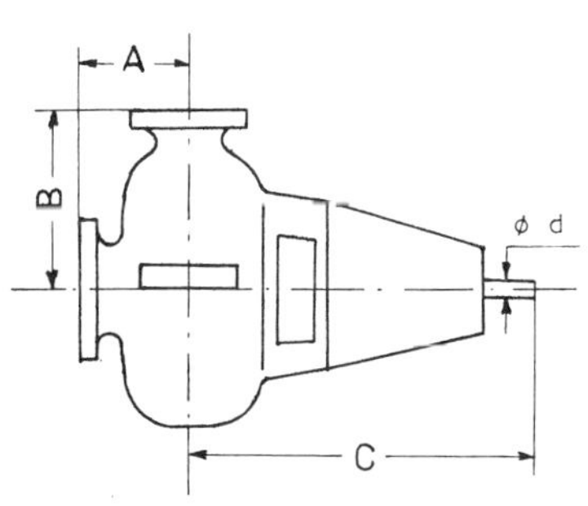


[1] Extra condition is:

$$\left(\frac{\Sigma F_x + F_z}{(F_x + F_z)_{max}}\right)^2 + \left(\frac{\Sigma F_y}{F_{y\ max}}\right)^2 + \left(\frac{\Sigma M_x + \Sigma M_y + \Sigma M_z}{(M_x + M_y + M_z)_{max}}\right)^2 \leqq 1$$

[2] With grouted baseplate and for steel pump structure.

[3] $F_y = \frac{2}{3} F_{y\ discharge} + F_{y\ suction}$

The 'moment only' shaft deflection was obtained by
first applying a force to a pipe 1 m from the pump
centre and subsequently in the same direction at
the pump centre itself, (see figure 2). The deflections were then subtracted.

The tests have been carried out on a number of
different baseplate pedestal assemblies supporting
a 2"x3"x13" pump having a 50 bar working pressure.
The tests were carried out both with and without
grout and also with and without a centering pin
under the pump.

Figure 3 shows the forms of baseplate which were
tested.
Apart from the grouting and pinning to the base or
not, the principal design differences are in the
construction of the pedestals:

A) Tapered box type pedestals bolted to the baseplate and with slide keys at the top of the
   pedestals.
B) U-channel pedestals welded to the baseplate
   side members.
C) 3 plate pedestal construction with 10 mm
   plate thickness welded to the baseplate side
   members.
D) 3 plate pedestal construction with 6 mm plate
   thickness welded to the baseplate side members.

The main concept of the baseplate construction
with fixed pedestals and a central pin under the
pumpcase was to create two triangle structures
linked at the attachment points by the pump casing
and the baseplate-pedestal construction.

The pumps were loaded with the moments as specified in API 610, 5th edition:
$\Sigma M_x$ = 4 067 Nm (= 3000 ft.lbs.)
$\Sigma M_y$ = 2 712 Nm (= 2000 ft.lbs.)
$\Sigma M_z$ = 2 034 Nm (= 1500 ft.lbs.)

The results are presented in table 4.

Pedestal types A and B gave the lowest shaft deflections which were only 20 % of the allowable
value 0.010 inch (= 0.25 mm).

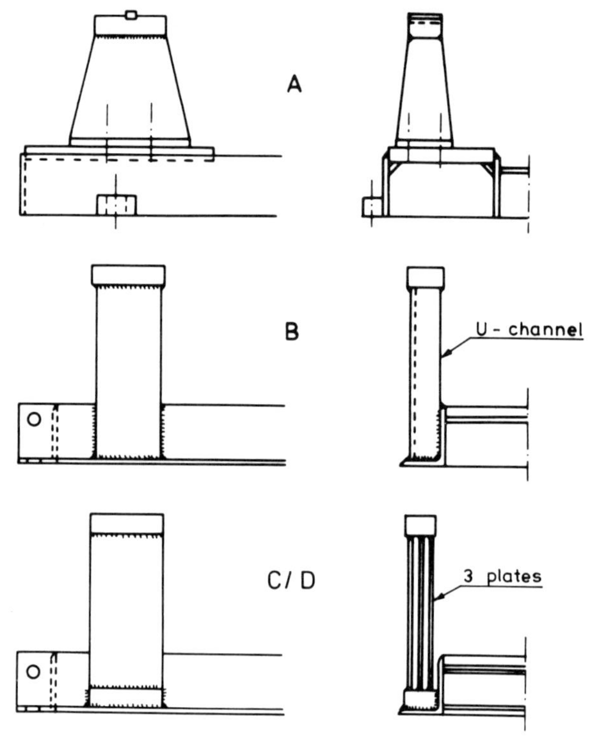


Fig 3   Tested baseplate forms

When the 6th edition API 610 forces and moments
are applied, the shaft deflections will be about
double these values and become about 35 % of the
new allowable figure.
The additional possibility of doubling the forces
and moments means reaching about 75 % of the
allowable shaft deflection value for this 50 bar
working pressure pump.

Table 4

| | Description of pedestal execution under a 2"x3"x13" pump (50 bar working pressure) | Grouted | Centering pin | Shaft deflection for $M_x$ = 4 067 Nm $M_y$ = 2 712 Nm $M_z$ = 2 034 Nm (API 610 5th ed.) mm | % ref. B | Shaft deflection to moment ratio (m/Nm x $10^{-9}$) $\frac{y}{M_x}$ | $\frac{y}{M_y}$ | $\frac{y}{M_z}$ | $\frac{z}{M_x}$ | $\frac{z}{M_y}$ | $\frac{z}{M_z}$ |
|---|---|---|---|---|---|---|---|---|---|---|---|
| A | Loose and bolted, |  | X | 0.240 | 462 | | | | | | |
|   | slide keys at top | X | X | 0.054 | 104 | 0 | 0 | −19.0 | 0 | 6.1 | −10 |
| B | Welded U-channel | X | X | 0.052 | 100 | 0 | 0 | −18.5 | 0 | 6.0 | 0 |
| C | 3 parallel plates |  | X | 0.240 | 462 | 0 | 0 | 24 | 44 | 16 | − 6 |
|   | 10 mm thick | X | X | 0.069 | 133 | 0 | 0 | 7.2 | 3.0 | 20 | 0 |
|   | | X | | 0.075 | 144 | 0 | 0 | 10 | 3.0 | 22 | 0 |
| D | 3 parallel plates | X | X | 0.078 | 150 | 0 | 0 | 8 | 3.0 | 24 | 0 |
|   | 6 mm thick | X | | 0.104 | 200 | 0 | 0 | 8.5 | 3.1 | 33 | 0 |

Although the pedestals formed by U-channels result in a stiffer structure, the 3 plate construction was preferred.  The maximum stresses in the
U-channels reach unacceptable values due to thermal expansion of the pump casing with liquid temperatures above 280°C. This will be explained further on in the paper.

The influence of the grout on the shaft deflection
was dependent on the design of the baseplate support frame and contributed by a factor 2.5 to 4.5.
The influence of a centering pin under the pumpcase was measured to contribute by a factor 1.1
to 1.3, see table 4.

### Shaft deflection calculation results

In the design stage of the 3 plate type pedestal
baseplate a calculation of the shaft deflections
was also performed. The calculations were carried
out using finite element modelling and were performed for a baseplate for a centerline support
steel pump (baseplate execution 'C').

The structure calculated was a non-grouted baseplate comprising a rectangular frame of L-profiles
with bridgeparts formed by U-channels, a coverplate and pedestals consisting of 3 parallel plates.
Essentially the same baseplate type was used for
the deflection tests.

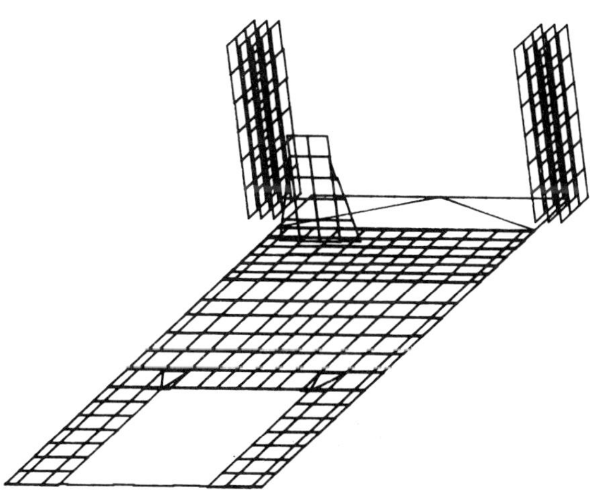

Fig 4   Finite element model

The L-members and U-channels were modelled as bars
and the coverplate, the pedestal plates and the
bearing housing support plate were modelled as
square and triangular plate elements, (see figure
4). The connections of pedestal plates and bearing
housing foot with the channels were modelled as
rigid bars.
The pump structure itself was assumed to be infi-
nitely stiff.
The loading on the pump was done with forces $F_x$,
$F_y$ and $F_z$ and moments $M_x$, $M_y$ and $M_z$ acting on the
intersection of Z-, Y- and Z-axes.

With the computer program NASTRAN, displacements,
stresses, reaction moments and forces were calculated.
The calculated results were compared with the
actual test figures of the same pump and baseplate,
which indicated a shaft deflection during test
of 0.24 mm when forced with maximum API moments
$M_x$ = 4 067 Nm, $M_y$ = 2 712 Nm and $M_z$ = 2 034 Nm.

The calculations indicated a shaft deflection of
0.21 mm, which is a difference of 0.03 mm compared
with the test figure 0.24 mm.

### Influence of pump casing pressure rating

The differences between the calculated and the
measured shaft deflection figure is thought to be
due to the assumption in the calculation that the
pump structure is rigid. For the 50 bar pump on
the grouted baseplate about 50 % of the total
shaft deflection is apparently caused by the flexibility of the pump structure itself and its connections to the pedestals and baseplate.
For another (4"x6"x10") pump having 30 bar working pressure on a non grouted baseplate with
3 plate (10 mm thickness) pedestals, calculations
and tests were also compared. The calculated shaft
deflection was 0.20 mm, whilst the test figure
was 0.28 mm, indicating a difference of 0.08 mm,
which is also thought to be caused by the flexibility of the pump structure and its connections to
the baseplate.
Apparent only 0.03 mm of the measured 0.11 mm
shaft deflection figure for the grouted situation
is caused by the baseplate and its pedestals.

A substantial increase of the baseplate stiffness
may have no significant influence on the total
shaft deflection. At the moment of writing this
paper we are continuing the calculations with alternative assumptions for the flexibility of the
pump structure and the different connections.

It is clear that more severe specifications concerning shaft deflections can only be answered by
stiffening the pump structure at the same time as
stiffening the support and baseplate structure.

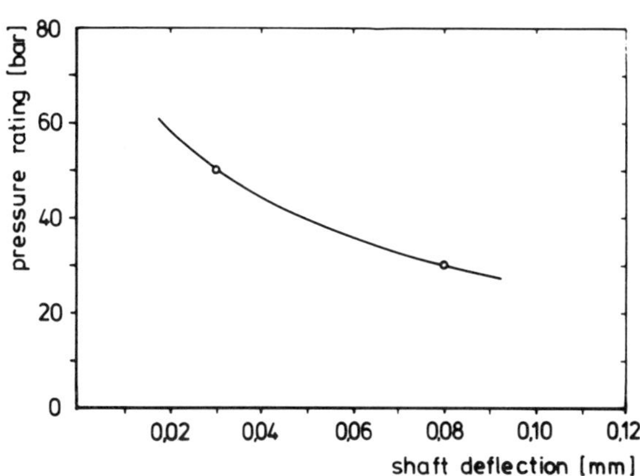


Fig 5   Effect of pressure rating

The pressure rating of the pump casing has an important influence on the shaft deflection. The foregoing analysis results in 0.08 mm and 0.03 mm deflection for a 30 bar and 50 bar pump respectively. Roughly doubling the pressure rating results in reducing the shaft deflection with about a factor 4, (see also figure 5).

Mounting a bearing housing support foot which is very stiff in Y- and Z-directions but flexible in X-direction, has a decreasing influence on the shaft deflection. This is advised as an effective measure for stiffening pump support structures. It is furthermore strongly advised to machine the surfaces of the bearing housing foot and the base-plate top.


## THERMAL AND VIBRATION BEHAVIOUR

### Pedestal stress measurements

When a centreline rigidly connected pump handles a high temperature liquid, the pedestals deflect sidewards due to the temperature expansion of the pump casing. To illustrate this, a temperature difference of 320°C for a stainless steel pump with 700 mm distance between the pedestals means an expansion distance of 4 mm. On hot pumps the deflections can easily be seen by the eye.

For the baseplate execution for the 50 bar pumps, shown in figure 3, execution A, the pump is able to slide horizontally along keys between the top of the pedestal and the pump; this means that no stresses are introduced in the pedestals by thermal pump case expansion.

In the execution B, C and D mentioned in figure 3 however, the pump is positively bolted down, so the pedestals are forced to deflect, which introduces stresses in the pedestals.
Strain gauge measurements have been carried out on 2 types of pedestals:
- the U-channel type pedestal;
- the 3 plate pedestal.

The tests were carried out by loading the top of one pedestal with the pump mounted and clamped to the test pedestal, whilst on the other side the pump was mounted with a small clearance to allow it to slide over the pedestal.

The outcome of the measurements on the U-channel pedestal was that the allowable stress (200 N/mm$^2$) was reached at the highest load point when at a deflection equivalent to a steel pump house at a temperature of 270°C, corresponding to a pumped liquid temperature of about 280°C. The highest stress occurred in the top weld between the U-channel and the baseplate side-member.

These measurements illustrate the limited use of the U-channel for high temperature pumping. An improvement is possible by turning the U-channel 180° so that its back is welded to the baseplate side member. This however was judged to be insufficient, so an alternative design was created in the form of a 3 plate pedestal.
This pedestal was also tested and gave acceptable stiffness properties as shown before, whilst the measurements of stress due to thermal pump expansion resulted in a maximum stress of only 88 N/mm$^2$ at the bottom of the plates at a deflection corresponding to a temperature of 400°C for a stainless steel pump casing.

### Pedestal temperature measurements

To evaluate the vertical thermal expansion of an uncooled pedestal, temperature and displacement measurements were carried out with the same base-plate on which the pedestal stiffness was measured. The pump casing was filled with oil which was heated to 400°C.
The total height of the pedestal from the bottom of the baseplate is 430 mm and this height increased by 0.42 mm, without forced air circulation. This height increase is reached when the average temperature difference with ambient is 81°C. When the temperature expansion coefficient is assumed to be 12.10$^{-6}$ m/m°C, the corresponding average pedestal temperature is about 100°C.
The temperature of the pedestal plates was measured at the inside and outside plate resulting in the temperature distribution as presented in figure 6.

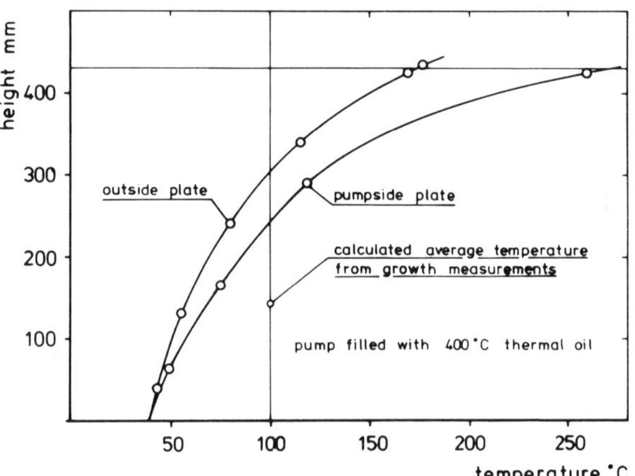


Fig 6   Temperature distribution, three plate pedestal

The vertical shaft deflection was also measured and amounted to 0.13 mm.
When an air circulation is present, the temperature of the plates will be lower.
The measured influence of a certain air stream along the pedestals was a height decrease of about 0.03 mm which means that the vertical shaft deflection is about 0.10 mm (= 0.004 inch) maximum. Cooling of the pedestals is judged to be unnecessary. This means that a cooling water system is not necessary; an advantage which should not be underestimated. From the measured deflection figures it is also possible to specify an alignment offset of the pump and motorshaft during (cold) mounting, so that in service there is a correct shaft alignment.

Normal clamping of a pump to its pedestals gives good contact between the metal surfaces of the pump and pedestal and consequently a good heat transfer. By introducing a thin temperature insulation plate between pedestals and pump an extra decrease of pedestal temperature can be achieved.

<u>Vibration measurements</u>

Bearing housing vibrations were carried out on the 50 bar and 30 bar pumps. The maximum vibration values were measured at shut-off condition but surprisingly there was no significant vibration difference between the grouted and non-grouted condition over the total flow range.

On the 4"x6"x10" 30 bar pump the natural frequencies were also measured whilst the pump was mounted on its baseplate. The measurements were carried out in non-grouted and grouted conditions and also with and without connected suction and discharge piping. The results were frequencies in the range 50 Hz - 55 Hz and 72 Hz - 140 Hz in X-direction and Z-direction respectively, thus not interfering with known exciting frequencies.

## CONCLUSIONS

The main conclusions from this paper are the following:

1) The final draft 6th edition API 610 standard specifies a pump baseplate assembly stiffness which is about twice the stiffness figure in the 5th edition.

2) The measured shaft deflections of 50 bar end-suction pumps on standard grouted baseplates are about 20 % of the allowable API 610 5th edition figures.

3) Grouting of standard baseplates gives a stiffness increase of 150 % to 350 %.

4) For the 3 plate pedestal type baseplate, connecting the bottom of the pump casing to the baseplate with a centering pin improves the stiffness by 10 % to 30 %.

5) The stiffness of the pump structure itself and the pump fastening play an important role in the stiffness behaviour of a pump installation on a baseplate.

6) With high temperature pumps, which are clamped to the pedestals and comprise no slide keys, attention has to be paid to the stress in the pedestals and their welds to the baseplate.

7) The described 3 plate pedestals need no cooling at liquid temperatures up to at least 400°C.

8) Future work on baseplates for refinery and petrochemical pumps should include the following subjects:
   - Further stiffness analysis of the pump structure itself to obtain a correlation between stiffness calculations and measurements.
   - Design of baseplate and pedestal assemblies with increased stiffness.
   - Design of free standing baseplates, i.e. baseplates which are not fixed to the floor with anchor bolts but are free to slide.

## ACKNOWLEDGEMENTS

This paper is published by permission of the Board of Stork Pompen B.V., Hengelo (O), The Netherlands. The author wishes to express his gratitude to Mr D.J. Burgoyne - Engineering Manager - for his help and advice.

## REFERENCES

1) American Petroleum Institute, Centrifugal Pumps for General Refinery Services. API Standard 610, Fifth Edition, Washington, March 1971.

2) American Petroleum Institute, Centrifugal Pumps for General Refinery Services. API Standard 610, proposed Sixth Edition (Final Draft), Washington, January 2, 1980.

3) International Organisation for Standardization ISO, Draft Proposal ISO/DP 5199, Second Draft Proposal for technical specification for centrifugal pumps: class II, July 1979.

# An engineer's guide to double seals

K R V LOCK, CEng, MIMechE and M L STOKES, BSc(Eng), CEng, MIMechE
Crane Packing Limited, Slough, Berkshire

SYNOPSIS   Engineers are giving greater consideration to the use of double seals for rotary shaft sealing duties.  The reasons for this growing interest include new demands for plant safety, the need for avoidance of solids to extend seal life, increased working pressures and the rising cost of product losses due to leakage.  This paper is presented as a guide to the use of double seals, with discussion on back-to-back and tandem seal installations with and without liquid sealants.  Comments on sealant selection, circulation methods, power requirements and seal housing design are included. The relative merits of pressurised and atmospheric sealant circulating systems are explained to assist the engineer in designing and selecting a complete double seal installation.

## INTRODUCTION

The mechanical seal has gained wide acceptance as a device to control leakage.  The arrangement of a single seal which derives its lubrication and cooling from the liquid being handled, remains the first choice and its selection is mainly determined by its pressure and temperature capability, and its corrosion resistance.  There are, however, occasions when the application imposes such demands on a single mechanical seal that it becomes incapable of surviving for any length of time.  In these instances a working environment needs to be created for the seal to keep wear and leakage within acceptable limits.  There are other situations where as a result of the inevitable leakage from the single mechanical seal, an unacceptable environment is created around the equipment.  The solution to these problems involves more than one seal.  This paper sets out some guidelines for engineers to obtain maximum sealing effectiveness from double seal installations.

## DOUBLE SEAL ARRANGEMENTS

Double seals are best understood by concentrating on the basic single seal, the 'major' seal, and mentally adding a second 'minor' sealing device either inboard, on the product side, or outboard, on the atmosphere side, of the major seal.  This distinction between the two possible positions for the minor seal is important.  It determines the performance characteristics of the arrangement and, therefore, the installation best suited to a particular application.

When a minor seal is interposed between the major seal and the product the arrangement is commonly referred to as 'back-to-back'.  Various seal arrangements which fall into this category are shown in Figure 1, with the major seal

identified by section lining.  A liquid or vapour, known as the sealant, has to be introduced into the space between the two seals at a pressure greater than that of the product to lubricate and cool both seals.  At the minor seal this pressurised sealant overcomes the product pressure to form the interface film or boundary film in the case of a lipseal.

When the minor seal is added to the atmosphere side, the major seal continues to be cooled and lubricated by the product, with minor seal operation usually supported by a lower pressure flush injected between the two seals. This arrangement is known as the tandem double, and examples are given in Figure 2.  Once again the major seal is identified by section lining.

## DOUBLE SEAL REQUIREMENTS

Seven basic duties are recognised as requiring double seals in one form or another.

A   High concentrations of solids in suspension - slurries.

B   Gases and vapours - dry running applications.

C   Highly viscous products - not suitable as the interface film.

D   High Pressure Velocity factors - pressure breakdown arrangements.

E.   Pharmaceutical and vacuum distillation duties - atmospheric contamination.

F   Hazardous products - safety applications.

G   Seals requiring performance monitoring.

Each of these duties will now be considered in detail.

A Solids in the product

In certain cases single mechanical seals
with hard faces and product circulation via
cyclone separators can operate successfully on
abrasive bearing products. However, when the
solids concentration becomes very high, it is
impractical to separate out the solids. Unless
the solids can be excluded they will not only
penetrate the faces and give rise to high wear,
but may also compact within the seal so
destroying its flexibility. A clean sealant is
therefore used to provide a satisfactory
interface film. This sealant is separated from
the product by a second seal which suggests the
arrangement in Figure 1a, the conventional
back-to-back double.

If product dilution by the sealant is
acceptable the double seal can be simplified to
that shown in Figure 1d, where the minor seal
is a flexible lip which allows sealant to pass
into the pump, but restricts product flow in
the reverse direction. Sufficient sealant flow
to cool the major seal must be allowed for.
One advantage this has is that the lip can
accommodate pressure transients in the product
whereas the back-to-back double may not.

B Gases and vapours

Apart from a few applications where the
loads on the seal are very low, mechanical seals
cannot operate dry. In those situations where
seals operate in vapour, as might be the case in
an agitated vessel, a sealant must be provided
to lubricate and cool the faces. This automatically leads to the selection of the back-to-
back double seal or the concentric faced version
which is a composite double seal (Figure 1b).

The concentric faced version is particularly suitable for agitated vessels since, by
virtue of its clearances, it can tolerate the
shaft movements usually associated with
equipment of this sort. Because both the vessel
pressure and the sealant pressure act in one
direction, the total thrust must be taken into
account when sizing the thrust bearing for the
agitator assembly.

C High viscosity products

Viscosity is one of the major factors
influencing the thickness of the interface film.
An increase in viscosity with a subsequent
increase in film thickness can lead to high heat
generation due to the shearing of the fluid film.
Furthermore, when the film is thick the possibility of leakage from within the film is
greater. Add to this the extra loads imposed
on the drive mechanism within the seal and it
can be seen that the environment is not ideal
for best performance. Assuming normal speeds of
1500 to 3000 rev/min a limit of 2500 cS is
considered acceptable for single seals. However,
when this viscosity limit is exceeded it is
essential to use a double back-to-back seal with
a clean liquid sealant forming the interface
film.

D Pressure breakdown

One of the basic factors defining the operating range for face materials is the Pressure
Velocity (PV) factor. By arranging an acceptable pressure differential across two rubbing
faces, PV can be kept within limits. The seal
installation shown in Figure 2a, the conventional
tandem double, is particularly suitable since
adjustment of the flush pressure controls the
pressure differentials across both seals. By
this means product pressures above the limiting
pressures of a single mechanical seal can be
accommodated.

Of the materials currently available
tungsten carbide and silicon carbide have the
highest PV capability, being in excess of 1000
bar m/sec. Material selection could well decide
the maximum pressures for a seal installation
and in extreme cases it may prove necessary to
install triple seals in order to accept a high
product pressure. As seals capable of accepting
140 bar plus are readily available, it is
doubtful if multiple seals will be required
except for a few specialised applications.

With the emergence of controlled leakage
hydrostatic seals it is now practical to use
these in conjunction with a conventional face
seal to break down pressure.

E Atmospheric contamination

Whilst the general object of any seal is to
control the escape of pressurised liquid there
are occasions when double seals are required to
prevent the atmosphere from contaminating the
product. One such example is pharmaceutical
applications where the interspace between the
seals is maintained under sterile conditions. A
fairly common application would be an agitated
vessel at 2 to 3 bar pressure where the sealant
may well be 3 to 4 bar saturated steam with
shaft speeds in the region of 200 to 400 rev/min.
Another example is vacuum distillation on a
refinery where if air is drawn into the pump
through the seal an extremely hazardous situation
may result, particularly with a pump on a reflux
duty. In circumstances such as these a sealant
in excess of the product pressure must be
circulated between double seals. The back-to-
back arrangement is the one most commonly used.

F Safety

In view of current legislation it is no sur-
prise that safety figures are high on the list of
double seal requirements. In general the back-
to-back double will meet most applications where
safety is a consideration. This installation
involves a pressurised sealant and the equipment
to provide it, the complexity and cost of which
may well be out of proportion to the level of
safety desired. The single seal with an external
throttle bush (Figure 2d), which in essence is a
tandem seal, may well prove adequate. The
selection of the minor seal depends upon whether
the product involved is toxic, creates a fire or
explosion hazard and whether the leakage is in
the liquid or vapour phase. Risk analysis is
assuming considerable importance in plant design
and it is this which can lead to variations in

the seal installation.

Consider for example the simplest tandem seal arrangement, Figure 2d. It has over the years proved very successful, and provided the leakage remains in the liquid phase, with no question of toxicity, it is adequate. Awareness of leakage is all that may be required to enable remedial action to be taken.

However, should there be a phase change on leakage or product solidification through decomposition or salt deposition it may prove necessary to provide a continuous flush of steam or water. A more effective device than the bush may need to be incorporated (Figure 2e). Rubber or PTFE lip seals can be used according to the nature of the product. Figure 2c shows an alternative easily installed arrangement where flushing is continuous, but at low pressure. Here the minor mechanical seal is externally mounted and subject to internal flush pressure.

In safety applications where, following mechanical seal failure, the minor seal must contain the full system pressure, albeit stationary, a continuous sealing lip may cease to be effective. The pressure actuated, free moving double lip seal (Figure 2f) can prove advantageous. Being normally non contacting, there is no wear on the lip whilst the main seal is operative. The lip seal is very pressure sensitive and will seal when the interspace pressure rises to between 0.2 and 0.5 bar, and yet is capable of withstanding 100 bar when the shaft is stationary.

The most complex form of this arrangement is the tandem double mechanical seal using, in one instance, a low pressure flush and in the other no flush at all. Where leakage from the seal is in the vapour phase, such as LPG, the mechanical seal with a gas back-up seal (Figure 2g), is ideally suited. This requires no complex flush system or additional equipment and as such does not rely upon the integrity of an external system in order to maintain overall effectiveness. Whilst experience has shown the gas back-up seal to be very effective on gas, its performance on liquids is not so clearly indicated.

Thus, when the leakage remains in the liquid phase or if it is in any way toxic the tandem mechanical seal (Figure 2a), using a low pressure flush, offers a high degree of safety. Provided the flush flow or circulation is maintained it will not be necessary to implement the shut down procedure, as would be the case in all the safety devices previously mentioned.


G    Monitoring of seal performance

Unscheduled shut downs on any plant are very expensive and so the demand for reliability and consistent operation is increasing. Performance monitoring is being applied in particular to sealing devices. One way to achieve this with seals is to contain, collect and monitor the leakage from seals. Whether the method used for leakage detection is a simple float operated leakage collection chamber or a sensitive excess flow valve used with a pressure switch, the prerequisite is a secondary seal in

tandem to contain the leakage. Any one of the devices mentioned, from the bush to the mechanical seal, can be employed according to the method and level of control required.


POWER AND HEAT GENERATION

Having decided upon the double seal arrangement to use for any application, the overall power requirement and, if necessary, the total heat load must be taken into account. Three values must be obtained.

a)   Power absorbed by both seals.

b)   The viscous shear of the liquid within the seal chamber.

c)   Total heat soak into the seal chamber.

The values for a) and b) will be required for the sizing of the prime mover, bearing in mind that the starting torque may be 10 times that of the dynamic figure. All three values are added together to provide the total heat load to be dissipated by the sealant or flush system. The tandem seal offers some advantage in this respect since the main seal can be cooled by the product being handled.


Power absorbed

Most seal manufacturers provide curves for the power absorbed by seals. For general consideration they are adequate, since they are invariably derived from tests and include the viscous shear. When size and speed are either very low or very high it is better to resort to calculation in which case the viscous shear must be treated separately. The average coefficient of friction varies little for carbon opposing any other seat material, assuming it to be lubricated, and a dynamic coefficient of 0.03 is a reasonable value.


Viscous shear

Viscous shear is dependent upon several factors, such as the characteristics of the liquid, particularly viscosity, and the external lines of the seal. To these factors are added those of speed, size and clearance over the seal making calculation very complex. Taking two of the most common liquids, oil and water, the curves in Figure 3 can be used to assess the power loss through viscous shear in the general size range of seals operating at normal speeds.


Heat soak

When a pump handles a hot liquid it may be necessary to apply cooling in order to create a desirable environment for the seal. The flow of heat will be related to the temperature difference between that of the product and that which is allowable in the seal chamber. The limiting temperature may not result from the seal but from the sealant used. The sealant temperature should not be allowed to rise above its initial boiling point otherwise vapolrisation with a resultant high wear will occur across

the running faces of the atmospheric side seal.
It necessarily follows that surface heat transfer coefficients and the various materials
within the pump must be known for accurate heat
flow estimation.

SEALANT AND FLUSH SELECTION

There are several factors which need to be
examined when selecting the sealant or the flush.

    a)  Chemical compatibility with the product.

    b)  Initial boiling point.

    c)  Decomposition and oxidation rate.

    d)  Flash point.

    e)  Pour point.

    f)  Viscosity.

As examples, a water sealant with concentrated
sulphuric acid would generate heat, and oil with
liquid oxygen would give rise to a totally
unacceptable situation. Oils and other hydrocarbon sealants decompose very rapidly above
$150^{o}C$ which could lead to loss of flexibility on
the atmosphere side seal. Finally if the sealant
is allowed to rise to a temperature above that
of the initial boiling point problems due to
vaporisation across the outer seal could be
encountered. Sealant temperature thus becomes a
very important factor and Table 1 gives the
maximum recommended temperatures for some of the
more common liquids.

Table 1

Recommended maximum sealant temperatures

| Sealant/flush | Temperature $^{o}C$ |
|---|---|
| Water | 80 |
| Methanol | 50 |
| Ethylene Glycol | 150 |
| Water/Glycerine (50/50 mixture) | 80 |
| Mineral Oil | 150 |
| Heat Transfer Media | 300 |

SEALANT SYSTEMS

It has been stated that back-to-back double
seals require a clean liquid sealant which has
to be provided at a pressure and flow to suit
the particular application. There are in
principle two systems,

    a)  the totally enclosed pressure system,

    b)  the atmospheric pumped system.

The former is shown diagrammatically in
Figure 4 where in its simplest form flow is
induced by a thermosyphon action. The flow can
be more positively induced by incorporating a
pumping device, such as a pumping ring or scroll
integral with the seal, or by a separate pump.
As heat loads increase a cooling coil can be
inserted in the vessel. A hand pump can be used
to maintain pressure in the most basic system
but for reliable operation this must be replaced
by a constant gas pressure source acting through
some form of pressure regulator or by an automatic topping-up and pressurising pump system.

The atmospheric pumped system, again in its
simplest form, is shown in Figure 5 where a
circulating pump draws sealant from the reservoir and supplies a pressurised flow. Cooling
and pressure control are achieved in the return
line. Variations on this design can be made
according to the heat loads, whether the sealant
is an oil, water or any other aqueous solution.
Atmospheric pumped systems can be made to serve
a number of seal installations by supplying a
quantity sufficient for all seals and controlling
each seal unit separately.

INTEGRAL PUMPING DEVICES

Reference has been made to integral pumping
devices, of which there are several designs.
Two such devices are described, these being the
pumping ring and the pumping scroll. The pumping
ring is a disc in which radial holes are drilled
so as to form a simple impeller rotating with
the seal, and the second is a stationary helix
either machined in the housing or a separate
insert having a relatively close clearance over
the seal and using the seal rotation to induce
the flow. It is the latter which is shown in
Figure 7. The scroll offers the benefit of
adding little heat to the sealant through
turbulence.

Each can induce flows of up to 4 l/min at
3000 rev/min and 1.5 l/min at 1500 rev/min
assuming water to be the sealant. Quantities of
this order are sufficient to lubricate and cool
the majority of double seal applications.

SEALANT CONTAMINATION

Whichever sealant system is used it must be
recognised that there will be some contamination
of the sealant by the product being handled.
There will come a point at which the sealant
will become saturated with the product depending
upon absorption rates. Unless the sealant is
periodically checked, complete sealant satura-
tion may give rise to corrosion or, in those
instances where the contaminant vaporises,
problems with the outer seal. Some provision
must, therefore, be made to decontaminate the
sealant either by periodic changing or by
automatic venting. The atmospheric pumped
system automatically degasses a sealant on its
return to the reservoir where it can be vented,
if harmless, or drawn off to a safe point, if
in any way harmful so maintaining the quality of
the sealant.

The situation is not so simple in the
totally enclosed pressure system as the

contaminant may well stay in solution. In
addition, if a gas is used as the pressure
source, this may be dissolved in the sealant
under pressure. Unless it is intended to shut
the system down and depressurise for venting
some provision may have to be incorporated to
allow venting whilst still under pressure. One
way is to install a secondary vessel above the
main pressure vessel and connect the two together in such a way that they can be isolated, and
the secondary vessel used for both filling and
venting.

In selecting and designing the sealant
system to be used in conjunction with a double
seal one should first consider the integrity of
the seal and then build into the whole system
the control and support necessary to maintain
the integrity of the whole installation.


MAINTENANCE

Whenever double seals are mentioned the
problems associated with maintenance immediately
spring to mind. All too often the seal
installation is dictated by the pump design and/
or a desire to select a pump which is capable of
accepting every possible available sealing
device. Such an approach often leads to an
installation being more complicated than it need
be. In the field of agitated vessels a considerable advance was made with the introduction of
the concentric faced double seal. This has been
designed as an integral unit forming its own
housing. In this way accumulation of tolerances
and working length setting have both been taken
into account. It results in simplifying the
initial assembly to the agitator and greatly
assists the subsequent maintenance.

Cartridge style units have many advantages
in that the sealing components are not handled
more than is necessary and once assembled remain
protected. Pressure testing can be conducted
before final assembly into the pump or mixer
thus avoiding the tiresome need to dismantle and
rebuild as a result of inadequate initial
installation care.

Figure 7a illustrates the cartridge nature
of the concentric faced seal together with a
suggestion for both the tandem double (Figure 7b )
and the back-to-back double (Figure 7c ) where
the pump design ends at a flat back cover. The
seal, seal housing, disposition of the circulation connections, flow devices and all other
aspects necessary for a good seal environment
become the responsibility of the seal maker.
For the pump maker there may also be an advantage since he can ring the changes without
altering the basic design of the pump. This may
of course be speculation, but as double seals
are used more and more to meet obligations
under current legislation it may be worth
considering.


CONCLUSION

In conclusion it may be said that an
engineer faced with a rotary shaft sealing
application should first consider a single
mechanical seal. However, when it is clear that
a double seal is the only answer this paper serves
as a guide to the selection of the appropriate
arrangement. It is hoped that the summary chart,
Table 2, will go some way to assist in the
selection.


ACKNOWLEDGEMENT

The authors would like to thank the Directors of
Crane Packing Limited for permission to present
this paper.


REFERENCES

1.    LISS, D. B. 'Circulation Systems for
      Multiple Seal Arrangements', ASLE Annual
      Meeting - 1974 Cleveland, Ohio.


2.    LOCK, K. R. V. 'Safety Applied to Mechanical
      Seals', 5th Technical Conference of the
      B.P.M.A. - 1977 Bath.

### Table 2 Summary guide to double seal selection

| SEALING APPLICATION | DOUBLE SEAL ARRANGEMENT (Refer to Figures 1 and 2) | | | | | | | | | | | SEALANT / FLUSH | | | | |
|---|---|---|---|---|---|---|---|---|---|---|---|---|---|---|---|---|
| | 1a | 1b | 1c | 1d | 2a | 2b | 2c | 2d | 2e | 2f | 2g | Steam | Oil | Water | Solvent | Other |
| Solids in Suspension | ● | | | ● | | | | | | | | | | ● | | |
| Vapours | ● | ● | | | | | | | | | | | ● | ● | ● | |
| Liquified Gases :- Petroleum | | | | | | | | | | ● | ● | | | | | |
| Ammonia | ● | | | | ● | | | | | | | | ● | | | |
| Chlorine | | | | | | ● | | | | | | | | | | ● |
| Dry Running Pumps | ● | ● | | | | | | | | | | | ● | ● | | |
| High Viscosity | ● | | | | | | | | | | | | ● | ● | | |
| High Pressure X Velocity | | | | | ● | | | | | | | | ● | ● | | |
| Contamination :- Biological | | ● | | | | | | | | | | ● | | | | |
| Explosion Risk | ● | | | | | | | | | | | | ● | | | |
| High Toxicity | ● | ● | | | ● | | | | | | | | ● | ● | | |
| Low Toxicity | | | | | | | ● | | ● | | | | | ● | | |
| Product Change :- Decomposition | | | | | | | | | ● | | | ● | | | | |
| Salt Deposition | | | | | | | ● | | ● | | | | | ● | | |
| Polymerisation | ● | | | | | | | | | | | | | | ● | |
| Atmospheric Icing | | | | | ● | | | | ● | | | | ● | | ● | |
| Hydrocarbon Pipeline | | | | | | | | | | ● | | | | | | |
| General Hydrocarbon Duties | | | | | | | | ● | | | | | | | | |
| High Shaft Speeds | | | ● | | | | | | | | | | | | | |
| Agitated Vessels | | ● | | | | | | | | | | | | | | |
| Pumps | ● | | | ● | ● | ● | ● | ● | ● | ● | ● | | | | | |

FOR EACH APPLICATION THE MOST LIKELY SELECTIONS ARE INDICATED

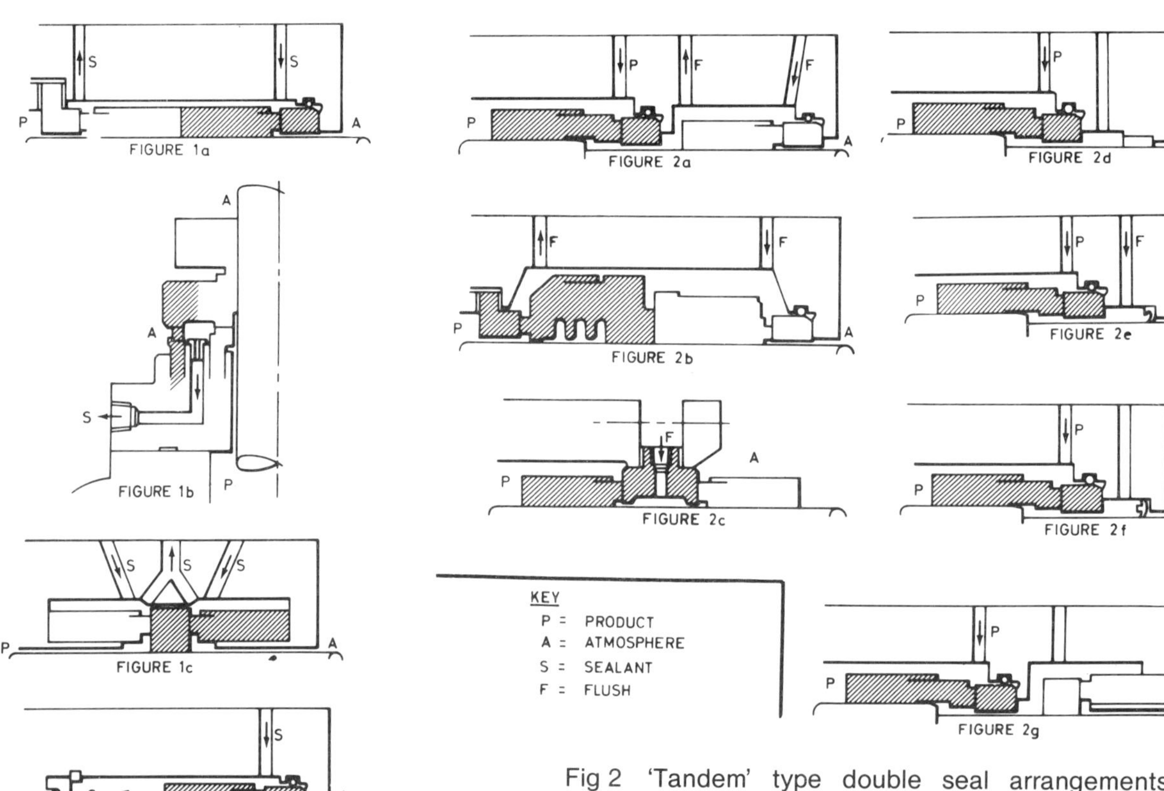


Fig 1 'Back-to-back' type double seal arrangements, requiring a supply of pressurised sealant
(a) Conventional back-to-back mechanical seal arrangement
(b) Concentric faced version of Fig 1a for agitator sealing
(c) Stationary mounted version of Fig 1a for high shaft speeds
(d) Simplified version of Fig 1a with the minor mechanical seal replaced by a flexible lip seal

Fig 2 'Tandem' type double seal arrangements, usually requiring a low pressure flush
(a) Conventional tandem mechanical seal arrangement
(b) Derivative of Fig 2a with an internally pressurised PTFE bellows seal as the major seal
(c) Face-to-face arrangement where the minor seal is internally pressurised by the flush
(d) Basic version of Fig 2a with a throttle bush as the minor seal
(e) Basic version of Fig 2a with a lip seal as the minor seal
(f) Basic version of Fig 2a with a pressure actuated double lip seal as the minor seal
(g) Basic version of Fig 2a with a gas back-up seal as the minor seal

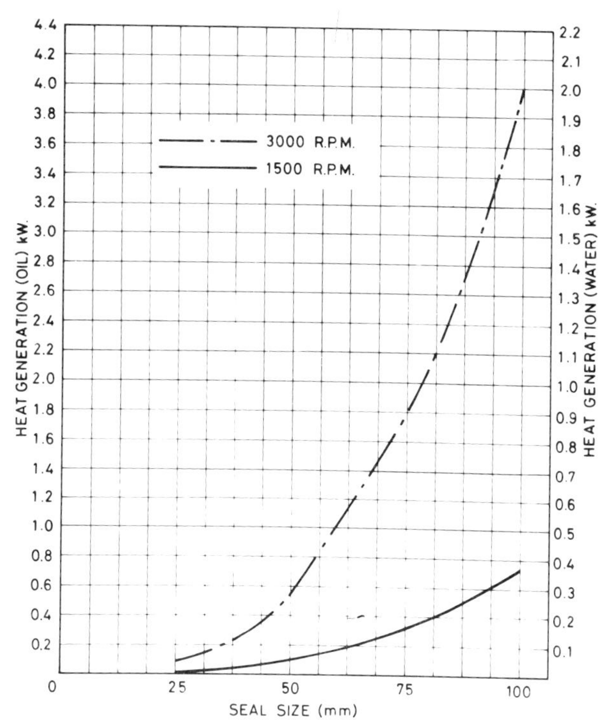



Fig 3   Viscous heat generation for oil and water

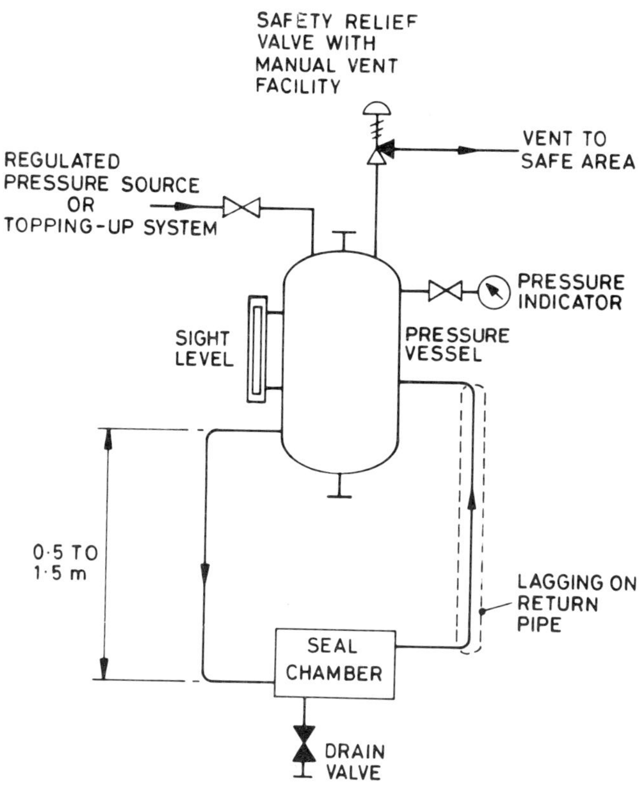



Fig 4   Totally enclosed, pressurised sealant system

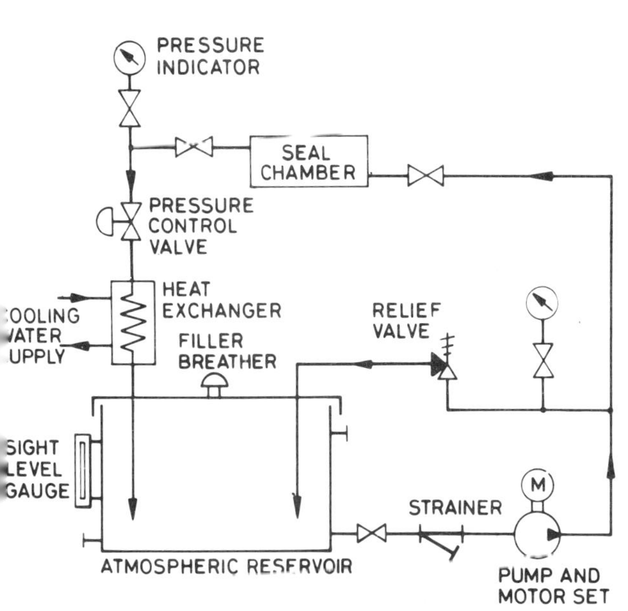



Fig 5   Atmospheric, pumped sealant system

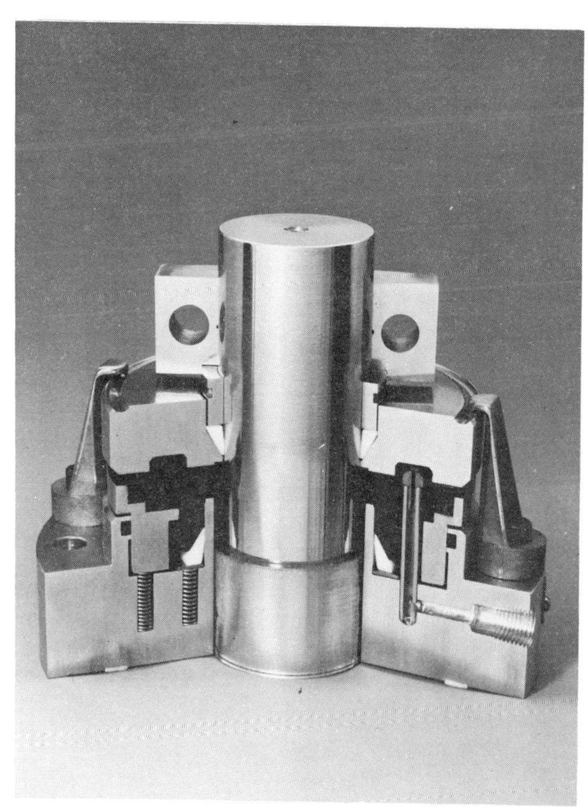

Fig 6   Crane Packing Type 151 concentric faced
double seal (sectioned)

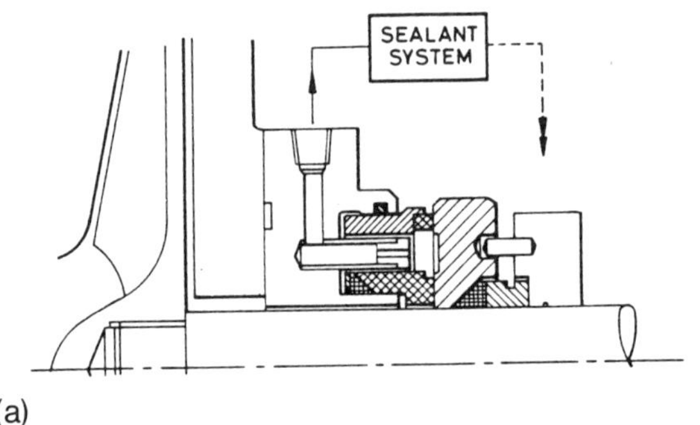


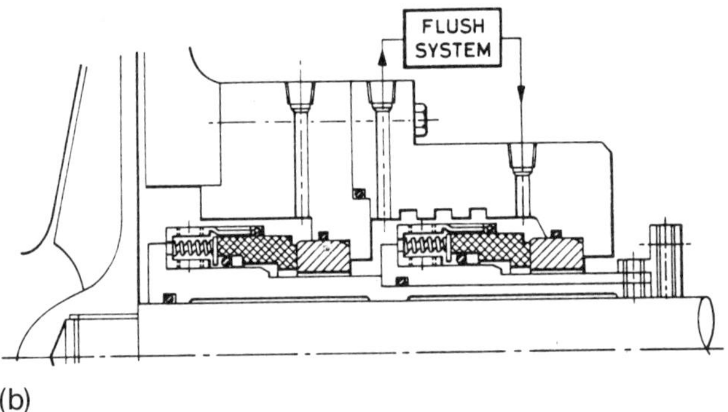


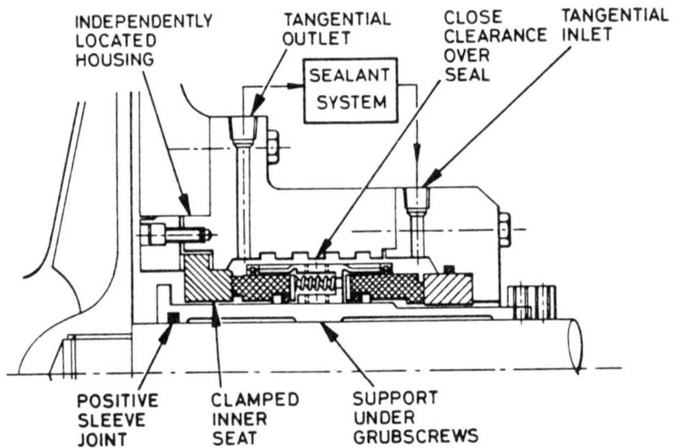


Fig 7 Cartridge style double seals
(a) Concentric faced arrangement
(b) Tandem arrangement
(c) Back-to-back arrangement

# Progress in mechanical drive steam turbines for process and energy cycles

I R STOPPS, BTech and C A COULTER, BSEE
General Electric Company, USA

SYNOPSIS Mechanical drive steam turbines for use in modern industrial plants must recognize variable speed operation, great diversity in rated speed and output, and widely differing steam conditions. The impact of these variables, and the considerations of reliability, efficiency, and producibility are discussed as they relate to the products of one of the world's largest turbine builders.

## DESIGN CHALLENGES

Mechanical drive steam turbines designed for industrial service driving centrifugal, axial, and reciprocating compressors frequently operate under conditions in which the design challenges are quite different from those encountered in producing units to drive generators for power production.

The primary application difference relates to the varying speed of operation under which industrial turbines must operate while producing either full load torque or a substantial portion of this torque. This is in contrast to the constant speed turbines required for generator drivers. This application difference leads to unique design challenges especially for the rotor, bearing, blade, and control requirements.

A variable speed turbine rotor must be designed to avoid resonance, or to be insensitive to vibratory excitation if resonance must be encountered. Such a rotor has a compact, husky appearance, with short shaft extensions beyond the journal bearings, and has sophisticated balancing provisions. (See Fig 1). Reliability dictates that a solid forged construction be used.

The journal bearings, which constitute a significant element in the total rotor system, must be chosen carefully to present acceptable stiffness and damping characteristics in order to assist in reducing the potential vibration amplitude. In view of the wide diversity of operating speeds which exists in the industrial turbine field, it is not surprising to find that tilting pad journal bearings, which offer exceptionally high damping characteristics, are required. (See Fig 2).

The design of the steam path elements also differs from the practice of constant speed turbine designs - again as a result of possible operation near resonance. The designer of a constant speed turbine may 'tune' the blading to avoid resonance, whereas

the designer of an industrial set may have to take more radical steps to avoid resonances, or may have to accept resonances and utilize a blading design which exhibits only relatively low service stresses. Thus, these elements in a variable speed turbine usually present a robust, massive appearance in comparison. (See Fig 3).

The control system of a constant speed set must be sensitive at only one operating speed, whereas the control system of an industrial turbine must exhibit acceptable characteristics of sensitivity, stability and speed of response over a speed range that may vary from 10 percent of the normal speed to 105 percent of normal speed. This challenge may be more severe when an extraction governing capability is involved. Not uncommonly, the variations of extraction rate are broad, and have no bearing at all on other turbine operating parameters.

In addition, high speed units frequently have low inertia rotors, so very fast response is mandatory for acceptable governing. Thus, the control system problems encountered in producing industrial turbines are frequently more demanding than those exhibited by other types of turbine equipment.

In addition the start-up sequence for mechanical drive turbines differs from the start-up routine used, for instance, in central station service. For a central station, the start-up sequence is established to best suit the turbine generator system. However, in the case of turbines which serve in complex process and energy cycles, the start-up routines are frequently dictated by process demands, with the result that load and speed changes may introduce severe transient operating conditions. This consideration may affect the design of variable speed units.

Mechanical drive steam turbines are used in a variety of process and energy applications. These include base load liquefied natural gas plants, ammonia plants, ethylene plants, coal gasification plants, steel mills and refinery applications to name a few. Each has its own unique requirement with respect to the design of the unit.

These different process and energy applications lead to an almost unique requirement for each and every turbine design except when there is plant duplication. Thus inlet pressures range up to 140 kg/cm$^2$, inlet temperatures up to 540$^o$C, and inlet flows up to 450 tonne/h. Turbine types vary from straight condensing, straight non-condensing or with extraction and/or admission at one or more points. Thus exhaust pressure can range from condensing to as high as 45 kg/cm$^2$. The power of these turbines typically varies from a few thousand kilowatts to nearly 100 000 kilowatts with speeds from 2500 rpm to over 14 000 rpm.

The effects of these variables dramatically affect the turbine construction. Obviously, high pressures tend to demand thicker walls, heavier flanges and heavier bolting. This tendency to size increase is mitigated by the greater density of higher pressure steam. High temperatures tend to require that more elaborate materials be chosen in the set design. This can affect producibility and cost, and carried to the extreme, can prevent a cost-effective turbine design from being realized.

A low ratio of inlet to exhaust pressure tends to reduce the number of stages of a given machine. Higher speeds, also tend to reduce the number of stages in a turbine and, in addition, to reduce the diameter of the rotating element.

When steam flow to the low pressure end of the turbine becomes excessive for a single row of low pressure blading, the designer resorts to multiple flow paths. Double flow and triple flow exhausts are common and even more extensive multiple flow designs have been considered. (See Fig 4).

Extraction requirements also have a profound effect on the mechanical design of the turbine. Obviously, an extraction flange is required, and most commonly in the industrial turbine field, extraction valve gear is also required to control the pressure existing at the extraction flange. (See Fig 5). The additional valve gear requires a separate operator, whose movements must be related to those of the valve gear at the inlet in order to obtain the required control of load, speed and extraction pressure.

As industrial plants become larger, to benefit from economies of scale, the ratings of the mechanical drive turbines required also increase. A dramatic trend toward increased turbine rating has required a rapid progression of technology and dictated the production of progressively larger units on which the plant operators may confidently rely.

The turbines shown in Table 1 are typical of the ever widening scope of activities which have been served by the turbine producers in the recent past. The progression was not always consistent: some 'landmark' turbines were ahead of their time in terms of one or more operating requirements.

Therefore, a policy has been adopted by the author's Company in which prototype designs are subjected to systematic laboratory and test examinations before being released for field service. The goal of this policy is the production of machinery of superior reliability and thermodynamic performance.

The wide variety of operating parameters cited above might well result in a disorderly array of equipment designed to serve in different installations were it not for the stabilizing effect of a series of standardizing specifications issued by several bodies concerned with various aspects of turbine design and operation. Perhaps the most notable of these is the American Petroleum Institute, whose specification 612 sets forth many demands relating to the design and operating requirement for turbines designed for use in process and energy cycles. The American Society of Mechanical Engineers has also been instrumental in issuing many specifications to guide turbine designers and users. These range from material and component acceptance tests to explicit specifications governing the performance testing of complete turbines. In Europe corresponding specifications have been issued by VDI, DIN and others.

The National Electrical Manufacturers Association has also issued specifications which set forth standardizing policies relating to permissible variation in steam conditions, governing accuracy, response and stability, permissible forces and moments to be tolerated by piping connections and other items relating to final installation and operating conditions.

Also a great many national bodies in different countries have developed refined specifications relating to the materials and processes involved in the manufacture of steam turbines. Perhaps the most well known material specifications are those issued by the American Society For Testing Materials. However, similar and often corresponding specifications have been developed by many other organisations.

In addition to these standard specifications, each process designer in describing the anticipated operating conditions for each driver, invariably develops an extensive series of stipulations to which the turbine equipment supplier must adhere when designing the apparatus which he will supply. The result of this field of specifications is that each turbine application will be served by machinery which has been 'custom' designed to well established and recognized design rules.

These diverse and sometimes competing specifications may appear to be unduly complex and unmanageable but they serve a very necessary purpose in maintaining a common appreciation of operating and maintenance requirements of industry and thus permit turbine builders to produce closely corresponding equipment.

RELIABILITY

Since the plants in which most industrial turbines are situated are extensive and incorporate vast quantities of elaborate process equipment, it is apparent that an enormous investment is required to construct such a facility. Furthermore the personnel required to staff the operation tend to become highly specialized experts with particular knowledge of the techniques essential to that plant's functioning.

Thus it is necessary to keep the product flowing properly in order to amortize the investment and to keep the experienced workforce.

Downtime is extremely expensive and can typically be upwards of 500000 dollars a day. Obviously a reputation of failure free operation is a major asset for the producer of any critical item of process equipment. Another similar consideration relates to the ability of the equipment manufacturer to produce a qualified staff of installation and service personnel when required.

The reliability of turbine drives has traditionally had the highest priority in industry and in this author's Company it has been the guiding light whenever design changes are contemplated.

In view of the extremely high cost of forced outages it might seem surprising that little or no equipment has standby spare units installed. The plant's first cost, physical size, and interconnection complications would make spare units untenable in most cases. Thus, since these turbines are at the heart of the process, reliability is of paramount importance.

Within this context however, improvements and innovations can and must be advanced. This does not create a dilemma but does force improved technology. Reliability can be clearly evaluated only on the basis of proven experience, and it is therefore important to focus on experience levels for components which contribute to high reliability. Any extension of demonstrated experience for a given component must be fully tested to maintain reliability standards. It is the policy of the author's Company to prototype test all new industrial turbine developments to demonstrate effectiveness for the intended operation and to reveal any characteristics which could affect operational reliability of the turbine equipment. In order to do this type of testing, vast investments are required in research and development laboratories. It is necessary to test materials over extended periods of time - for instance 20 years - to predict life expectancy. It is also necessary to do full scale testing of aerodynamic and thermodynamic models. The prototype tests may involve producing special equipment with which to conduct the test or designing the test element to be compatible with existing test equipment. For example, prototype testing of new rotating blades designed for high pressure stages involves an existing test turbine casing into which a specially instrumented wheel, equipped with the blades under test, is fitted. This unit drives a second turbine which is arranged to serve as a dynamic brake (See Fig 6).

Modifying the steam flow through the braking turbine changes the torque developed in the system and thus controls the load imposed on the blades under test. The instrumentation yields information relative to the centrifugal stresses, vibratory stresses, natural frequencies, effective damping of the blade, vibrational amplitude, thermal gradients and temperature distribution, effects of variation of nozzle constructional spacing, and dovetail load distribution.

Prototype tests of this nature are deemed to be most indicative of component capability, because in a few weeks of well organized test operation, the behaviour characteristics of the component can be intimately studied over a much broader range of operation than would be indicated by a few years of field service, since field service usually is related to a narrower scope of operation. Furthermore prototype testing reveals numerical values of the parameters, thus leading to a superior evaluation of component capability.

Prototype testing has a fundamental aim of proving the acceptability of the design prior to its production release. However a second benefit is also realised: optimization of the design that results from reviewing test behaviour. Thus the component which ultimately reaches the customer as part of a new turbine can be compared to a 'second generation' design, were prototype testing not practised.

Turbine reliability may be enhanced by the use of modern electronic control gear. Governing systems utilizing solid state elements can be designed with redundant back up circuits, self diagnostic features, and independent multiple power supplies. These features are incorporated for reliability, but other advantages develop which are related to the virtual elimination of mechanical links, levers and pin joints, which historically have proven to be liabilities in substained accurate and sensitive control systems.

As well as being more reliable, the sophisticated electronic control system exhibits a much longer speed range, faster response and greater stability than other governing systems. It can also be more easily adapted to a micro-processor based plant monitoring and control system.

Further assurance of reliability is made by in-process quality control testing of turbine components. This testing may tend to become so routine that it escapes recognition but it is an everyday necessity to all producers of technically demanding equipment. Planned tests such as the hydraulic pressure testing of turbine casings are handled in an almost unnoticed manner. But unplanned testing, such as can occur when a machinist uncovers a suspected casting imperfection, can become more involved. The quality control tests must be handled in an efficient and reliable manner to ensure that fully acceptable components pass through the production cycle and adhere to the established shop cycle.

Test work is sometimes integrated into the manufacturing cycle and is an in-process operation on a sub-assembly. For example, on electronic control boards an accelerated life test is an established routine. Such tests, while important in maintaining the reliability record of the turbine manufacturer, are frequently not obvious to a casual observer. Many other standard tests are run on each complete production turbine requiring specially designed test stands, each of which has provision for steam supply, steam exhaust, oil supply, instrumentation connections and other facilities to simulate operation in the final installation.

As mentioned earlier, when the unit is installed in the field, any planned or forced outage must be dealt with efficiently and effectively. Therefore not only must a turbine manufacturer have at his disposal a highly trained service organisation, but field diagnosis techniques must be at their fingertips. One example of this is the HIBAL programme developed by the authors' Company which permits easy balance of turbine rotors in the field.

During the initial test period in the manufacturing plant, rotor behaviour is measured as a result of applying a series of carefully planned unbalances. From this study, a series of 'influence co-efficients' are derived for the particular rotor system, and these co-efficients may be used at any subsequent time to define corrective balance weight additions required to modify the operation of the rotor whose dynamic performance may have been degraded by carry-over deposits or erosion.

All the data necessary to utilize HIBAL is obtained from the non-contacting probes at each end of the turbine, and from the 'key phasor' probe. The key phasor provides an index from which the phase angles of the shaft displacement peaks, as measured by each of the four vibration probes, are determined.

Each rotor has a series of readily accessible balance planes into which balance weights may be inserted. Special ports in the turbine permit modifying the balance weights without disturbing the turbine structure or requiring a prolonged shutdown. Balancing by material removal from the rotor body is not permitted. Thus it is always possible to restore the rotor to a condition corresponding to that of its initial configuration when, for example, rotor reblading may be required. This operation merely requires the simple removal of any balance weights that have been added subsequent to the original rotor production.

Another example is the author's Company's ability to use telemetry techniques to measure blade vibration on turbines operating in a process environment. In this case, strain gauge signals are transmitted from the rotor by a diminutive battery operated radio transmitter. The signals are received on an antenna mounted on the bore of the casing and recorded for subsequent analysis by appropriate equipment. This is just another example of the testing that is carried out to achieve the highest standard of reliability. (See Fig 7).

EFFICIENCY

Within the past decade the changing economic pattern has emphasised the value of machine efficiency for almost all types of operation. Nowhere is this more true than in relation to the operation of steam turbines in industrial plants where the cost of fuel, added to substantial increases in material and labour costs, has contributed markedly to the massive increases in plant product cost.

In order to mitigate to some extent the impact of these rising costs on the final product cost, leaders in the industrial world are striving to achieve increased efficiency in every facet of their operations. These efforts have several immediate effects on their desires regarding the steam turbine drivers: turbines of improved mechanical and thermodynamic efficiency are sought; improved thermal cycles involving higher steam temperatures and pressures are being investigated and progressively larger units leading to improved plant performance are becoming profitable.

It is the authors' Company philosophy to design turbines for maximum efficiency without sacrificing operating reliability. Thus although maximum efficiency has been sacrificed to achieve ultimate reliability in certain cases, there have been more instances of improved efficiency as a result of careful development studies. In fact current generation turbine designs have improved 10 points in efficiency over earlier configurations. It is worthwhile to explore some of the changes that have been made in efficiency features.

There is now a strong influence to peak speed turbine designs: that is to incorporate sufficient turbine stages so that each stage operates at a velocity ratio that will afford maximum stage efficiency. There is an extensive use of various leakage control devices to improve efficiency. Special root and tip spill strips are often fitted to the high pressure stages of modern turbines to inhibit steam bypassing the intended flow path. (See Fig.8). Shaft packing of more elaborate design is utilized in today's turbines, as compared to those built in past years.

Many of the turbines which serve industrial applications now operate with short energy ranges in which exhaust pressure is high or in which a large portion of the throttle steam is extracted at comparatively high pressure. In such cases, inlet losses are very costly to overall turbine performance. For this reason, oversized trip throttle valves and oversized inlet control valves and valve chests are employed to hold steam velocities to low values and thus to limit the flow induced pressure drops to low values.

The turbine blading itself has also been the subject of design refinement to improve efficiency. Both stationary and rotating blade shapes involved in the production of industrial steam turbines have been re-designed in the past decade. The new stator blade shapes as well as more effectively converting pressure energy to velocity energy, produce lesser nozzle edge wakes that are the primary source of rotating blade resonance excitation. The new rotating blades have section shapes designed by a revolutionary holographic transform method and present surfaces to the steam stream which more faithfully follow natural flow lines, with the result that energy wasting flow separation is minimized. (See Fig.9). The rotating blade shape philosophy is extrapolated in the cases of stages near the condensing end of the turbine to produce blades whose basic shape, chord, and inlet and exit edge angles continuously change with blade section radius in order to recognize more effectively the flow variations that develop at progressively greater radii of rotation.

Bearing losses, particularly on high speed sets, have been recognized as major consumers of shaft power.

The greater stability and load distributing advantages of tilting pad thrust and journal bearings are well known. However only in recent years has the true magnitude of thrust bearing losses, notably in high speed operation, been recognized. This item has been intimately studied in an elaborate test programme conducted by the author's Company. As a result of this test programme, an advanced design of tilting pad thrust bearing is now being produced which by virtue of higher permissible unit loads, results in lesser bearing losses.

Turbine casings too, have been designed with reliability and efficiency in mind. Double shell construction to minimize pressure stresses, freely expanding nozzle structures, and careful attention to rapid changes in wall thickness are directed towards trouble free, long life service. The higher internal pressures which may be tolerated by these structures, permit achieving better pressure ratios on the early stages of high pressure machines and thus contribute to improved performance levels.

Exhaust hood losses have also been subjected to intensive study. Recent designs of exhaust hoods have minimized the use of mechanical strutting in the high velocity flow path. Instead, flow guides incorporating a diffusing flow path are fitted to result in an efficient conversion of steam velocity leaving the last stage blade to a pressure head which is controlled by the turbine exhaust condition. These more efficient exhausts are also designed to afford more rigid support to the low pressure end journal bearing.

While the overall performance of a given turbine is indicated most ideally in the well conducted loaded performance test involving water brake load absorbers, calibrated torque measuring couplings, carefully monitored steam supplies, and precise auto recording instruments, the individual performance characteristics of many of the critical turbine elements can best be evaluated by precise laboratory tests.

The performance verification of many of the turbine elements described in the previous paragraphs was obtained through a series of carefully prepared tests that were run in the aerodynamic laboratory of the author's Company. This facility has a vast array of equipment especially constructed to study the aerodynamic performance of turbine elements. Flow tests utilizing either air or steam can be used to evaluate shape changes in valve chests, extraction nozzles, exhaust hoods or other major frame elements. (See Fig.10). Individual stage performance tests can study actual operating changes resulting from stage clearance variation, the incorporation of different blade contours, band arrangements, or any number of diverse stage design modifications. These improvements have not only been proved by component test programmes but also by full load operation under field service conditions.

<u>FLEXIBILITY</u>

The wide variations in applications, and operating parameters which have been cited previously, coupled with the requirement to produce efficient turbines for each application, demands that the manufacturer be capable of supplying machines with widely varying physical configurations. Yet each turbine must be designed and produced in an economic and timely manner. This complex objective can be achieved if there is a carefully planned flexibility in the design and manufacturing processes. Such flexibility may be attained through the use of 'group technology'.

Group technology is a method of simplifying the design and manufacturing processes by segregating corresponding turbine elements into building blocks or groups, and subsequently applying to each member of the group similar technological treatment. In the authors' Company, the building block concept is well advanced, and turbine designs are routinely processed using these structured components. Since over four million different turbine designs are possible, each unit can be in effect, a 'custom' designed machine utilizing standardized components.

The major results of such structuring is manufacturing economies normally associated with large scale production, but now applied to small lot production. This grouping of technology further benefits by a much greater vertical integration of the design and production process which is enormously enhanced by computer aided design and manufacturing practices.

Turbine hardware has been developed so that successive sizes of a given component can be defined in terms of application capability ranges, physical sizes, resonant frequency characteristics, or other parameters, and these data have been committed to computer memory banks. The result is that as well as analysing the thermodynamic requirements of a potential new proposition, the computer is able to select appropriate, fully designed components from its hardware list and identify the best turbine design for the application.

The building blocks considered are shown in Figure 11. All of these components have been designed with common interfaces so that interchangeability exists to a high degree.

The application of the computer in the engineering phase has been dramatic. The design process has been automated both for proposals and for detailed requisition engineering once a project commences. The design programme creates a turbine which is optimized in accordance with steam conditions, compressor horsepower and speed. The programme performs the required design calculation with greatly improved productivity - in the order of 3 to 5 times over previous manual calculations. In addition, the quality of design and the optimisation of hardware has improved and information is available to the customer more rapidly.

The philosophy of computer aided design work extends into the drafting room through the use of interactive computer drafting. (See Fig 12). Each of the product structured components is fully designed and stored in the memory banks of a computer that has been developed especially for such work. The components that have been defined in previous programmes can be called and appropriately positioned with respect to one another. The computer accumulates necessary dimensional information and may subsequently issue a fully detailed hard copy drawing of individual turbine components or of a total assembly. The productivity improvement over conventional drafting is in the order of 3 to 1 and the quality and accuracy of drawings is greatly enhanced.

Each of these computer aided design routines has built-in flexibility to permit intelligent manipulation or control of the design under study. For instance, it may be desirable to study the effect of changing stage diameters or the number of stages before an extraction point. Capability exists for the computer engineer to insert a special turbine design and to evaluate its effect on steam consumption, cost, resonant speed or any of a number of other parameters. Similarly in the drafting phase the design draftsman may describe a special item of hardware to be incorporated to marry two initially incompatible items, to control overall turbine length, or to accomplish some other desired objective.

The use of structured components has an equally impressive effect in the manufacturing process. This is particularly true in the case of the author's Company, which has a major commitment to numerically controlled machine tools. Essentially all large machines have numerical controls as a result of a 10 year investment programme and because the geometry developed by computer graphics is in digital form, a direct link exists with the numerical control programming effort. The improved accuracy and repeatability of this method of manufacture provides high quality in the end product and the standardization of turbine components allows the re-use of proven hardware. (See Fig 13 ).

The mechanisation of design and manufacturing operations has resulted in many other advantages. The productivity of the individual worker has markedly improved, and the quality and uniformity of the product has also benefited from automation. Production cycle times have been reduced as a result of both standardization of components and increased productivity. Production costs benefit directly from these items, but in addition the accuracy of cost projections and control have also been improved. As noted briefly above, there is a direct customer benefit that derives from the more rapid availability of firm customer drawings. The extensive use of highly structured components also permits more accurate assessment of manufacturing cycle time, so that machine tool



   C58/81

loading schedules can be predicted with far greater accuracy than was possible before the era of computer aided design (CAD) and computer aided manufacturing (CAM) techniques.

The CAD-CAM era has brought and will bring many significant changes to the industrial scene. In order to utilize these techniques, significant investments are necessary to install new equipment, and further investments are required to secure the services of technically trained leaders and to upgrade established personnel to utilize the new techniques. In addition it is recognized that some of the new equipment will be subject to rapid obsolescence as technological advance continues to create a new environment.

Further developments over the next few years are certain to bring forth or inspire changes that will make an already interesting field almost entrancing.

Fig 1 Typical compact, robust rotor for variable speed service showing balancing provisions and short shaft extensions

Fig 2 Tilting pad journal bearing

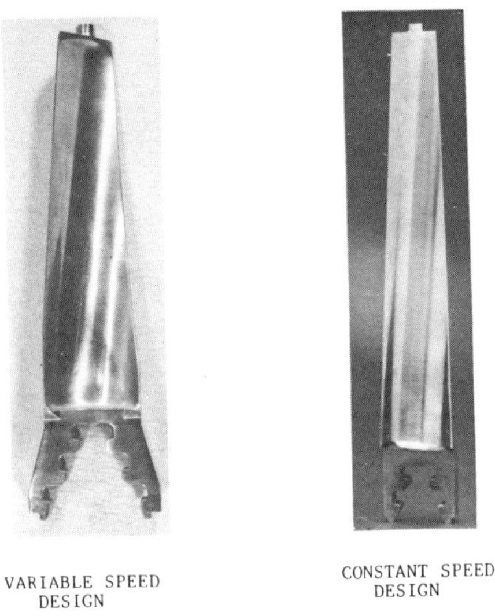



Fig 3    Comparison of variable and constant speed blades for similar power ratings

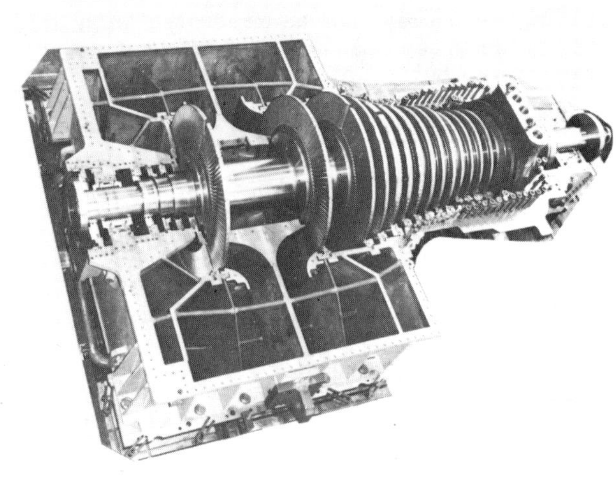

Fig 4    Mechanical drive steam turbine in assembly showing double-flow exhaust configuration

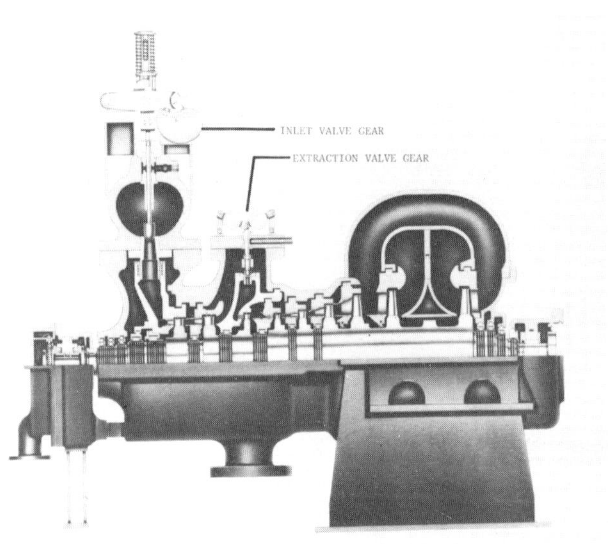



Fig 5    Cross-section of high speed turbine with automatically controlled extraction

Fig 6    Blade vibration test rotor

Fig 7    Rotor stage installed strain gauges and battery operated radio transmitters for recording blade operating parameters by telemetry

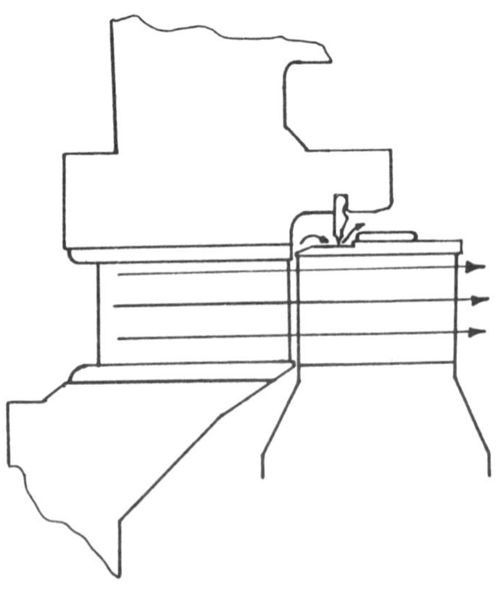

Fig 8    Diaphragm with installed tip spill strips to inhibit steam bypass

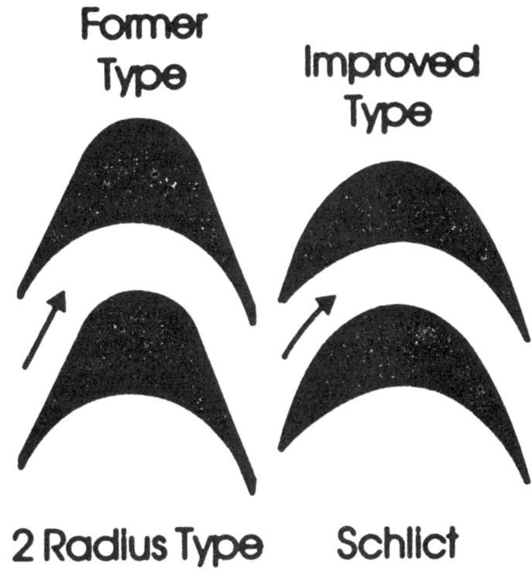



Fig 9 Comparison of high efficiency Schlict rotating blade contours and former blade type

Fig 10 Transonic test facility for stage performance determination

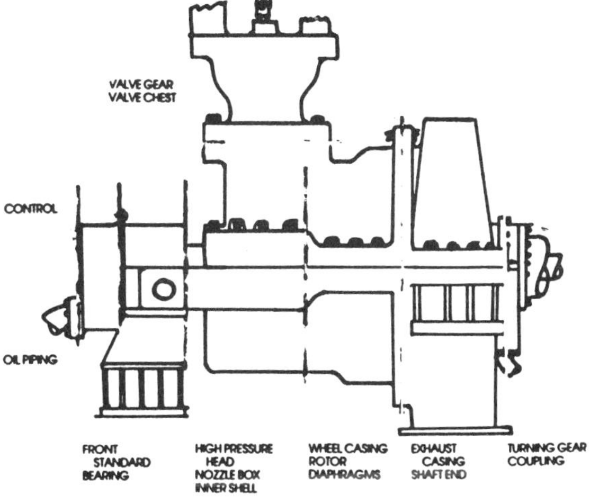



Fig 11 Component structuring showing major elements and interfaces

Fig 12 Interactive computer drafting

Fig 13 Numerically controlled machine: an essential element in computer aided manufacture

# C59/81

# Gas expander power recovery in fluid catalytic cracking

T A DZIEWULSKI, BSc, CEng, FIMechE
Foster Wheeler Energy Limited, Reading

SYNOPSIS  Typical Fluid Catalytic Cracking gas expander power recovery system is described.  Methods of cleansing the flue gas to remove catalytic fines and to improve catalyst economy are outlined. Expander turbine staging and configurations are reviewed.  Thermodynamic performance sensitivity to various factors is shown.  Matching an expander to a FCC system and computer evaluation of system performance are discussed.  Effects of erosion on blade life and efficiency are described, together with the effect of high gas temperature on blade rupture life.  Expander blading design features, overspeed protection and computer evaluation of train rotor system response to upset conditions are discussed.  Auxiliary drivers, train configurations and expander materials are described.  Future trends in FCC power recovery development are outlined.

## Fluid Catalytic Cracking Power Recovery System

Since 1963, when the first full scale Fluid Catalytic Cracking (FCC) power recovery installation was commissioned at Shell Canada's Oakville refinery, using a gas expander (Ref. 1), there are now in service or being installed throughout the world, over 30 FCC power recovery units. Until the introduction of the expander, the FCC flue gas pressure energy was simply dissipated across a control valve and an orifice chamber and wasted.  FCC expander saves this energy by converting it into mechanical power.  With the increase of fuel and power costs, FCC power recovery has become economically very attractive, offering return on investment in two to four years.

The FCC unit in a petroleum refinery converts vacuum gas oils into high octane petrol, petrochemical raw materials and other light products. A typical FCC power recovery system is shown in Fig. 1.  The catalyst, fluidised by aeration, is circulated continuously through the reactor. During the cracking reaction, the catalyst becomes coated with compounds of coke, sulphur and nitrogen.  To reactivate the catalyst, a portion of it is continuously diverted to the regenerator where the coating compounds are burnt off with the aid of air supplied by an air compressor. The compressor, which is part of the power recovery train, is, in this system, driven by a gas expander, i.e. a gas turbine.  The expander turbine recovers the pressure energy in the hot flue gas (the products of combustion in the regenerator), which would otherwise be wasted. Regenerator flue gas pressure available for expansion was about 1.75 bar abs. in earlier FCC units, but modern systems operate at higher pressures, up to about 3.8 bar abs.  The flue gas temperature for earlier FCC units was in the range of 600 $^{o}$C to 675 $^{o}$C.  At these temperature levels the flue gas contained some 8 to 11 percent CO, which was burnt in a CO boiler with supplementary firing to generate high pressure steam.  Modern FCC units however are usually designed for complete CO combustion to $CO_2$, which requires regenerator continuous operation at 730 $^{o}$C to 760 $^{o}$C.  On these units a waste heat boiler is used to extract remaining energy from the hot exhaust gases leaving the expander.  In doing work in the expander, the enthalpy of the gas drops, with a corresponding gas temperature drop of about 110 $^{o}$C to 200 $^{o}$C.  Since the effluent gas is usually required to meet the environmental air pollution standards, the flue gas from the waste heat boiler is ducted to an electrostatic precipitator, to reduce the catalyst content or via scrubbers if $SO_x$ removal is required, before discharging the gas to atmosphere.

## Flue Gas Temperature Control and Afterburns

During normal operation, flue gas temperature is controlled by injecting water or steam into the FCC regenerator, to maintain the gas temperature at the required level.  Afterburns commonly occur during start-up, shut-down and fluid-bed process upsets.  The regenerator suffers from sporadic, but fairly frequent coke afterburns, which occur initially in the regenerator, but can spread downstream to the third stage separator and the expander inlet.  FCC systems designed for incomplete CO combustion also suffer from sporadic CO afterburns, when as a result of the fluid-bed process upset, the flue gas which contains some 8 to 11 percent of CO at high temperature, receives some unburnt excess oxygen and ignites spontaneously, producing flue gas temperatures of up to 1000 $^{o}$C.  The CO afterburns are less frequent, but are more hazardous.  In a complete CO combustion system, the flue gas contains only minimal quantities of CO (in the order of four hundred ppm ), the danger of CO afterburns is, therefore, practically eliminated.

If unchecked, the afterburns have a most
detrimental effect on the life of the downstream
equipment. Preventive emergency measures are,
therefore, used to stop afterburns. Water is
injected instantaneously into the regenerator
and into the piping upstream of the expander to
quench the flame. Coke afterburns are thus
fairly easily and effectively controlled.
However, CO afterburns, in an incomplete CO
combustion system, may still occur downstream of
water injection nozzles, subjecting the expander
to excessive temperature. Other emergency
measures include shutting off the supply of flue
gas to the expander and by-passing it to the
waste heat boiler and secondly reducing the air
supply to the regenerator by venting off the
compressor discharge to atmosphere.

## Cleansing of Flue Gas

The 'dirty' flue gas from the regenerator
contains catalyst fines, which unless removed,
would erode the expander blades rendering them
unserviceable in less than 1000 operating hours.
To obtain a more reasonable blade life, say
40 000 hours, particle content must be reduced
to about 100 ppm. The bulk of the heavy fines
is centrifuged out by primary and secondary
cyclones located within the regenerator vessel.
Flue gas is then ducted to an external third
stage multi-cyclone separator which is the most
critical item in the power recovery system, for
the final clean-up operation. The Shell Oil
Company has pioneered and developed the design
of a third stage separator for over 20 years.
The majority of FCC power recovery installations
use Shell separators, although there are several
installations which utilise other makes. From
the third stage separator the cleansed gas is
ducted to the expander (Fig. 1). The catalyst
separated in the third stage separator is concentrated in a blowdown gas underflow of about
2 to 4 percent, via a critical-flow nozzle which
controls the flow rate. The underflow gas containing separated catalyst passes on to a scrubber
where the catalyst is separated by water injection and the water catalyst slurry piped to
disposal. The underflow gas from the scrubber
is passed to the waste heat boiler.

Current designs of third stage separators
aim at the following performance:

Particle size greater than 12 microns - 100%
                                                removal

Particle size greater than 10 microns - 99%
                                                removal

Particle size greater than 5 microns - 91%
                                                removal.

Fine particles of 5 microns or smaller follow
the gas stream without impinging on the expander
blades and thus have little erosive effect. It
must be mentioned that it is very difficult and
costly to sample the gas entrained catalyst in
order to measure the efficiency of the separator.
Instead, it is more practicable to assess the
effectiveness of the separator by measuring the
expander power loss due to blade erosion.

The latest separator designs are based on maximum continuous gas temperature in the separator
of about 740 °C, allowing the use of stainless
steel internals. Water or steam is injected into the regenerator to cool the gas upstream of
the separator. Although an increase in gas
temperature would yield big gains in power
recovered, it offers no process advantages and
would necessitate metallurgical upgrading of the
regenerator and the separator, which presently
is uneconomic.

## Catalyst Economy

In order to economise on catalyst usage, a
catalyst classification and recovery system may
be used (Ref. 3). It reduces catalyst make-up
rates by returning a selected portion of the
fines to the regenerator. The underflow from
the third stage separator, containing the removed catalyst fines, is ducted on to a fourth
stage (final stage) separator. This separator
classifies the fines so that the coarser particles of about 25 microns and larger are separated,
whilst the finer particles pass with underflow
to the scrubber for disposal. The coarse catalyst is periodically emptied from the separator
and transported to the regenerator for recycling,
so that about one third of fresh catalyst make-
up may thus be saved. In order to properly
optimise the catalyst recovery system, it is
necessary to analyse the entire FCC flow system
using a computer model. In addition, it is
advisable to carry out site tests of the catalyst
flow behaviour to establish the optimum catalyst
withdrawal and addition rates.

## Number of Expander Stages and Expander Configuration

The number of expander stages selected for a
given application, depends on the isentropic
enthalpy drop available, but in the final analysis it is a design decision affecting expander
isentropic efficiency and gas velocities. The
isentropic efficiency affects the amount of
energy recovered per stage, whilst the gas
velocity affects the rate of erosion of the
blading. Optimisation of the number of stages
required, must therefore consider both the
design and economic factors. The design factors
include the isentropic enthalpy drop available
from the flue gas, the enthalpy converted by the
stage into work, stage velocity ratio, rotational speed, rotor diameter, blade height, stage
reaction plus some minor design variables
(Ref. 4). The economic factors are chiefly the
expander overall efficiency, the blade life and
the total capital cost.

Single-stage FCC expanders have been used,
with some efficiency penalty, for isentropic
enthalpy drops as high as 260 kJ/kg of flue gas.
With the advent of two-stage and multi-stage
expanders, the isentropic enthalpy drop for a
single-stage expander for peak efficiency, should
today be set as low as 130 kJ/kg of flue gas.
For higher isentropic enthalpy drops, two or
more stages give better efficiency and lower gas
velocities. For axial-inlet overhung-rotor
configuration, the number of stages is presently

limited to two (Fig. 4 and 5), but in the future
it will most probably extend to three stages.
Three axial-flow stages with overhung-rotors
have been used on aero-engines and aero-engine
derivative power turbines. The limit for this
type of configuration may possibly be four
stages. Single-stage expanders have been used
almost exclusively for the last 15 years for the
FCC power recovery. Currently, two-stage and
multi-stage expanders are entering service.

Presently, there are two basic expander
configurations used: an overhung rotor with an
axial inlet, limited to a maximum of 2-stages
(Fig. 4 and 5) and a between-bearings rotor with
radial-inlet configuration which permits multi-
staging (Fig. 6). These two configurations
represent two basically different schools of
thought. An axial-inlet expander permits
straight inlet ducting with resultant good
distribution of catalyst particles across the
stator nozzle annulus. A low, high-tuned,
foundation can be used. The disadvantages
include a relatively high cascade gas velocity
and consequently higher erosion rates. A typical
single-stage FCC expander, with an expansion
ratio of 2.5 and inlet temperature of 725 $^{o}$C has
a stator nozzle outlet velocity of about 460 m/s
and a rotor blade exit velocity of about 475 m/s.

The between-bearings rotor configuration
expander uses two radial inlets and two radial
exhausts, to balance out the differential thermal
expansion of the inlet casing and to obtain a
reasonable distribution of catalyst fines across
the stator nozzle annulus. Such configuration
requires a high, low-tuned, foundation. Future
development may bring single inlet and single
outlet arrangements to remove the disadvantages
of the 'two-inlets' 'two-outlets' configuration.
The main advantages of a multi-stage type of
expander are its lower gas velocities and hence
lower blade erosion rate and higher efficiency.
For example, a four-stage expander operating at
conditions similar to those quoted above for a
single-stage expander, has gas velocities
approximately half those of a single-stage
machine. Lower velocities increase blade life,
but since more stages are used, the cost of
replacement blading is increased. However, a
multi-stage expander efficiency, as compared with
a single-stage machine is some 4 to 6 points
better.

Expander Thermodynamic Performance Sensitivity
to Gas Conditions

Deviations from the design (rated) conditions in expander mass flow, inlet and exhaust
pressure and inlet temperature, seriously affect
expander thermodynamic performance. Reduction
in mass flow, also causes a reduction of expansion ratio, which together result in a loss of
expander power, approximately in accordance with
a parabolic relationship between the mass flow
and power. Fig. 3 shows a typical performance
map for a single-stage FCC expander. Thus, with
gas inlet temperature constant at say 700 $^{o}$C and
assuming constant exhaust pressure, if the mass
flow is reduced to 88% of the rated (100%) flow,
the absolute inlet pressure will be reduced from
100% to 90%, and the expander power output will
be reduced by 20%. This illustrates how sensitive the expander is to deviations from the rated

conditions. The effect of changes in gas inlet
temperature on expander performance can also be
seen from the map.

Matching of Expander to the FCC System

The expander is designed and sized for given
rated conditions of gas mass flow, inlet and outlet pressures, inlet temperature and gas data.
Because of the serious consequences to the thermodynamic performance due to deviations from the
rated conditions, it is essential that the rated
conditions are calculated very accurately at the
design stage. In a correctly designed power
recovery system, the expander inlet valve,
(regenerator pressure control valve), should, at
the rated operating conditions, be fully open
with about 0.14 bar pressure drop across it.
The expander stator nozzle throat acts as a
fixed orifice in the system, controlling the
expander expansion ratio. Should the expander
stator nozzle throat area be mis-matched, say be
too large, then, for a given mass flow, the
expander inlet pressure will automatically drop.
This, in turn, will result in an increase in
pressure drop across the pressure control valve
(which maintains the regenerator pressure), thus
wasting gas pressure energy. The power recovered
by the expander will be reduced as a result of
the reduction of expansion ratio.

It is essential, therefore, that before the
expander design is finalised, the initial
process design data is checked by calculation
which includes the effect of the actual piping
layout on system pressure losses and to ensure
that expander stator nozzle and rotor blade
throat areas are sized correctly. It is also
highly desirable to carry out field tests during
commissioning, to verify the expander operating
conditions. To achieve optimum efficiency it
may be necessary to carry out a matching operation, similar to that frequently employed on
turbo-charged diesel engines, where the expander
is matched to the system by installing and testing stator nozzles with different throat areas.
For an FCC expander this may be a difficult
exercise to perform in practice. There is a
need, therefore, for expander manufacturers to
consider ways and means of making this facility
available in the future.

Computer Evaluation of FCC System Performance

In addition to establishing the rated thermodynamic performance of the complete expander
train, it is essential to establish the effects
on performance due to the variable system
components, using a computer programme to verify
the process and equipment performance. The
study should include the performance of the
regenerator, third stage separator, piping
system, air compressor, expander and the motor/
generator. Each of the components should have
its normal, part-load and abnormal performance
cases included in the programme. The interaction of all the variables produces variations
in expander performance and determines the power
balance of the complete train. Thus the regenerator performance determines the flue gas
conditions and catalyst flow. The air compressor
performance is affected by ambient temperature
and, for example, its inlet guide vanes. The
expander gas inlet conditions and mass flow are

affected by the upstream equipment, cooling
water injection, the amount of by-pass and the
afterburns. The system start-up procedure
should also be evaluated using the computer
programme. Programmes currently available are
based on steady-state conditions, but development work is going ahead aimed at producing
programmes which will simulate transient operating conditions.

Effect of Erosion on Blade Life and Efficiency

Blading erosion due to the catalyst fines,
which are present in moderate quantities during
normal operation and in large quantities during
process upsets, is inherent in FCC operation.
The mechanism of erosion, which is very complex,
has not been investigated extensively. Erosion
rate depends on the velocity and size of the
particles, i.e. on their kinetic energy. The
velocity of small particles is higher than that
of the larger particles and both of these deviate from the gas velocity vectors. Light
particles deviate less than the heavier particles, from the direction of the gas stream. In
general, the erosion rate may be taken as
proportional to the square of the gas velocity.
Thus if gas velocity is halved, the kinetic
energy of the particles and their erosion effect
will be a quarter of the original. The rate of
erosion also depends on the cascade gas flow
pattern, the angle of impingement, particle
abrasiveness and gas temperature (Ref. 5). The
blade and cascade design criteria which govern
these factors are difficult to establish, except
for the velocity effect which is clearly the
choice of the designer. It is essential to
avoid any local concentration of the particles,
by obtaining an even distribution across the
gas flow annulus. G.H.H. expander (Fig. 6)
incorporates a patented design feature (ELIN-
UNION patent) whereby a stellited step is
provided before each blade row, on both the
stator and the rotor, which diverts and distri-
butes dust particles across the flow annulus.
FCC expander operation experience shows certain
blade erosion trends and patterns. The convex
surface of rotor blades is not affected by the
erosion. The catalyst particles are centrifuged
towards the tip of the rotor blade on the
concave surface of the blade. The erosion
usually starts at the blade tip trailing edge
area, progressing downwards towards the blade
root section. The blade leading edge is also
attacked, mainly by the heavier particles, with
the erosion being greater near the blade base.

Reduction of expander efficiency due to
erosion of rotor blades shows a relationship
between expander power output and rotor blade
throat area (Ref. 6). For up to 20 percent
erosion (i.e. 20 percent increase in rotor
blade throat area), the reduction in power
output is approximately linear with the increase
in throat area. Above 20 percent erosion, the
power reduction becomes more rapid, approaching
a point where blade replacement becomes necessary. Another way of assessing the degree of
erosion is the loss of blade weight. It is
usually accepted that a 10 percent loss in blade
weight due to erosion signifies the end of its
service life. Some single-stage expander
installations have achieved five year blade
erosion life, but the average life may be as low

as two or three years. Multi-stage expanders
are likely to achieve better results. Blade
erosion life is critically affected by major
process upsets when vast quantities of catalyst
pass through the expander. The effect of the
erosion of blades with operating time on
expander power output loss must be allowed for,
when establishing the power balance design
criteria for the expander train.

Effect of Temperature on Blade Rupture Life

Expander blading is subjected to sporadic
temperature excursions above the rated temperature, during transient operating conditions of
the FCC system or during afterburns of CO or
coke, which are of random duration for each overtemperature incident. Exposure to overtemperature
shortens the blade rupture life considerably. Blade rupture strength is governed by the
Larson-Miller relationship (Ref. 7) where for a
given stress, the time to rupture is related to
temperature by the equation:

$$\sigma = \text{function of } T (\log t + c)$$

where $\sigma$ = rupture stress in bars

T = blade temperature in degrees
Kelvin

t = rupture time in hours

c = material constant (for high temperature alloys c = 20)

therefore

$$\sigma = \text{function of } T (\log t + 20)$$

Using the above relationship, the blade rupture
life may be calculated by means of a computer-
ised monitoring system which continuously
computes the time-temperature effect. On some
FCC expander installations, the overtemperature
effect on blade life has been more critical than
the erosion effect. Overtemperature, such as
that produced during afterburns, shortens the
blade rupture life dramatically. For example,
an afterburn at 815 °C lasting only 5 minutes,
reduces the blade rupture life by about 10 000
hours. Design rupture life of expander blading
is usually 100 000 hours.

Expander Blading Design Features

FCC expander turbines are axial-flow type.
Because of relatively low expansion ratios,
reaction type design is used as it offers the
advantages of higher efficiencies and lower gas
velocities, in comparison with an impulse type.
Free vortex flow design is usually used, resulting in a twisted rotor blade. Expander turbines
are designed for 50 percent reaction at blade
mid-height, with the degree of reaction increasing from blade root to tip section. The rotor
blades have a conventional gas turbine aerofoil
profile with a low aspect ratio, i.e. a relatively long chord which gives a sturdy blade. The
blades are unshrouded and are attached to the
disc by a fir-tree root. In order to minimise
the effects of erosion, the blade inlet and
trailing edges are thickened, which reduces the
aerofoil efficiency slightly. The concave surface is usually flame coated with tungsten
chromium carbide, to resist erosion. The stator
nozzle blades and outer shroud are also similarly coated.

To prevent rotor blade fatigue failure due to vibration, it is essential to verify blade suitability with regard to their natural vibration modes and excitation frequencies. The blade natural vibration frequencies including frequency scatter, should be established by actual shaker tests, with the blades mounted in the disc. Blade frequencies have a range or scatter due to the manufacturing tolerances and coupling effects. The scatter is particularly wide for short blades. Rotor blade steady state (centrifugal and gas bending) and alternating (vibratory) stresses should be evaluated using Goodman or Soderberg diagrams, to establish the safety factor. Operation in resonance with nozzle blade passing frequency or other sources of excitation should be avoided. Such operation is only acceptable with very high safety factors.

## Expander Overspeed Protection

All turbo - type prime movers require provision of an emergency quick-closing, tight, trip valve, which shuts off the supply of power fluid within such a time as to prevent the overspeed runaway of the rotor. FCC expanders present some problems in this respect. The great majority of existing FCC expander installations use an emergency shut-down system which is not fast enough to shut-off the inlet flue gas in the event of coupling or shaft total fracture between the expander and the compressor. Should this happen, speed acceleration of the expander to a runaway self-destructing speed would occur. Such emergency shut-down systems assume that the risk of expander coupling or shaft total fracture is a very remote possibility. Historically, this solution was partly forced by the lack of suitable,large (1 to 1.5 m diameter),quick-closing trip valves. Today, such valves with the required one to two seconds closure time, are available. However, a quick-closing tight shut-off provided by the trip valve, creates a number of problems in the FCC system. For example, the regenerator and third-stage separator design pressures may be exceeded and may, therefore, require a provision of pressure relief valves. The reliability of such valves in high temperature service and dirty gas environment is questionable and their use is normally avoided. The alternative of increasing the design pressure of the system, is very costly. The emergency trip valve in the expander inlet line, is therefore, designed to prevent full closure, in order to allow depressurisation of the system, without the use of relief valves. This type of shut-down system stops the expander train in ten to fifteen seconds. It takes advantage of the fact that the air compressor and the generator, when driven above synchronous speed, act as a brake, assisting in slowing down the train. When the expander trip valve closes on shut-down, the by-pass valve opens. The relatively slow shut-down operation is most suitable for FCC units where a slow shut-down is required to prevent catalyst from slumping or a reversal of catalyst flow in the standpipes. Catalyst flow reversal may result in an ingress of hydrocarbons into the regenerator and consequently excessively high temperature being generated. On expander trip, steam is immediately injected into the regenerator to keep the bed fluidised. On restarting, slumped catalyst in the regenerator is refluidised by the air.

The decision on whether to use a quick-closing tight trip valve, or a slow-closure type, for emergency shut-down, is not a simple one and presents a dilemma. Another possible solution, from the safety viewpoint, would be to design the expander casing, such that on fracture of the rotor disc or blading, the disintegrating parts would be contained.

## Computer Evaluation of Train Rotor-System Response to Upset Conditions

A computer study is necessary to establish the speed acceleration-time relationship of the train rotor-system due to various upsets, such as:
Total fracture of expander coupling or shaft, putting the generator on line, generator trip out, generator short circuit, steam turbine trip and compressor inlet guide vanes failure to open.

The speed acceleration-time curves are needed to determine the required closure time of the expander trip valve and opening time of the by-pass valve.

## Expander Power Output

Power recovered by an expander may be calculated from the equation:

Power output in kW =

$$(GRT_1/102)(k/k\text{-}1) \, \eta_o \left[ 1-(P_2/P_1)^{k-1/k} \right]$$

Where  $G$ = gas mass-flow in kg/s

$R$ = gas constant in mkg/kg°K

= 847.84/molecular weight

$T_1$ = gas inlet temperature in degrees Kelvin (°K)

$k$ = specific heat ratio of flue gas (Cp/Cv)

$\eta_o$ = overall efficiency (isentropic and mechanical)

$P_1$ = gas inlet pressure in bar abs.

$P_2$ = gas exhaust pressure in bar abs.

For a single stage expander, an isentropic efficiency of 80 percent is a good approximation. With a mechanical efficiency of 98%, this gives an overall efficiency of about 78%. Multi-stage expanders have isentropic efficiency some 4 to 6 points higher, with a mechanical efficiency of about 98%. Fig. 2 shows a typical single stage FCC expander specific power output in kW per $10^5$ kg/hr of flue gas mass-flow as a function of gas inlet temperatures and pressures. Presently, the commercially available FCC expander power range is about 3500 kW to 25000 kW for a mass-flow range of about 60 000 kg/hr to 550 000 kg/hr.

## Expander Auxiliary Drivers and Train Configurations

An FCC expander train requires an auxiliary driver, for start-up or for power make-up. The start-up driver accelerates the whole train to

near operating speed by overcoming the inertia
of the rotor system and the expander windage
losses. Initially, the expander does not
contribute any power, until the whole FCC system
comes on stream. A variety of train configurations is feasible (Fig. 7). Depending on
expander operating speed, speed changing gear
may be required, but is avoided if possible.
Configurations 'a' and 'b' use steam turbine
as start-up and power make-up driver. They have
been used on earlier FCC installations where the
expander power was insufficient to meet the air
compressor requirements. The arrangement 'c',
is used on small FCC units, where an induction
motor/generator with a speed increasing gear is
capable of providing the necessary torque for
start-up. However, such an installation, with
full-voltage start-up, requires an electric
system capable of accommodating high starting
currents. For modern high temperature high
pressure FCC units, where the expander recovers
more power than is needed by the air compressor,
configurations 'd', 'e', 'f', 'g' and 'h' are
used, i.e. a steam turbine plus a motor/gener-
ator. The steam turbine is normally used only
for start-up and bringing the train up to a
speed when the motor/generator can be started.
The motor/generator then assists in the start-up.
When the train achieves the full speed, the
steam turbine is put into an idling mode. It
may then require cooling steam. Some installations utilise the steam turbine to generate
additional electric power. Configurations 'i'
and 'j', with an expander driving a generator,
are used for retrofit installations which
require power recovery but do not want to dist-
urb the existing air compressor train, or when
the plant energy balance makes such an arrangement desirable. It has the advantage of making
the FCCU totally independent of the power
recovery system. However, since such an
arrangement has low rotor inertia and as there
is no compressor in the train to act as a brake,
generator load-shedding presents considerable
overspeed problems. In such an arrangement,
expander gas inlet must be provided with a quick-
closing emergency trip valve. In addition, a
careful study of the behaviour of the rotor-
system under load-shedding conditions is essential. As an alternative solution to this
problem, the use of eddy current brake coupled
to the expander/generator train has been
suggested (Ref. 11).

A proper optimisation of the expander train
configurations and auxiliary drivers, subject to
usual economic considerations, must evaluate a
variety of factors, many of which may be unique
to a given installation and site conditions. It
is also important to take into account the fact
that the auxiliary driver may be required to
operate for long periods at part-load and
consequently at low efficiency levels.

## Motor/Generators

Majority of FCC power recovery installations
use motor/generators of an induction type. The
motor/generator works either as a motor or a
generator, depending on whether the power
delivered by the expander is smaller or larger
than the power required by the compressor train.

An induction motor/generator, connected to a
sufficiently large electrical system which sets
frequency, will generate or absorb power as a
function of slip, i.e. the difference between
actual speed and synchronous speed. Excess
power from the expander or power make-up steam
turbine, is absorbed by the generator when it is
driven above synchronous speed. When driven
below synchronous speed, power is automatically
delivered to the train, when the motor/generator
operates in the motor mode. Under such operating conditions, the steam turbine may be used
without its own speed control. Synchronous type
motor/generators are being introduced currently
into FCC power recovery trains. Their application requires special attention as they need
excitation control.

## Expander Materials

Gas flow-path materials must be suitable
for elevated temperature service and have good
erosion resistance properties. With gas temperatures above 700 $^{o}$C, additional problems of
hot corrosion and sulphidation may be encountered. Typical FCC expander materials for
service above 700 $^{o}$C are shown in Table I.

## Future Trends

The development of computer programmes for
FCC system analysis will help to improve the
understanding of the steady state and transient
operating conditions, resulting in overall
improvements. The flue gas separator continuing
development will result in cleaner gas and more
stable operation, which will extend the expander
blade life. Power recovery is currently being
applied to FCC plants as small as 15 000 BPSD,
with corresponding gas expander sizes as low as
3500 kW. Future increases in energy costs are
likely to lower the minimum economic size for
power recovery applications even further. The
use of expanders, which requires cleansing of
flue gas, yields environmental and ecological
benefits. Therefore, the present and future
anti-pollution legislation will provide an
additional incentive for power recovery applications. The development of the expander itself
will produce further improvements in reliability,
blade life and efficiency. A variable stator
nozzle or a change-out stator nozzle assembly
may be found necessary for performance matching
purposes and for part-load operation. Two-stage
and three-stage overhung-rotor expanders will
compete on future FCC plants with multi-stage
between-bearings expanders, offering interesting
alternative design philosophies.

At present, there appears to be no process
advantage in raising the flue gas temperature
above the level required for full CO combustion.
Third stage separator and expander, suitable for
this temperature level are still yet to be
developed, subject to economic viability. This
will require better materials for the third stage
separator and more effective cooling methods of
expander disc and blading.

In the near future, the economic break-through in power transistors and microprocessors which are used to produce variable voltage and variable frequency, will make available economically attractive variable-speed electric motor/generators with special characteristics. This will change radically the FCC power recovery train composition. Thus, start-up steam turbine will not be needed. The electric motor/generator will be capable of operating at high speed, up to about 6000 RPM, thus eliminating the need for a speed-increasing gear and generate electric power at 50 Hz. It will require only low starting current, will provide variable speed for start-up and supply braking effect on load shedding.

With the continuing rise in energy costs, FCC power recovery, which is already well established, will continue to offer considerable economic advantages, a reliable service and will find more extensive application to new and retrofit installations.

REFERENCES


1. WILSON J.G. Energy recovery pays off at three Shell refineries. Oil and Gas Journal, April 1966, p. 77.

2. KRUEDING A.P. Cat cracker power recovery techniques. Chemical Engineering Progress, Vol. 71, No. 10, p. 56.

3. ARNOLD B.W. and GWYN J.E. Optimum cat fines recovery benefits FCC units. Oil and Gas Journal, May 28, 1979, p. 72.

4. RENARD D.J. Application of turbomachinery to FCC process. Proceedings of The Elliott Company, Bulletin H-40, Sept. 1979, p. 13.

5. KOSTERS C.H. Solid particle erosion. Proceedings of the Elliott Company, Bulletin H-40, Sept. 1979, p. 33.

6. STETTENBENZ L.M. Minimising erosion and afterburn in the power recovery gas expander. ASME paper, 70-Pet-6 1971.

7. LARSON F.R. and MILLER J. A time-temperature relationship for rupture and creep stresses. ASME trans. 1952, Vol. 74, p. 765.

8. BALFOORT J.P. Power recovery systems and hot gas expanders. Proceedings of the third turbomachinery symposium, Texas A. & M. University, 1974, p. 1.

9. BETKE W.C. Evaluation of turbomachinery for the FCC Process. Proceedings of the Elliott Company, Bulletin H-40, Sept. 1979, p. 7.

10. STETTENBENZ L.M. The power recovery gas expander in the fluid-bed catalytic-cracking cycle. Proceedings of the 47th annual meeting of the Western Gas Processors and Oil Refiners Association, Newport Beach, California, U.S.A., Oct. 5, 1972.

11. GEARY C.H. Power recovery and the eddy current brake. Proceedings of the Elliott Company, Bulletin H-40, Sept. 1979, p. 39.

12. DZIEWULSKI T.A. and BEWS J.H. Recover power from FCC units. Hydrocarbon Processing, Dec. 1978, p. 131.


TABLE 1

| | SINGLE AND TWO STAGE EXPANDERS | MULTI-STAGE EXPANDERS (G.H.H.) | |
| --- | --- | --- | --- |
| | | DIN Specification | Equivalent U.S. Specification |
| Stator Nozzles | Haynes Stellite 31 or Waspaloy | DIN 1.4981 X 8 Cr Ni Mo Nb 1616 | AISI 347 with 16% Ni & 1.8% Mo |
| Rotor Blades | Waspaloy | DIN 2.4969 Ni Cr 20 Co 18 Ti | Nimonic 90 |
| Rotor Disc | Waspaloy | DIN 1.4980 X 5 Ni Cr Ti 2615 | A 286 |
| Shaft | AISI 4340 | DIN 1.4922 X 20 Cr Mo V121 | AISI 420 with 0.2% C, 1% Mo and 0.3% V |
| Inner Inlet Casing | AISI 304 or 347 | DIN 1.4981 X 8 Cr Ni Mo Nb 1616 | AISI 347 with 16% Ni and 1.8% Mo |
| Outlet Casing | AISI 304 or 321 | DIN 1.4581 G X 7 Cr Ni Mo Nb 1810 | ACI-CF-8C and 2% Mo |

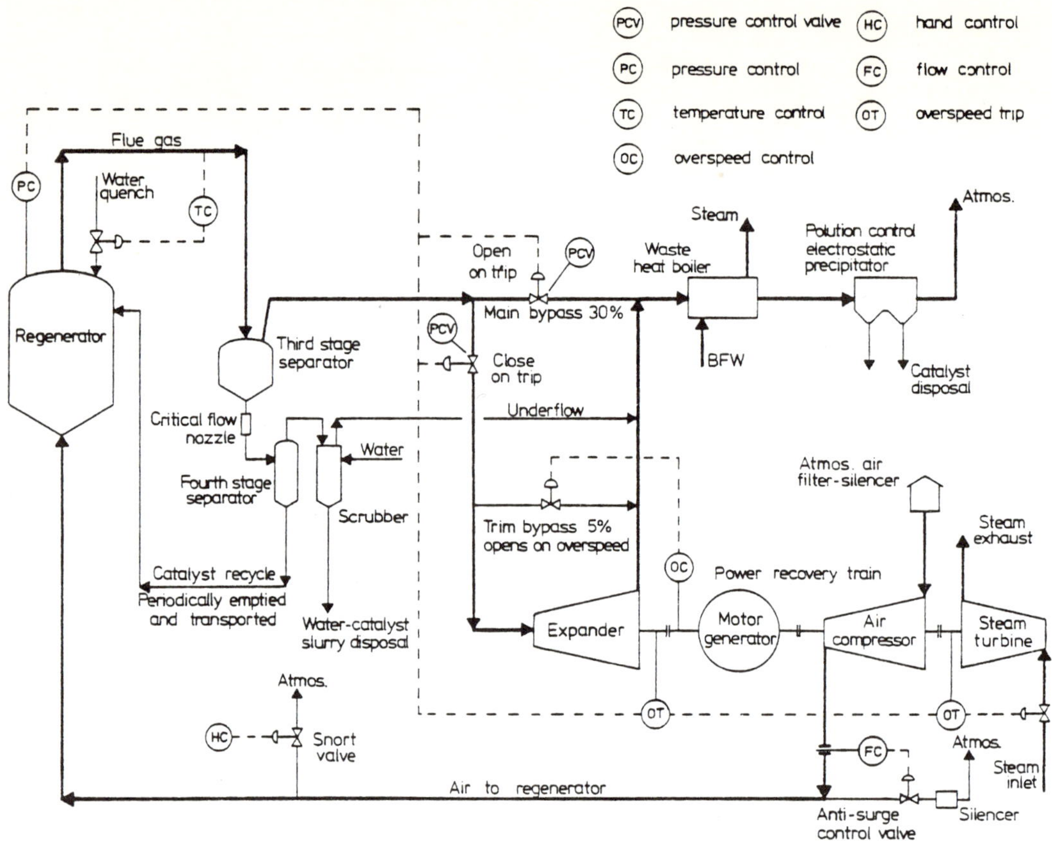


Fig 1   Typical FCC power recovery system

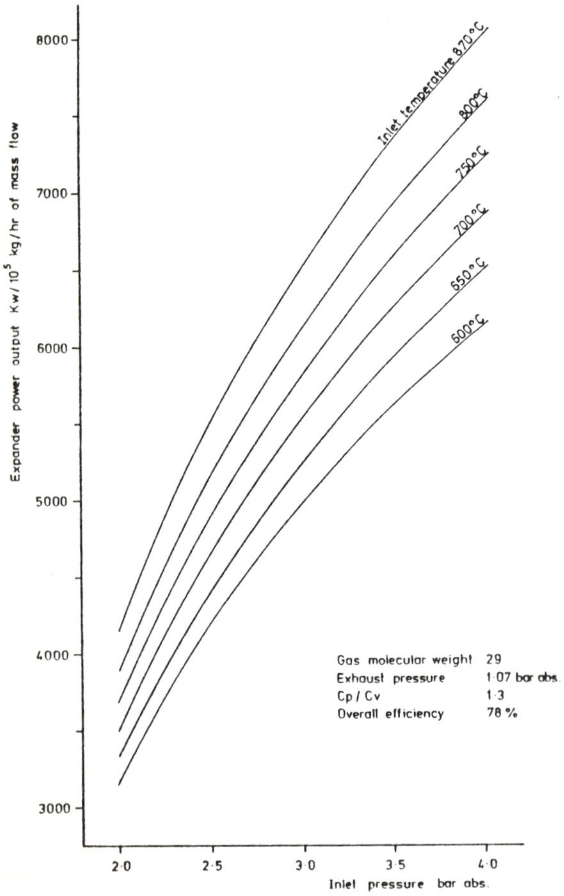

Fig 2   Expander specific power output versus inlet pressure and temperature

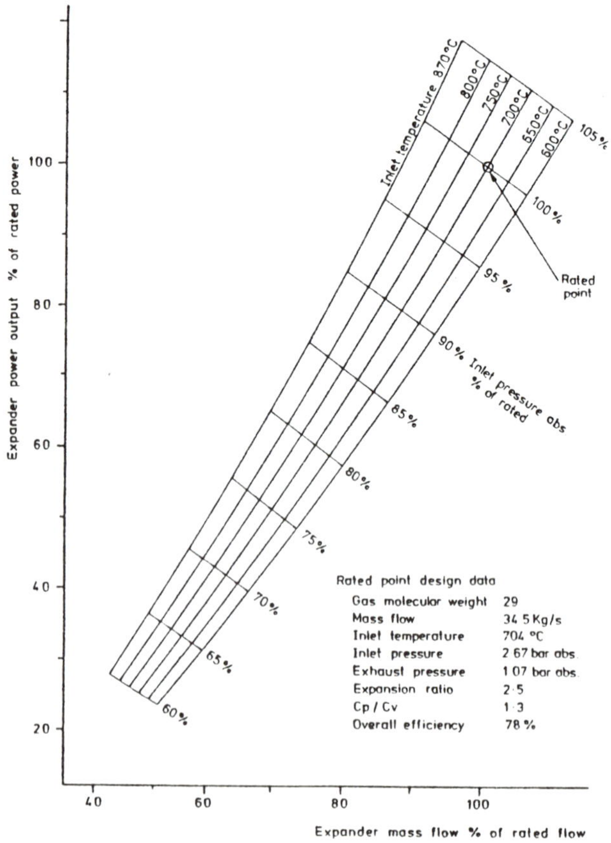

Fig 3   Relative expander power output versus relative mass flow, relative inlet pressure and inlet temperature

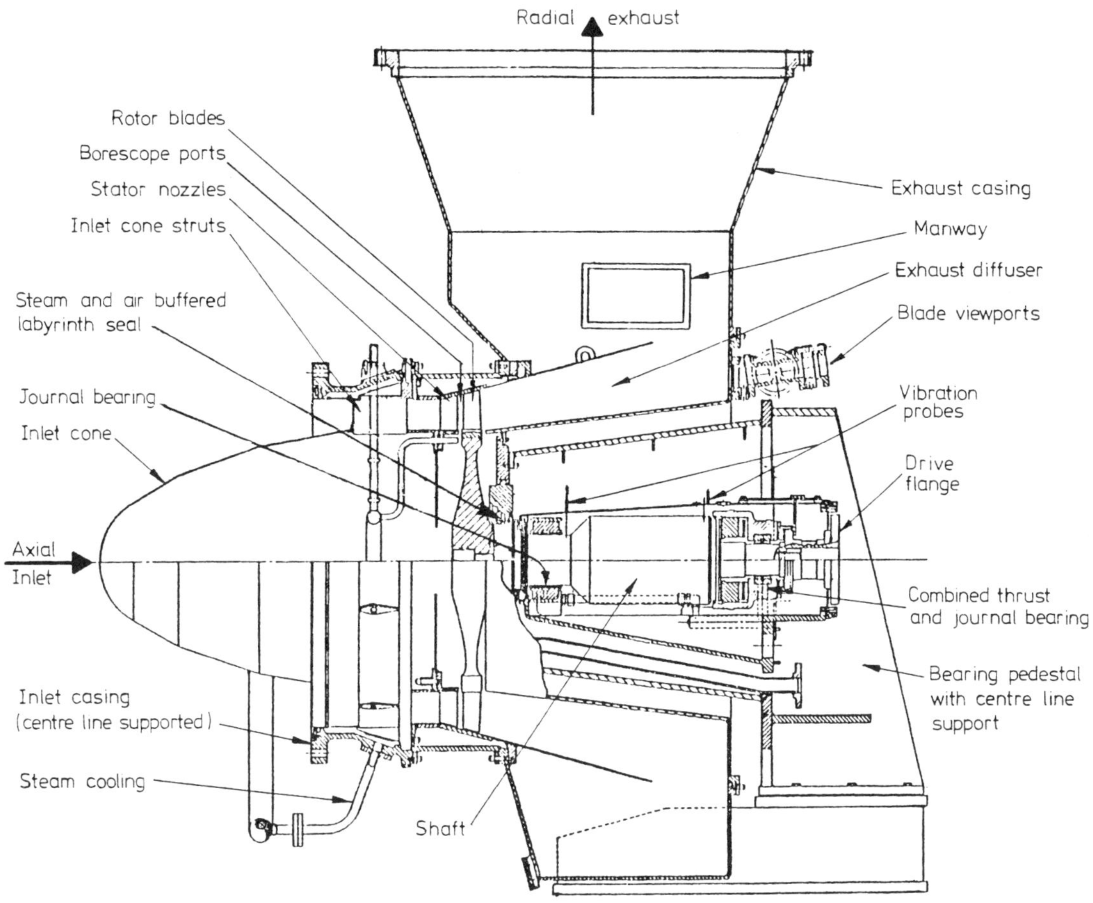



Fig 4   FCCU single-stage expander (Courtesy of Ingersoll-Rand)

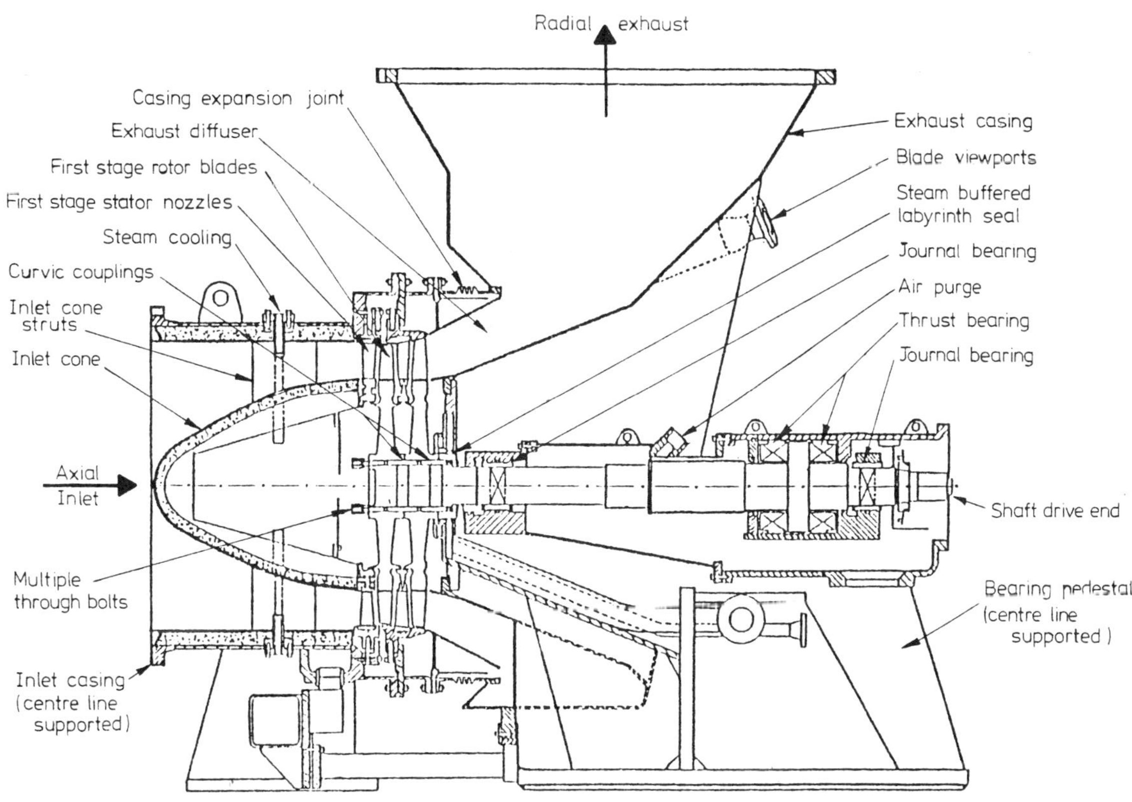



Fig 5   FCCU two-stage expander (Courtesy of Elliott Company)

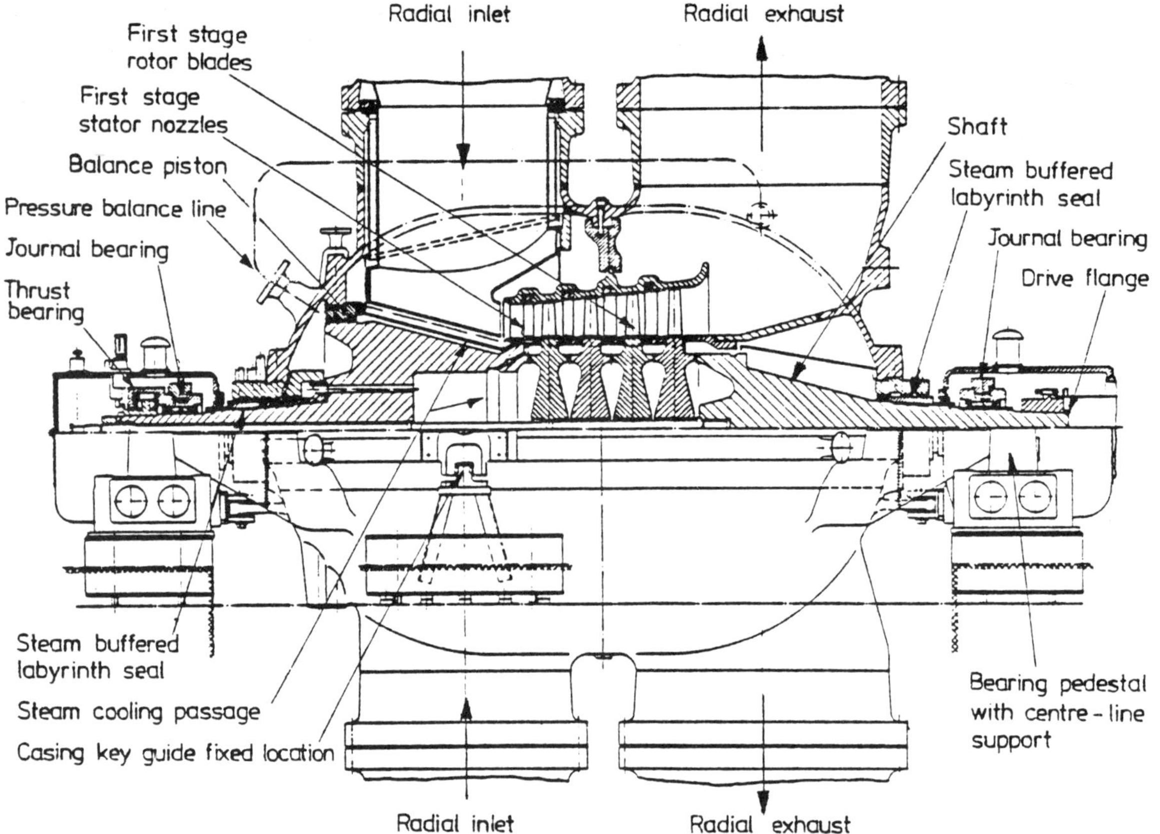


Fig 6  FCCU multi-stage expander (Courtesy of MAN- GHH Sterkrade)

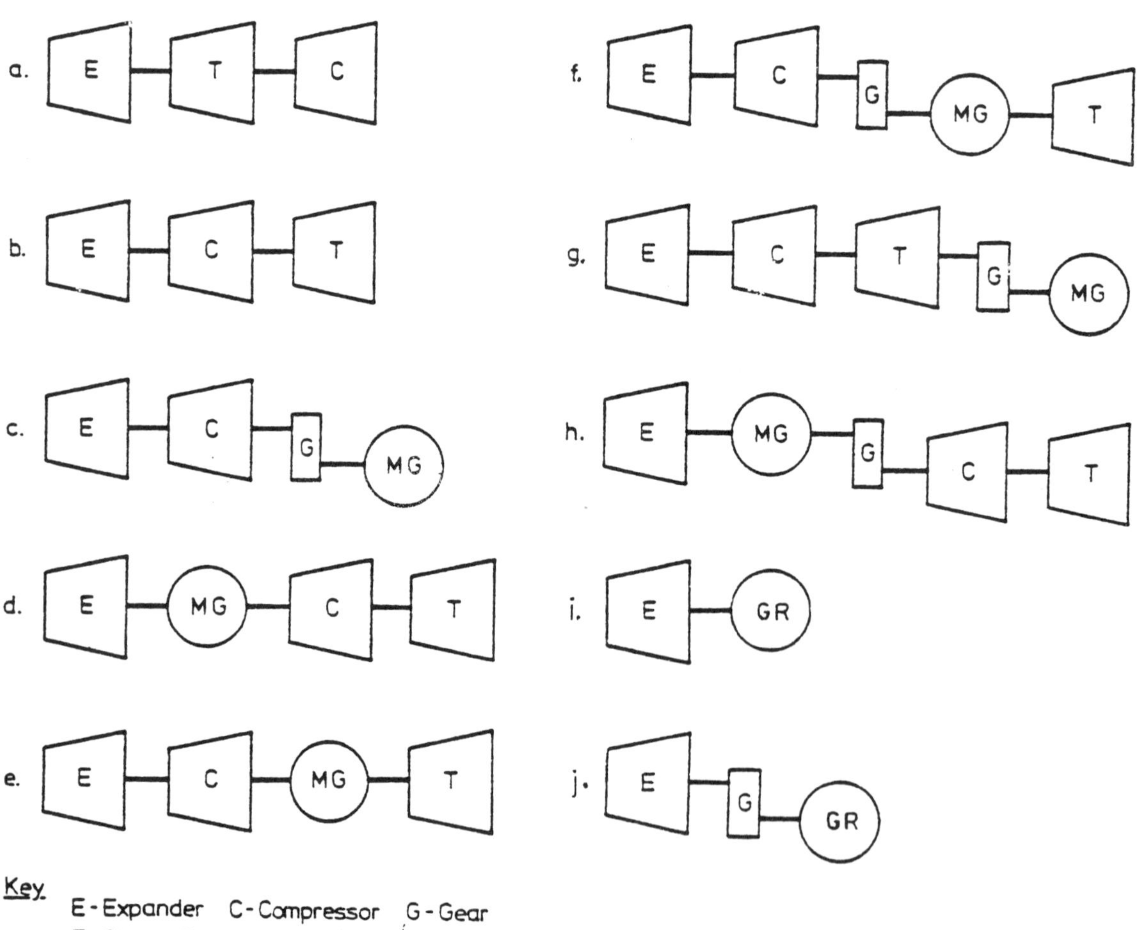


Fig 7  Typical FCCU train configurations

# Single stage turbines for mechanical drive in the oil, petrochemical and related industries

R J COLES, CEng, MIMechE and G H PLATT, BSc, CEng, MIMechE
APE-Allen Limited, Bedford

SYNOPSIS    After discussing when turbine drives should be considered and possible applications, further comments are made on specification, ancillary equipment, steam conditions, performance, materials, testing, reliability, maintenance and future developments.

## INTRODUCTION

The oil, petrochemical and process industries are very important users of steam turbines for driving pumps, fans, blowers, reciprocating compressors and centrifugal compressors. There is a wide variety of applications for both general purpose and special purpose machines. This paper is restricted to single stage turbine drivers which are usually general purpose machines.

For the majority of situations electric motors are the obvious choice to drive the smaller mechanical units so why use a steam turbine? As with other industries steam turbines should be considered as drivers when one or more of the following conditions exist:

(a)   when heat/steam is produced by the process

(b)   when low pressure steam is required by the process or for driving auxiliary equipment

(c)   where a variable speed driver is specified

(d)   where high speed direct drives are required by the driven machine

(e)   where no reliable electrical supply is available

(f)   in explosive or hazardous areas

(g)   in hot damp areas.

Thus some of the more usual processes using steam turbines are refining, ammonia, ethylene, methanol, sulphuric acid and acetic acid. Many other plants and processes may use steam turbines depending on factors listed above. (Ref. 1)

## APPLICATIONS

Within these plants where do we find applications for steam turbines? Process compressors, pumps, circulators and fans can be conveniently driven by steam turbines if the process itself provides or uses steam. This is a form of combined heat and power which gives a very efficient cycle as shown in Fig. 1. Appendix 1 shows a typical economic justification for a single stage turbine drive. (Ref. 2)

Many boiler auxiliaries such as feed pumps, fans and fuel pumps are driven by steam turbines as it is often essential to have an alternative source of power in case of electrical failure. This is easily arranged if steam supply is nearby. Utilities such as cooling water pumps and compressors can be driven by steam turbines as complete failure of these services can lead to total shutdown of the plants with consequent loss in production. These are just two examples of where single stage steam turbines can be used on standby duty. In this type of application it must be possible to start the turbine driven unit automatically and in a very short time. This can easily be achieved with the overhung wheel single stage turbine as shown in Fig. 2. With correct design and operation this type of machine will start from cold even with very high inlet conditions of over $10^7$ Pascals, 500°C and will be up to speed and on load normally in less than 30 seconds. For the more usual steam conditions met by general purpose turbines, adequate start-up time may be achieved by the between bearing single stage design.

The turbine can be used to drive more than one piece of equipment or as an alternative or additional driver through a clutch to a gas expander or electric motor. This is a useful way of reducing first costs on machinery or for conveniently improving efficiency of an existing motor driven unit. Where the steam turbine is used in conjunction with a gas turbine or gas expander in this way the set-up is similar to combined cycle installations except that the steam turbine may not take its steam from heat provided by the gas turbine exhaust. There is no reason why a combined cycle set-up should not be considered for small size units. For example, it would be possible to provide approximately 375 kW and to provide exhaust heat to process from a steam turbine taking waste heat from a 1 220 kW gas turbine. The same would be true of any waste heat source that can be used to utilise steam.

A further application that should be considered where one of the conditions outlined applies is for a vertical spindle drive. Single stage turbines have been used certainly up to 2500 kW to drive vertical spindle pumps with resultant reductions in space and foundation requirements. Since vertical spindle electric motor driven pumps are very common, this may also mean that the pump itself can remain the same design for a turbine drive.

A recent trend has been to look at the possibility of using very low pressure steam through a turbine to drive a conveniently situated pump or other piece of rotating equipment. Steam just above atmospheric pressure, say $1.5 \times 10^5$ Pascal's, can be used in a condensing turbine in order to recover more energy from the steam. The main problem with this is that, because the steam volumes are large at these low pressures, the turbines themselves become very large per horse power and consequently relatively expensive. However, it is a useful application where there is an excess of low pressure steam.

SPECIFICATION

What is the specification for suitable single stage turbines? Typical sectional arrangements of suitable turbines are shown in Figs. 2 and 3. The turbine would normally be supplied complete with control system having speed governor and overspeed trip mechanism, an oil system complete with pumps, filters, coolers and interconnecting pipework as necessary, any speed reduction gearing and instrumentation. Normally it would only require to be supplied with steam, cooling water and perhaps electrical power for auxiliary drives and instrumentation. Most users and contractors will call for API* 611 if it is a general purpose duty and API 612 if it is a special purpose duty. These specifications themselves set out in which category the turbine is. The use of API has been a problem for European turbine makers for many years. In fact, provided the turbine is of proven design and meets the intent of API there should be no difficulty so long as the engineer analysing machines realises this and applies engineering logic. Other API specifications are called up for associated equipment such as API 613 for gears, API 614 for oil system, API 615 for noise and API 670 for vibration monitoring instrumentation. NEMA** and ASME*** codes are also used along with user and conctractor's specifications. All these specifications are based on conservative grounds and on experience and are tending to become more and more onerous. If care is not taken, therefore, in applying these specifications, a grossly over engineered and expensive arrangement will result.

---

*     American Petroleum Institute

**    National Electrical Manufacturers'
      Association

***   American Society of Mechanical Engineers

Most single stage turbines fall in the general purpose category and are basically specified to API 611. Provided that the turbine meets the intent of this specification then the machine should be acceptable to both user and contractor. The principal requirement is for proven machinery. Other points worth mentioning are:

(a)   The turbine should have ease of access for maintenance purposes. This can be achieved by either a horizontally split or vertically split casing. In the case of the horizontally split machine, the steam joint has to be broken but then the whole machine, wheels, bearings and glands are available for inspection. With the vertically split case, it is only necessary to break the steam joint if the wheel is to be inspected as the glands and bearings can easily be worked on by removing the gland and bearing cover.

(b)   The turbine should have mounting arrangements such that expansion can be accommodated without affecting alignment. Centreline or near centreline support is therefore desirable or some other means of achieving the same result. With the overhung turbine this can be achieved by the use of a centring ring arrangement.

(c)   Provided that the terminal points are matched to the pipe standard, the manufacturing standard within the turbine itself may be to any recognised and acceptable specification.

(d)   Materials are referred to as various United States standards. Again providing the materials used are equivalent or better than these then they should be acceptable.

STEAM CONDITIONS

Steam conditions covered by API 611 specification are up to $4.13 \times 10^6$ Pascals, 400°C. Theoretically if the steam conditions go above these then API 612 specification should apply. In fact it is no problem for the single stage turbine made by some European makers (shown in Fig. 2) to go up to inlet pressure of $13.5 \times 10^6$ Pascals, and temperatures to 530°C. This is because the barrel type construction gives optimum conditions for low hoop stress. Provided the machine is designed from the start to cope with these conditions there should be no difficulty. Materials must be carefully selected to have high temperature characteristics and care must be taken to cope with increased thrust loads and gland sealing problems. On the other hand, no minimum conditions are specified in API. It has already been said that this is an area where increasing uses for steam turbines are being found and because of wetness problems it is probably more difficult to cope with these higher conditions. Here again, care will need to be taken to ensure that the wetness of the steam is not too great and again careful selection and high grade material is very important. The process requirements usually dictate the steam conditions to be used so it is best left to the turbine maker to provide a machine to meet the prevailing conditions.

PERFORMANCE

Provided that the turbine is able to run at its optimum speed, the wheel efficiency of a single stage turbine can be over 60%. These sort of efficiencies are achieved in practice and are not purely theoretical figures. Many single stage turbines have been used in the past to drive pumps, fans etc. directly at the speed of the driven unit. We see many machines, therefore, running at 3600 revolutions per minute (rpm) or 3000 rpm and even down to 1500/1800 rpm with resulting poor efficiencies of as low as 20%. The principal reason for this has been to simplify the overall mechanical arrangement and to eliminate intermediate gearing, but another influence has been that API 611 limits turbine speed to 6000 rpm. If a good efficiency is required, the turbine maker should be free to select the optimum turbine speed even if it is as high as 8 to 10000 rpm. Gearing can then be used to reduce the speed down to the driven unit speed. Clearly, if the driven machine speed itself is high (say above 6000 rpm) there is little point in putting a gear into the train as a reasonable efficiency will, in any case, be obtained. If a simple arrangement is required then the direct driven unit clearly has the advantage.

Multi-stage turbines are now being considered for smaller and smaller powers in an attempt to improve efficiency. For small powers however, the various losses in the turbine form a greater percentage of the rated output of the machine. It is possible that these losses may cancel any improvement in efficiency in a multi-stage machine with the result that a single stage turbine may prove to be more efficient.

EQUIPMENT

Turning now to the equipment provided with the turbine, we make the following observations:
(a) Governors

There are many different types of governors and governing systems now available ranging from a simple mechanical type to all hydraulic type. These systems can cope with any requirements that may be needed, whether it be for constant or variable speed application and constant pressure or differential pressure output from the driven unit. Fig. 4 shows typical characteristic curves obtained from the same pump with different governing systems.

For many applications a mechanical governor to NEMA A specification and operating directly on to a control valve or valves is sufficient. This type of system is prone to wear on linkages and joints and in this respect an hydraulic type system has an advantage. However, the simplicity and ease of adjustment of the mechanical arrangement may be attractive.

In addition, the hydraulic system has an advantage of being able to operate a number of control valves if better part load performance is desired. This type of governor can be made to meet the more demanding requirements of NEMA B, C or D, and speed ranges of 20 to 105% can be achieved. There are also systems which use hydraulics to provide the actuating power, but which use a standard package type governor to provide the control.

Recent specifications are beginning to call for electronic governing. Whilst this certainly increases the ease with which a turbine can be integrated within the overall control system of a given plant, it must be remembered that the electronic signal still, in the end, needs to be converted to a mechanical movement to achieve control of the turbine.

(b) Control valves/emergency valves/hand valves

For most applications a single control valve is sufficient provided the turbine is operating somewhere near its rated point. If, for whatever reason, the turbine is rated at considerably more than its normal operating output then 2, 3 or 4 control valves may be desirable to achieve improved steam consumptions at normal loads. This is shown in Fig. 5. In this context we are talking about automatic valves and not hand valves. Hand valves can be fitted, but should ideally only be used to enable full output to be obtained from the turbine in the event of unforeseen low steam conditions or similar situations. Unfortunately what may happen in practice is that the hand valves are opened and are left open, thus increasing steam consumption at a given load.

Much is often made of a separate trip or emergency valve to shut the turbine down quickly in the event of a trip condition arising. There are two schools of thought on this matter. One school suggests it is better to have a combined control and emergency valve as with this arrangement the valve is continually moving and therefore less likely to stick in an emergency. The other school feels that a separate valve specifically designed as an emergency valve is more reliable. With an hydraulic system the usual method of closing these valves is by diverting the control oil to drain, so in any case both control valves and emergency valves will close. On a mechanical system the emergency valve is latched open and when triggered off, is shut by spring force.

(c) Oil systems

These vary considerably depending on lubricating oil and control oil requirements. The simplest arrangement is oil ring lubricated bearings which on lower speeds and reasonable steam conditions require no external oil system. However, if an hydraulic control system and pressure lubricated bearings are required, a standard oil system should comprise:

Reservoir ($2\frac{1}{2}$ - 3 minutes retention time)
Main oil pump (shaft driven)
Auxiliary oil pump (motor or turbine driven)
Oil filter

Oil cooler
Relief valves
Interconnecting piping
Instrumentation

This type of system is perfectly adequate for most applications, but for larger and special purpose drives it may be felt necessary to go to the more sophisticated arrangements such as a separate larger oil console, duplicate oil cooler, filters and separately driven main oil pump. It would appear that rarely, if ever, is it necessary to go to the API 614 oil system for the type of drive under discussion here as in this case the oil system can become almost as costly as the turbine itself.

If desired the turbine oil system could usually feed the driven machine with lubricating oil without too much problem. This can be achieved by fitting either a larger or additional main oil pump.

Again within the hydrocarbon industry it is usual to call up the API codes and other specifications such as NEMA and ASME. This means that such things as coolers and filters are often far more expensive than they need be.

## (d) Gearing/coupling

Since maximum efficiency is not usually achieved at the same speed as that required by the driven machine, it is often desirable to drive through speed reduction gearing. Standard type single or double helical gearing is usually acceptable provided the appropriate service factor is taken into account. Better in some respects is the use of integral gearing. This gives an extremely compact arrangement, thus saving installation space and keeping foundation costs down to a minimum. It also eliminates any alignment difficulties between turbine and gear shafts. In addition only four bearings are required for the turbine and gear, thus reducing maintenance requirements. Gearing to API 613 can be supplied in a separate gearbox, but this specification cannot be applied to integral gearing.

Couplings should be chosen carefully to allow for expansion and any malalignment between turbine and the driven machine. It is usual for the driven machine maker to supply the coupling and to provide the turbine manufacturer with a free issue half coupling to enable this to be fitted and balanced with the turbine.

## (e) Instrumentation/alarms/trips

It is possible to add almost any amount of instrumentation to a turbine and this is usually selected to suit philosophy of the plant, the particular area of operation, as well as the machine itself. As a minimum the turbine should be fitted with pressure gauges for live and exhaust steam and for control and lubricating oil, thermometers for oil temperatures and an overspeed trip mechanism. It is now common for rather more instrumentation to be fitted, in particular monitoring equipment for vibration and axial positioning of the turbine rotor. Discretion should be used when deciding how many of these sensing points are required, bearing in mind we are talking in the main of general purpose machines. For example, it may be thought that one vibration probe per bearing is sufficient rather than the two specified in API 670. Additional pressure and temperature sensing points can be added almost as required. Additional alarms and trips can also be fitted to suit customer preferences provided they are routed through a solenoid trip arrangement of some kind. Usually trip/alarm facilities are for low lubricating oil pressure, high lubricating oil temperature, high bearing temperature, high exhaust steam pressure and excessive inlet pressure and temperature or excessive vibration. All instrumentation should be selected to suit the location of the turbine in the plant with particular reference to the appropriate hazard zone.

## MATERIALS

Turbine materials should be carefully selected to meet the specified steam conditions. Typical materials would be:

Turbine casing: For live steam temperatures up to 450°C - cast steel. For high live steam temperatures - alloy steel having special high-temperature characteristics.

Turbine wheel disc: Forged steel having high-temperature characteristics.

Turbine shaft: Heat treated carbon steel; with integrally geared turbine having high gear ratios, the shaft is manufactured from chrome alloy steel, the gear pinion teeth being cut on an enlarged section of the shaft.

Turbine blading: Moving blades - 12% chromium alloy steel.

Valve spindles and cones: 13% chromium alloy steel.

Exhaust cover: Fabricated from carbon steel.

Gear casing (integrally geared machines only): Fine-grained cast iron.

Gear wheels (integrally geared machines only): Chrome alloy steel.

In addition to the effect of steam pressure and temperature note should be taken of other factors such as where operators are likely to be standing on the machine for maintenance purposes. Forces and moments from external piping are also a factor to be considered, but this can usually be resolved by the pipework designers making suitable provision for expansion to eliminate as far as possible all forces and moments on the turbine branches. Specific process demands must also be met such as the removal of copper bearing alloys if the turbine is to be used in an ammonia plant.

## TESTING AND COMMISSIONING

With regard to shop testing, obviously a full load, full speed test is the ideal. This is not often possible, however, as it is normal for the turbine to be delivered to the driven machine maker's works for lining up and despatch or for the turbine to be delivered direct to site. A no-load full speed test is therefore usually accepted to prove that the turbine is mechanically sound and to set the governor and tripping mechanism. If it is felt essential that a full steam test is carried out before despatch to site, there is usually a way around the problem although the solution may be costly. It may be possible for the turbine maker to test the turbine itself on a suitable brake in their test bay. This does not prove the complete unit however, and it would be better to test the combined turbine and driven machine. The practicability of this depends on the avail-ability of steam at the driven machine maker's works, or on some suitable means of loading the driven machine at the turbine maker's works. If, as is often the case, neither of these alternatives are available, then some suitable outside testing facility must be found, such as a government test unit, another manu-facturer's works, or a nearby power station.

Since in most cases, the turbine is only no-load tested in the works, any full tests will usually be carried out on site in the commissioning phase. Here the main difficulties are problems of obtaining steady conditions of steam pressure and temperature and load as well as the difficulty of measurement and the accuracy of installed instruments.

It is advisable for commissioning to be carried out if possible by the turbine maker's personnel. This can be done on a daily rate basis which ensures a quick and efficient job as a client is anxious to get the commissioning engineer off site as soon as possible.


CO-ORDINATION

Throughout the period from order to comp-lete commissioning it is essential that close co-operation is maintained between end user, engineering contractor, driven machine maker and the turbine maker. The practice of holding a vendors co-ordination meeting at the order stage with all parties present is very useful, as it eliminates immediately most difficulties and misunderstandings.


RELIABILITY

A steam turbine, as with most rotating equipment, is essentially a reliable piece of equipment. Major breakdowns are relatively rare, but if they do occur they are usually due to such things as thrust bearing failure, shed-ding of rotor blades or overspeed. Most of these difficulties are caused by some unfore-seen load or steam condition arising, or from very poor maintenance. These failures would result in the turbine being severely damaged and would necessitate a replacement rotating element or turbine wheel. However, since this is a rare occurrence, it is not usually worth-while to hold a spare wheel or rotating element in store unless the turbine is a critical piece of equipment, or if there are a number of the same type turbines installed on site.

More usual problems are associated with governor hunting or maladjustment, sticking control valves, excessive gland leakage, oil leaks, malalignment between turbine and driven machine, or blocked oil filters. In other words, the usual maintenance problems. As stated before, the turbine therefore should be arranged for ease of maintenance and clear and concise working instructions should be provided by the turbine maker. Spares need to be ordered with the turbine, but in most cases consumable spares would be adequate, i.e. bearings, gland packing, governor springs etc.

With reasonable maintenance and steam conditions a turbine should be able to operate continuously for up to four years without shutdown.


FUTURE DEVELOPMENTS

It is established that the steam turbine is already a reliable and efficient machine. What of the future? It is doubtful whether much improvement can be made in the efficiencies of single stage turbines. The trend may, therefore, be for smaller multi-stage machines. Mainten-ance standards become increasingly difficult to achieve, so that simpler machines may become more desirable. Standard turbines are already relatively simple, but can become grossly over-complicated by the application of specifi-cations. Computerisation and automatic control are already more common. This will no doubt lead towards electronic governing although it will not necessarily mean finer control on the turbine itself. Turbines may have to be used for wider applications than previously, in part-icular in areas where excessive low pressure steam is available or where waste heat is avail-able to raise additional steam. Further consid-eration may well be given to the use of single stage turbines in conjunction with gas turbines or gas expander turbines, either as part of a combined cycle arrangement or as a helper turbine.


REFERENCES

1.    COLES R.J. and GOLDSMITH T.J.    Steam turbines for the hydrocarbon processing industry. APE Engineering, 1979, No. 29, p. 17-21.

2.    WATKINSON D.P.  The steam turbine in CHP* schemes.  APE Engineering, 1980, No. 32, p. 12-14.

3.    APE-Allen Limited   Profit by energy conservation.  APE Engineering, 1980, No. 34, p 15-18.

---

*Combined heat and power.

Typical economic justification for single stage
turbine drive

1.   A typical scheme has gas fired boilers at
33 bar, dry saturated.  Previously steam was
reduced through pressure reducing valves to the
process pressure of 11 bar.  Proposed installation of a single stage turbine showed the
following estimated savings:

Conditions:

Set rating        930 kW
Steam inlet       33 bar, dry saturated
Steam exhaust     11 bar (4.8 per cent  wet at
                             full load)

2.   For an average process steam demand at
11 bar, dry saturated of 27 720 kg/hr the
turbine operates with an exhaust wetness of
3.6 per cent. requiring 28 750 kg/h dry saturated steam at 33 bar at the turbine inlet, to
give an output of 675 kW.

Enthalpy change through turbine = 93.6 kJ/kg
Boiler efficiency         = 80 per cent
Cost of gas               = 23p/therm (0.218p/MJ)
Cost of electricity       = 2.6p/kWh

Annual cost of gas to generate output 675kW:

$$\frac{\text{boiler steam flow}}{\text{boiler efficiency}} \times \frac{\text{enthalpy change}}{\text{in the turbine}} \times \frac{\text{gas}}{\text{cost}} \times \frac{\text{annual}}{\text{hours run}}$$

$$\frac{28\ 750}{0.8} \quad \times \quad 93.6 \quad \times \frac{0.218}{1\ 000} \times \ 8\ 000$$

$$= \ £58\ 664$$

Equivalent annual cost of obtaining 675 kW
from public electricity supply:

$$675 \quad \times \quad \frac{2.6}{100} \quad \times \quad 8\ 000 \qquad = \quad £140\ 400$$

Gross annual saving by using single stage steam
turbine instead of electric motor:

$$£140\ 400 \quad - \quad £58\ 664 \qquad = \quad £\ 81\ 736$$

3.   No additional staff are required to
operate the set as it can run unattended and
because of its simple rugged construction
maintenance costs will be low.  However, in
the calculations to assess potential savings,
running and maintenance costs should not be
more than the allowed          per year.

4.   A typical capital cost for a complete turbine installation including steam and water
piping, cabling, building and civil works
would be £100 000.

∴ Simple payback period = capital cost/annual
savings = 100 000/79736 = 1.25 years.

(Ref. 3)

A Mollier diagram of a typical large thermal power station which burns
the equivalent of 2.6kW to 3.3kW of fuel for 1kW power supplied to the
electricity supply network

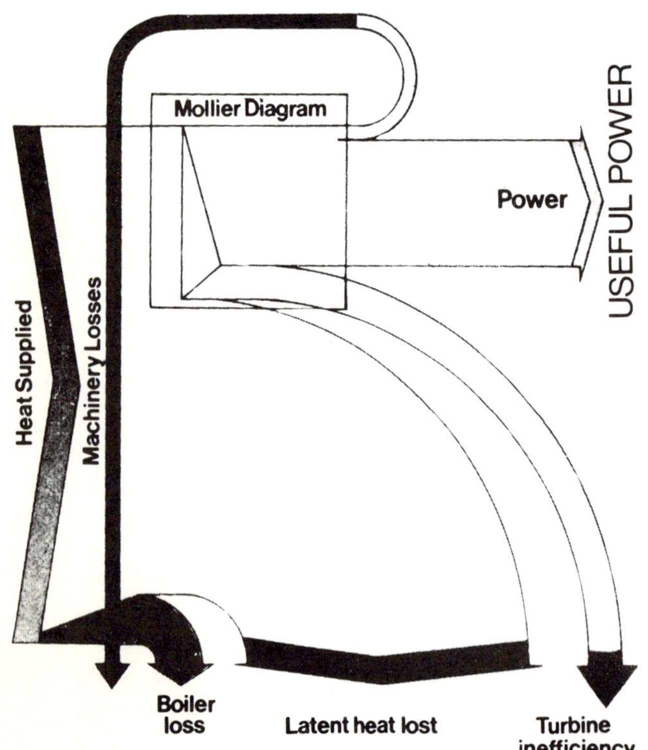


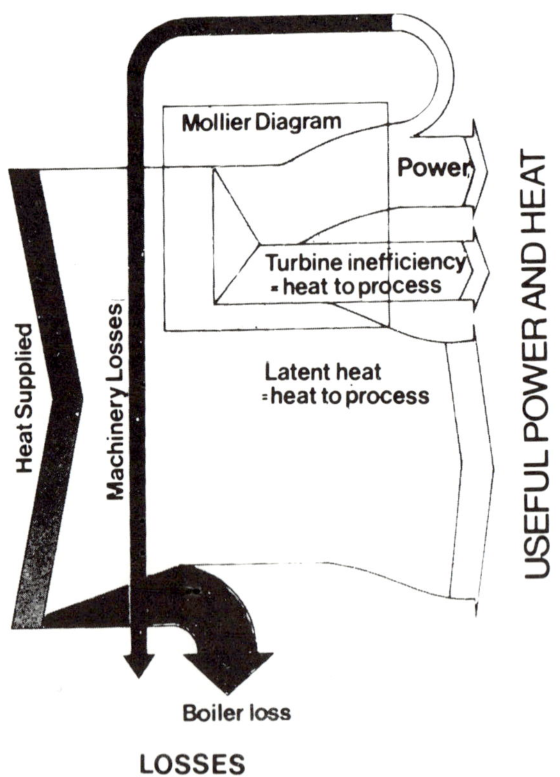


This diagram is of a typical industrial combined heat and power
scheme where overall losses are kept to a minimum, and approximately
1.2kW to 1.5kW of fuel produces 1kW of power and heat, substantially
increasing fuel conversion efficiency

Fig 1    Mollier diagram for power station and combined heat and power scheme

                                   

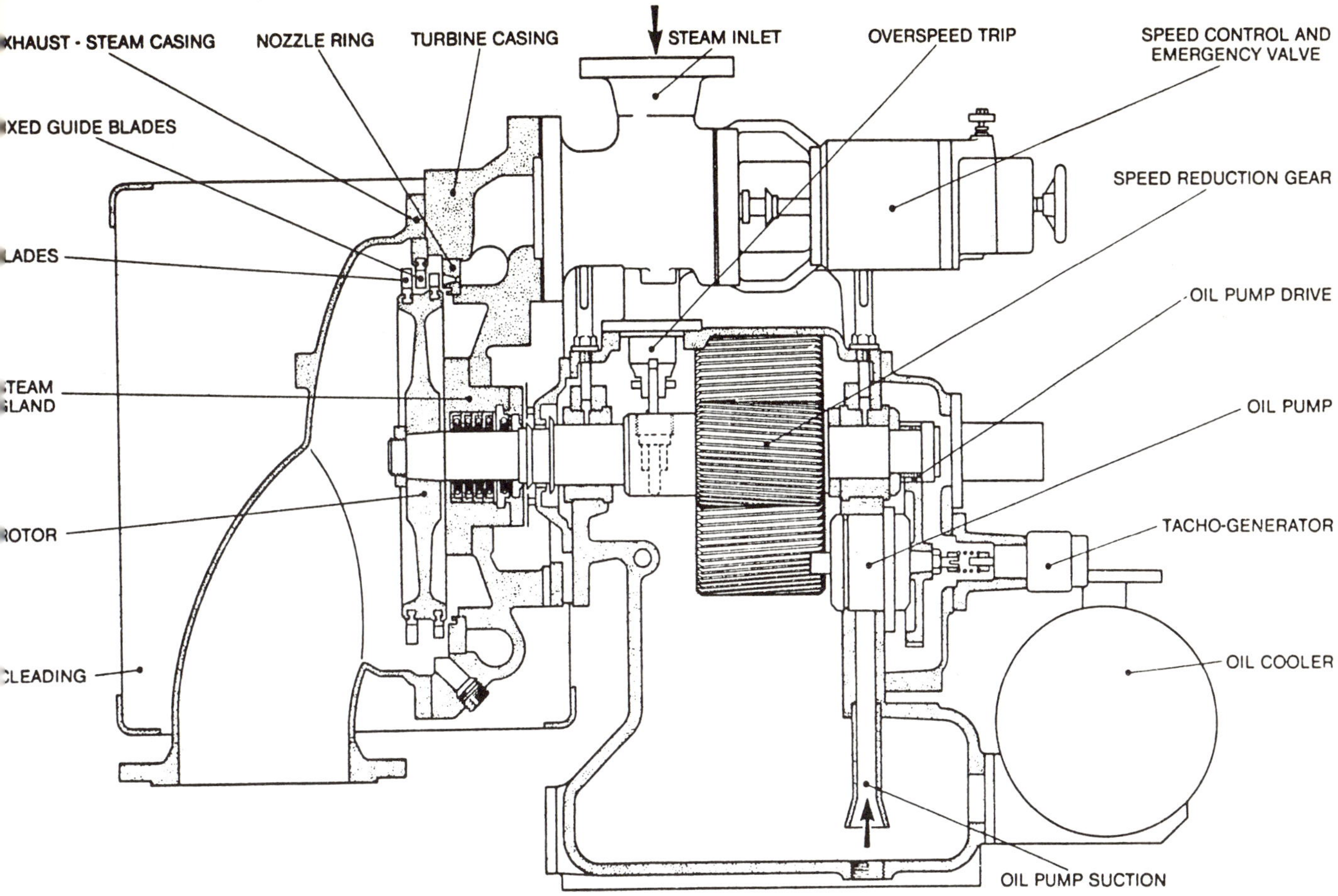

Fig 2    Typical section through a single stage turbine (overhung type)

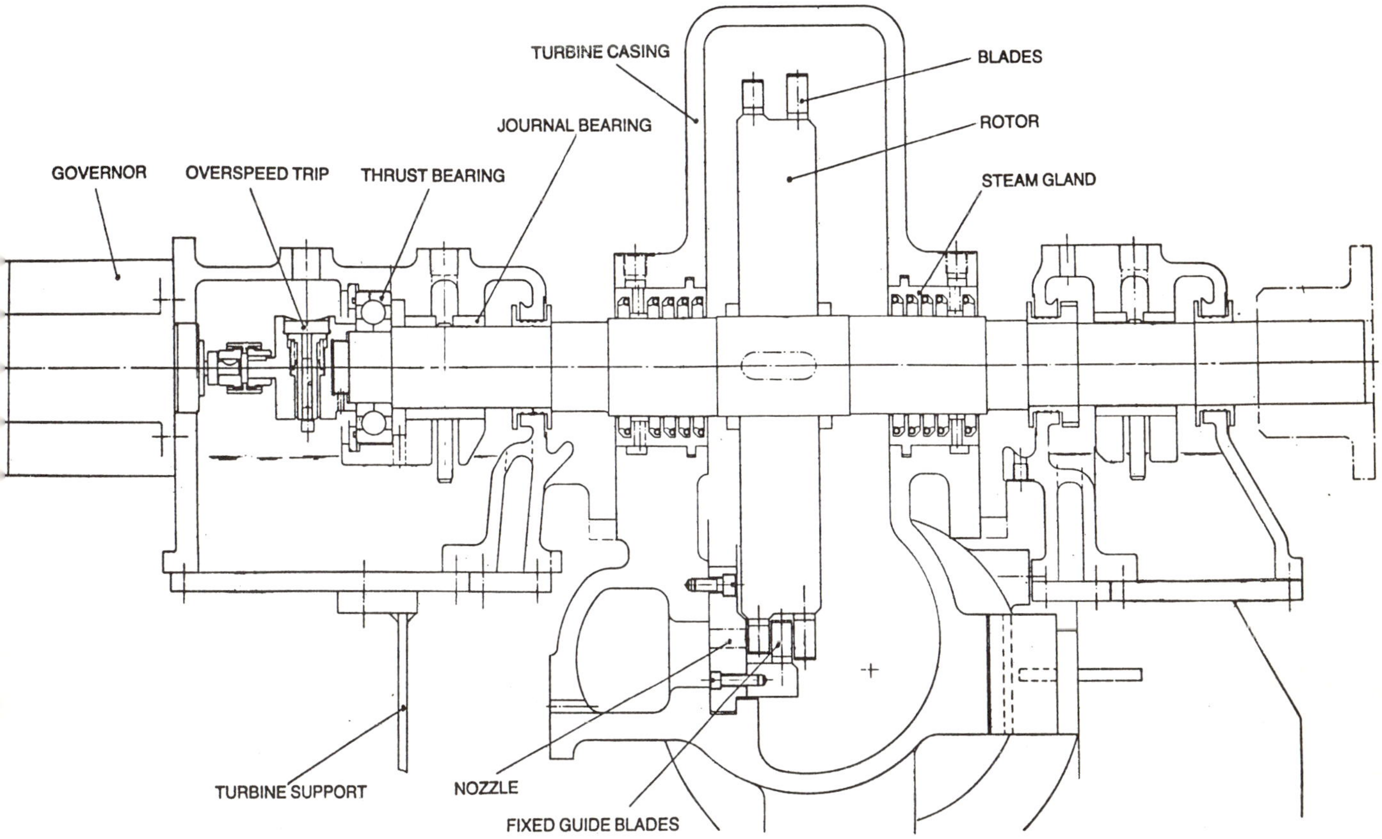

Fig 3    Typical section through a single stage turbine (between bearing type)

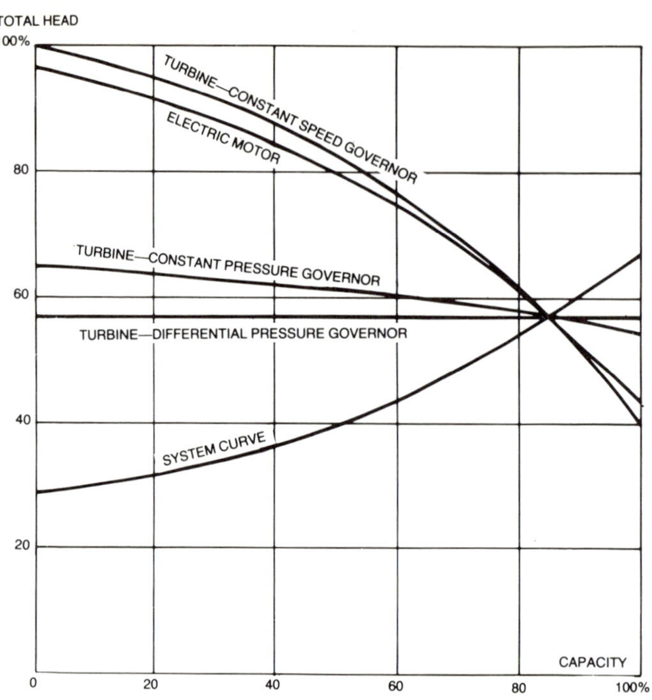


Fig 4    Pump characteristics with various governing systems

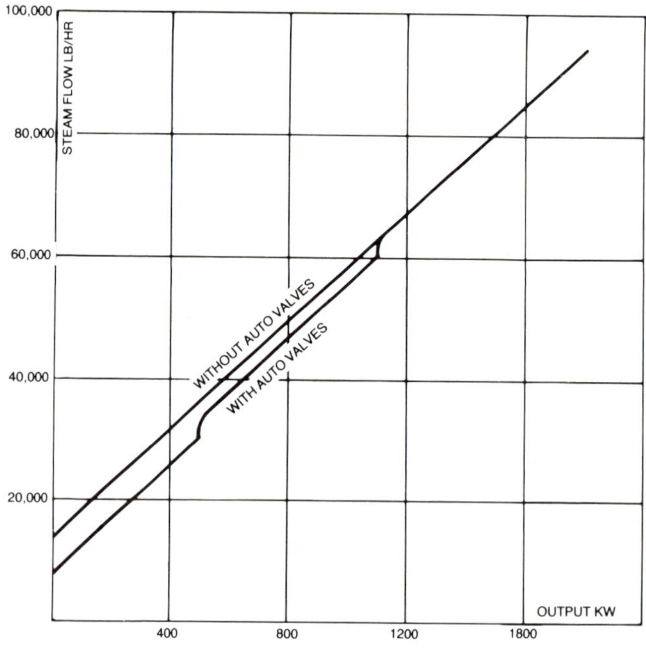


Fig 5    Steam flow versus output for turbine with or without automatic control valves